LOGIC AND ARCHITECTURE SYNTHESIS

LOGIC AND ARCHITECTURE SYNTHESIS

Proceedings of the IFIP TC10/WG10.5 Workshop on
Logic and Architecture Synthesis
Paris, France, 30 May–1 June 1990

Edited by

PETRA MICHEL
Siemens AG
Munich, Germany

GABRIELLE SAUCIER
INPG/CSI
Grenoble, France

1991

NORTH-HOLLAND
AMSTERDAM · NEW YORK · OXFORD · TOKYO

ELSEVIER SCIENCE PUBLISHERS B.V.
Sara Burgerhartstraat 25
P.O. Box 211, 1000 AE Amsterdam, The Netherlands

Distributors for the United States and Canada:
ELSEVIER SCIENCE PUBLISHING COMPANY INC.
655 Avenue of the Americas
New York, N.Y. 10010, U.S.A.

Library of Congress Cataloging-in-Publication Data

IFIP TC10/WG10.5 Workshop on Logic and Architecture Synthesis (1990 :
Paris, France)
Logic and architecture synthesis : proceedings of the IFIP
TC10/WG10.5 Workshop on Logic and Architecture Synthesis, Paris,
France, 30 May-1 June 1990 / edited by Petra Michel, Gabrielle
Saucier.
p. cm.
Includes bibliographical references.
ISBN 0-444-89023-8
1. Integrated circuits--Very large scale integration--Design and
construction--Congresses. 2. Silicon compilers--Design and
construction--Congresses. I. Michel, Petra. II. Saucier,
Gabrielle. III. Title.
TK7874.I3282 1990
621.39'5--dc20 90-28307
CIP

ISBN: 0 444 89023 8

Printed in The Netherlands.

PREFACE

The IFIP* *Working Conference on Logic and Architecture Synthesis* had the goal to provide an international forum for experts to meet and openly discuss the latest results on synthesis related matters. Researchers from academia, industry and governmental laboratories took the opportunity to share their knowledge and contribute to this three day event.

With this book we want to introduce a broader number of people to the essence of the working conference – the presented papers.

The papers cover the whole spectrum from high-level synthesis to technology mapping, including a historical overview on half a century of logic synthesis and asking the question whether high-level synthesis is practical at all. The reader undoubtedly will be left with the impression that though the field of synthesis has made considerable progress in the last few years, there is still a huge number of problems to be dealt with.

The editors hope that this book may have an inspiring effect on the readers to help solving the open problems.

Munich and Grenoble, August 1990
Petra Michel
Gabrielle Saucier

* IFIP: International Federation for Information Processing

ACKNOWLEDGEMENT

The editors want to express their gratitude to the members of the Programme Committee of the IFIP Working Conference on Logic and Architecture Synthesis for participating in the review process of the papers.

Dr. Gerd Venzl deserves a special thank for his efforts in putting together the final programme.

ORGANIZATION

Program Chairpersons:

P. Michel, Siemens AG, FRG

G. Saucier, Institut National Polytechnique de Grenoble, France

Program Committee

T. Akino	Matsushita Electric Ind. Co., Ltd.	Japan
F. Brglez	MCNC	USA
G. De Micheli	Stanford University	USA
A. R. Newton	University of California at Berkely	USA
P. Pype	IMEC	Belgium
W. Rosenstiel	FZI at the University of Karlsruhe	FRG
F. Somenzi	University of Colorado at Boulder	USA
G. Venzl	Siemens AG	FRG

CONTENTS

Low Level Synthesis I

Low Level Synthesis II

Technology Mapping

Case Studies

Keynote Talks on Logic Synthesis

Logic and Architecture Synthesis
P. Michel and G. Saucier (Editors)
Elsevier Science Publishers B.V. (North-Holland)

Half a Century of Logic Synthesis (The First Two Decades, Anyway)

E.J.McCLUSKEY

CENTER FOR RELIABLE COMPUTING
Computer Systems Laboratory
Departments of Computer Science
and Electrical Engineering
Stanford University

1. INTRODUCTION

Claude Shannon was a Research Assistant in Electrical Engineering at M.I.T. when he published his master's thesis; it described how to use Boolean Algebra as an abstract representation for switching circuits, [1]. The basic ideas in this 1938 thesis are now universally used for logic design. Specific topics in the paper include: how to represent a contact network (circuit containing only relay contacts and switches) as a Boolean algebra expression, how to use the postulates and theorems of Boolean algebra to transform this expression into a simpler expression (if such exists) from which a network with the same behavior and fewer components may be formed.

Another important algebraic contribution is the series expansion of Boolean functions:
$f(X_1,...X_n) = X_1X_2f(1,1,X_3...X_n) + X_1X_2'f(1,0,X_3...X_n) + X_1'X_2f(0,1,X_3...X_n) + X_1'X_2'f(0,0,X_3...X_n)$
This expression, its dual, and obvious generalizations are now called the Shannon expansions. They are extensions of the "Law of Development" presented in Section 24 of [2], the book by Cauturat (sic) cited as one of Shannon's two references on Boolean algebra. Shannon calls the expressions such as $f(1,1,X_3...X_n)$ "coefficients."

Series-parallel circuits (those containing only series and parallel connections) correspond directly to Boolean algebra expressions. Circuits that include other connections such as the bridge circuit of Fig. 1 do not have simple algebraic representations and are much more difficult to design. There is an extensive discussion of such circuits in Shannon's paper since they are so often the most efficient realization of a function. Since there is no direct analog for this problem in gate networks, in some real sense such networks are easier to design than contact networks. (There have been attempts to use non-series-parallel connections of transistors within complex CMOS gates, but there are many reasons to avoid such designs.)

Despite being titled "A Symbolic Analysis...," a major portion of the paper discusses synthesis; but most of the discussion is limited to totally symmetric functions. Shannon gives design methods for implementing such functions with contact networks. His networks and design methods seem to have been rediscovered in connection with MOS pass transistor networks, [4].

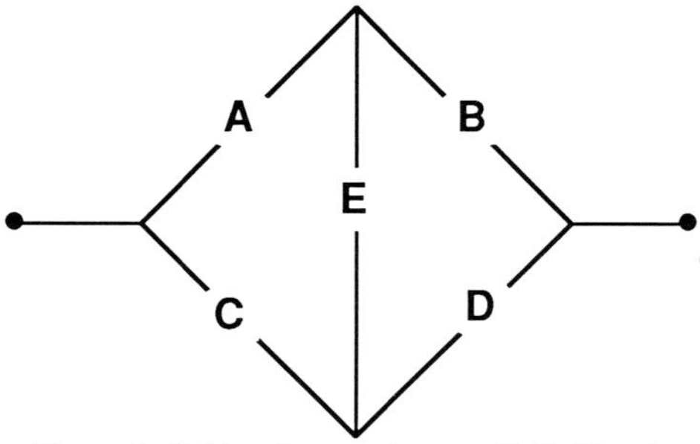

Figure 1. Bridge (non-series-parallel) Circuit

In his 1938 paper, Shannon summarized the synthesis status quo as follows:
"...For the synthesis problem the desired characteristics are first written as a system of equations, and the equations are then manipulated into the form representing the simplest circuit. The circuit may then be immediately drawn from the equations. By this method it is *always possible* to find the *simplest* circuit containing only series and parallel connections, and in some cases the simplest circuit containing any type of connection. ..." (italics added for emphasis)

The next paper that Shannon wrote on this subject was published in 1949. Some of the earlier optimism had vanished as demonstrated by the following quote from the later paper:
"...It is even more difficult to show that a circuit found in some way is the most economical one to realize a given function. The difficulty springs from the large number of essentially different networks available and more particularly from the lack of a simple mathematical idiom for representing these circuits..."
The 1949 paper also has a theorem proving that "Almost all functions" are not totally symmetric thus pointing up a limitation to the synthesis discussion in the 1938 paper.

As is typical in such situations, there is confusion as to who published the first paper describing a formalization for binary circuits. The 1955 book by Higonnet and Gréa, [5], credits Nakasima and Hanzawa* for the first paper relating Boolean algebra and contact networks. In his 1949 paper, Shannon also cites Nakasima's 1938 papers in Nippon Electrical Communication Engineering, but does not mention the 1936 and 1937 papers cited by Higonnet and Gréa. Shannon as well as Higonnet and Gréa list two 1939 papers by Piesch.

An SRI report by William Kautz, published in the IEEE Computer Transactions, [6], presents the early development of switching theory in the Soviet Union. The introduction starts with the following summary: " The origins of Soviet switching theory are usually attributed to Shestakov, whose dissertation in 1938 showed how Boolean algebra can be applied to the representation and transformation of contact networks. The first publications in a public journal appear to have occurred in 1941. Soviet authors who choose to discuss beginnings usually credit both Shannon's 1938 paper and Shestakov's dissertation. A 1910 book review by P. Erenfest is sometimes mentioned as the first recognition that the algebra of logic might be used as an analytical tool for telephone switching networks."

In any event, there was about a ten year lapse (probably due to the World War) and then very substantial progress in formal design techniques in the next decade. Most of the major

---

* I have included in the reference list only documents that I was able to inspect personally. Other work that is cited in these publications is mentioned but not included in the reference list. The interested reader is invited to consult the listed references to learn more about the other publications.

advances occurred at the Bell Telephone Laboratories, Harvard and M.I.T. Not too surprisingly, this work was influenced more by Shannon's papers than by any of the other early papers.

2. BELL LABORATORIES AND M.I.T.

The progress at Bell Laboratories resulted from the activity of their switching research department as well as their internal training courses. Two of the most notable achievements from the switching research department are the Karnaugh map, [7], and Moore's results on finite automata, [8]. As an outgrowth of the training at Bell Laboratories, one of the Bell instructors offered the Bell course on switching circuits as an M.I.T. course and then the course text was published, [9]. The enthusiasm and optimism for this course is evident in this excerpt from the foreword by John Mesar:
"... automatic digital computing systems. ... The abilities they do have, however, seem great enough to promise the elimination of much of the mental drudgery required by our modern world. It is this glimpse of a coming age of 'mechanized intelligence' that recently caught the imagination of professional people and caused the upsurge of interest in switching."

The switching circuits course became a standard M.I.T. course taught by Professor Samuel H. Caldwell, who modified the content and, in 1958, published his own text , [10], which became widely used when Caldwell's students introduced the switching theory course into the curriculum at other colleges and universities. Caldwell calls the expressions such as $f(1,1,X_{3}...X_{n})$ in the Shannon expansion "residues."

The majority of the material in the Bell Laboratories book describes circuits for specific functions; there are 12 chapters that discuss circuits for counting, translating, selecting, connecting, lockout, finding and hunting, timing, pulse generation, checking, registration, and calculation. The approach is very similar to the "design with MSI" techniques that were so popular in the 1970s. Caldwell specifically rejects this approach in favor of presenting a design methodology rather than specific circuits. In his preface, he says, "There is no harm in discussing a useful circuit, but there can be harm if the usefulness is the only reason for discussing it." He also rejects: " the extreme of treating switching-circuit theory as an exercise in mathematics." He comments further: " Most engineers are not familiar with the ideograms of symbolic logic. Some of these look to them more like the brands that identify western cattle ranches than mathematical symbols."

The Caldwell approach was made possible, in part, by the development of formal systematic techniques for minimization of Boolean expressions and for sequential circuit design. It's not too surprising that these techniques were developed by two of Caldwell's own doctoral students. The sequential circuit methodology comes from David Huffman's doctoral thesis which was submitted in 1953 and published subsequently by the Franklin Institute, [11]. Huffman summarizes his work as "An orderly procedure by which the requirements of a sequential circuit can be reduced to the requirements of several combinational switching circuits." Some aspects of the Huffman technique were presented earlier in England by Montgomerie, [12]. E.J. McCluskey's 1956 doctoral thesis presented "a systematic procedure for writing a Boolean function as a minimum sum of products." This research was started at M.I.T. under the supervision of Caldwell and Huffman, was completed at Bell Laboratories and published in the Bell System Technical Journal, [13]. The McCluskey technique was a simplification and extension of earlier work by Prof. W.V.Quine of the Harvard University philosophy department, [14]. Prof. Quine also published two other papers on Boolean minimization — one paper introduced iterated consensus, [14] and the other treated what have come to be called unate functions. In his 1952 paper, Prof. Quine makes the following observations:

" But there remains a problem which, despite the trivial character of truth-function logic, has proved curiously stubborn; *viz.*, the problem of devising a general mechanical procedure for reducing any formula to its simplest equivalent."

" It is noteworthy that the staff of the Computation Laboratory of Harvard University have found it worth while to set forth elaborate procedures for simplifying truth functions, and even to tabulate all the simplest equivalents of formulas involving four or fewer letters."

" ...we worked out a routine which could be depended upon to reveal a simplest normal equivalent, and indeed all the simplest normal forms."

3. HARVARD UNIVERSITY

At the same time that this activity was going on at Bell Laboratories and M.I.T., Harvard University had a completely independent activity aimed at designing logic networks. The independence is perhaps unbelievable with both Harvard and M.I.T located within walking distance of one another in Cambridge, Massachusetts. Nevertheless, there is no evidence of any collaboration between these two academic institutions during this time period. Dr. Howard Aiken was Professor of Applied Mathematics and Director of the Computation Laboratory at Harvard. The Mark series of computers was designed and built at this Laboratory; their work on formal design techniques came in response to their need for better tools to design their computers. The Harvard methodology is presented in a book published in 1951, [15]. This book is organized similarly to the Bell Laboratories book, [8], in that 6 chapters describe circuits for specific functions; but differs in using vacuum tubes (triodes and pentodes) as basic components rather than the relays of the Bell book. The Harvard book includes an appendix listing "minimal" realizations of all types of 4-variable functions. Aiken's preface gives interesting insight into the thinking at this Laboratory:

"...The lack of adequate mathematical methods for the investigation of the functional behavior of electronic control circuits represented the largest single obstacle to the rapid development of the subject and to the instruction of students looking forward to an active career in the field...."
"...completely concurred with the opinion of the lecturer as to the desirability of undertaking a detailed investigation of the general subject of electronic-control-circuit synthesis."
" As regards the mathematical approach to the subject matter of this volume it should be noted that several alternatives exist.....Again, Boolean algebra was employed by Claude E. Shannon in his discussion of relay circuits. It is believed, however, that the algebraic approach adopted in the present volume provides a particularly convenient vehicle of thought and has the considerable advantage of lying within the province of the average reader's previous mathematical experience."

A connection was established between the Harvard Computation Laboratory and Bell Laboratories just after the publication of the Harvard book, [15]. Bell supported the Harvard research in switching theory in return for reports which were submitted to Bell. Four of these reports are reprinted an an appendix in the book by Curtis, [16]. This book presents decomposition theory (based on the work by Ashenhurst and others at Harvard) and the preface says: "...it is intended as a guide for a one-semester advanced course in switching theory, and thus is meant to augment a prior course based on a text such as *The Design of Switching Circuits* by Keister, Ritchie and Washburn." This statement is another example of the gap between M.I.T. and Harvard — even though Curtis studied at M.I.T. as well as Harvard and includes the material from McCluskey's M.I.T. thesis in his book, he ignores the text by Caldwell.

4. ELSEWHERE

While most of the subsequent work on synthesis developed from the M.I.T, Bell, Harvard work, there are some important isolated books that deserve mention. The Higonnet and Gréa book was published in France in 1955 followed by an English edition in 1958, [17]. Also in 1958, not only did the Caldwell book appear but also books by Naslin, Moisil, and Phister. While the Higonnet and Gréa book was published in France, it probably should be included with the M.I.T. developments since the preface was written in Cambridge, MA and thanks Caldwell for introducing the authors to switching theory and advising on the book. Caldwell, Higonnet and Gréa worked together at the Graphic Arts Research Foundation on the Photon machine, a photographic type-composing machine. A revised version of this book, which includes a table developed by Ed Moore of Bell Labs of "best" 4-relay circuits, was published in English in 1958. In this same year, a text in French was published by Naslin, [18]; the presentation in this book depends heavily on the M.I.T. work of Huffman and McCluskey (see Neslins preface). Strangely enough, Naslin was unaware of the work of his countrymen Higonnet and Gréa until his text was completed (he states in his preface that he only learned of the Higonnet and Gréa book from a review in the Soviet journal *Avtomatika i Telemexanika.*) Also in 1958 the book by Phister, [19], appeared. This book emphasizes digital computer design rather than switching circuit or logical design theory, but it contains important material on design using flip-flops.

1959 saw the publication of a Romanian text on switching theory which was published in an English translation in 1969, [20]. This book is important because of the surveys of other work included in the prefaces and, in particular, for the discussion of the extensive work that was carried out in Romania. However, it does not appear to have had much influence on subsequent research outside Romania.

Finally, the book published by Gavrilov in Russia in 1950 should be mentioned. According to Kautz, "...this one publication is clearly responsible for setting down most of the important technical problems to be solved, the terminology, and the general viewpoint toward the subject to be taken by Soviet researchers to follow. This book was the most advanced single source on relay switching network design anywhere in the world for at least five years following its publication. Unhappily, it was never translated into English. A German edition appeared in 1953." Another opinion of the Gavrilov book appears in [5]: ".. qui ne parait pas apporter de contribution importante sur les précédents, autant que l'on puisse en juger en l'absence d'une traduction complète."

5. SUMMARY AND CONCLUSIONS

Most of the fundamental developments in synthesis took place in the decade following Shannon's 1949 paper, [3]. This is not to ignore the many important later contributions, which should be mentioned here, but which must be omitted due to time and space limitations. The interested reader should consult more recent publications such as [21] and [22] for information about recent results in synthesis.

ACKNOWLEDGEMENTS

This work was supported in part by the National Science Foundation under grant No. MIP-8709128 and the Innovative Science and Technology Office of the Strategic Defence Initiative Organization administered through the Office of Naval Research under Contract No. N00014-85-K-0600.

REFERENCES

[1] Shannon, C.E., "A Symbolic Analysis of Relay and Switching Circuits," *Trans. A.I.E.E,* pp. 713-723, 1938.
[2] Couturat, L., *The Algebra of Logic*, The Open Court Publishing Company, Chicago and London, 1914.
[3] Shannon, C.E., "The Synthesis of Two-terminal Switching Circuits," *BSTJ,* Jan. 1949.
[4] Mead, C. and L. Conway, *Introduction to VLSI Systems*, Addison-Wesley, Reading, MA, 1980.
[5] Higonnet, R.A. and R.A. Gréa, *Étude Logique des Circuits Électriques et des Systèmes Binaires*, Éditions Derger-Levrault, Paris, 1955.
[6] Kautz, W.H., "A Survey and Assessment of Progress in Switching Theory and Logical Design in the Soviet Union," *IEEE Trans. Comp.*, pp. 164-204, Apr. 1966.
[7] Karnaugh, M.," The Map Method for Synthesis of Combinational Logic Circuits, *AIEE Trans. - Com. and Elect.,* Nov. 1953.
[8] Moore, E.F., "Gedanken Experiments on Sequential Machines," in *Automata Studies*, Princeton Univ. Press, 1956.
[9] Keister, W., A.E. Ritchie and S.H.Washburn, *The Design of Switching Circuits*, Van Nostrand, 1951.
[10] Caldwell, S.H., *Switching Circuits and Logical Design*, J.Wiley, 1958.
[11] Huffman, D.A., "The Synthesis of Sequential Switching Circuits," *Journal Franklin Institute*, March, April 1954.
[12] Montgomerie, G.A., "Sketch for an Algebra of Relay and Contactor Circuits," *Jour. IEE*, Part III, pp. 303-312, 1948.
[13] McCluskey, E.J., "Minimization of Boolean Functions," *BSTJ*, Vol. XXXV, No.6, pp. 1417-1444, Nov. 1956.
[14] Quine, W.V., "The Problem of Simplifying Truth Functions," *The American Mathematical Monthly,"* Vol. 59, No. 8, pp. 521-531, Oct. 1952.
[15] Staff of the Harvard Computation Laboratory, *Synthesis of Electronic Computing and Control Circuits*, Harvard Univ. Press, 1951.
[16] Curtis, H.A., *A New Approach to the Design of Switching Circuits,* D.Van Nostrand, Princeton, N.J., 1962.
[17] Higonnet , R.A. and R.A. Gréa, *Logical Design of Electrical Circuits*, McGraw-Hill , 1958.
[18] Naslin, P., *Circuits a Relais et Automatismes a Séquences*, Dunod, 1958.
[19] Phister Jr. M., *Logical Design of Digital Computers*, John Wiley, New York, 1958.
[20] Moisil, Gr.C., *The Algebraic Theory of Switching Circuits*, Pergamon Press, Oxford, 1969. A translation of an earlier book in Romanian, *Teoria Algebrică a Automate*, Editura Tehnică, Bucharest, 1959.
[21] Brayton, R.K., et al, *Logic Minimization Algorithms for VLSI Synthesis*, KAP, 1984
[22] Saucier, G. and P.M. McLellan, *Logic and Architecture Synthesis for Silicon Compilers,* North-Holland, 1989.

Logic and Architecture Synthesis
P. Michel and G. Saucier (Editors)
Elsevier Science Publishers B.V. (North-Holland)

IS HIGH-LEVEL SYNTHESIS PRACTICAL?

R. Camposano

IBM Thomas J. Watson Research Center
Yorktown Heights, NY

With very few exceptions, high-level synthesis of digital systems is currently not used in industrial design practice. This paper explains the main causes for this and argues that there is reason to believe that in the near future high-level synthesis will indeed be integrated into practical design automation systems.

1. Introduction

The term high-level synthesis is mostly used to describe design activities above the well-known logic level. More specifically, high-level synthesis is the automatic synthesis of a *circuit structure* from a *behavior* at a *register-transfer level* and above [3]. In practice, the behavior is mostly specified in a sequential language such as a proper subset of behavioral VHDL [10], C or Pascal. A structure is given by a netlist consisting of memory elements (RAM's, registers), functional units (ALU's, adders, combinational logic) and interconnection (multiplexers, busses, wires).

The main advantages that high-level synthesis promises are a *shorter design cycle*, and the possibility to *master complex designs*. Since high-level descriptions are in general concise, they are easier to write, easier to understand, less error prone, and faster to simulate. But although automation at this level may increase productivity significantly, it also comes with a price: design automation often can not match the abilities of a skilled human designer, resulting in larger or slower hardware.

The task of high-level synthesis is rather formidable. A behavior is essentially a collection of operations. These operations are partially ordered by *data dependencies* and they must follow certain control sequences such as *alternation (if-then-else)* and *looping*. High-level synthesis decides at which *time* each operation is to be executed, i.e. the control step in synchronous systems, and which *hardware* (memory, functional units, interconnection) is to be used. The former task is called *scheduling*, the latter is called *allocation*.

Compare these goals with 'classical' logic synthesis. Logic synthesis is concerned only with optimizing combinational logic. After decades of research, only now are logic synthesis tools becoming standard industrial practice. So it is not surprising that high-level synthesis has not been used for real designs with one exception, digital signal processing applications [6, 9]. This paper explains the main reasons for this lack of practicality, not by describing the technical problems involved in high-level synthesis (which have been solved to some extent [3]), but rather by focusing on the missing parts.

The rest of the paper is organized as follows. The next section shows the need for high-level synthesis. Section 3 deals with the tools needed to bridge the gap between high-level synthesis and logic synthesis, namely finite state machine synthesis and retiming. Section 4 briefly discusses the effects of partitioning designs generated by high-level synthesis. Verification, which we believe to be very important for the acceptance of high-level synthesis, is also insufficiently developed as shown in section 5. Section 6 briefly discusses the problems involved in specifying behavior. The paper's conclusions state that much work is still needed to make high-level synthesis practical, that this work is under way, and that in some areas high-level synthesis already is or will very soon be practical.

2. The need for high-level synthesis

If high-level synthesis becomes used in practice, it will ultimately be for economic reasons. Sufficient applications for which the advantages explained in the previous section are relevant would trigger the need for high-level synthesis. These applications include:

- *Glue logic,* which usually consists of combinational logic and only a few storage elements.
- *Controllers and sequencers,* which generate signals to control some other circuit or machine. Typically, this is a finite state machine with, at most, hundreds of states.
- *Protocol engines,* which implement a bus protocol or a communication protocol. These applications may involve several communicating finite state machines and possibly asynchronous parts.
- *Digital signal processing applications,* often characterized by complex algorithms performed on regular data streams.
- *Processors,* which execute an instruction set, e.g., microprocessors or co-processors.

High-level synthesis is of advantage for all of the above applications, if design time rather than design optimality is the main criterion. As design automation becomes more mature, automatic design systems may of course outperform human designers, e.g., state assignment, scheduling of long sequences, etc.

If integrated circuits are used, the above applications are implemented as processors, ASIC's or PLD's. The market for integrated circuits in 1989 was approximately $19 billion in Japan and the Far East, $13 billion in the U.S. and $6 billion in Western Europe [1]. Table 1 shows that close to 40% of this market are potential applications for high-level systems. Even if only in a small fraction high-level synthesis is of real advantage, the economic need is easily demonstrated. Furthermore, printed circuit boards, often used to implement large systems, also include many potential applications for high-level synthesis.

Memories	35%
Processors	20%
Linear ICs	17%
ASICs	17%
Logic Families	8%
PLDs	3%

Table 1. IC Market Share in 1989 [1]

As far as future predictions go, these are difficult. E.g., in 1981 it was predicted that the U.S. IC consumption in 1989 would be approximately $28 billion [5]; the real consumption was only close to $13 billion [1]. Nevertheless, it seems safe to predict that the total IC consumption will continue to grow, and that the combined market of ASICs, PLDs and processors will also grow. Since ASICs, processors and PLDs compete for applications (ASICs are in a sense squeezed in between processors and PLDs), it is not clear how the market share of them will evolve.

3. Finite State Machine Synthesis and Retiming

A design generated by high-level synthesis must be further refined. Typically, the combinational logic is optimized by logic synthesis and then the design is finished by a layout tool. However, the step between high-level synthesis and logic synthesis is too large, resulting in multiple problems [13]. E.g., designs generated with high-level synthesis may show a large amount of redundancy, the memory elements may be placed in the wrong position with respect to the logic, or the states of a finite state machine may not be properly encoded.

This gap is however being closed by active research in the areas of *finite state machine synthesis* [7] and *retiming and resynthesis*[12, 14] . FSM synthesis can be used for optimizing sequential control logic. Typical issues are state encoding, state minimization and partitioning of finite state machines. FSM synthesis can not be used for data-path circuits, because it requires some sort of state enumeration; attempting to enumerate states in a data path with hundreds of latches is impossible in practice. The correct technique to optimize data paths seems to be retiming and resynthesis, which allows to move latches and to perform optimizations across and including latches. The relation of these techniques to high-level and logic synthesis is depicted in fig. 1.

As these synthesis techniques become more practical, high-level synthesis will also become more practical and better integrated into existing design systems.

4. Partitioning

In order to be useful, high-level synthesis must be able to deal with very large systems. The generated RT-level designs thus are often too large for logic synthesis and

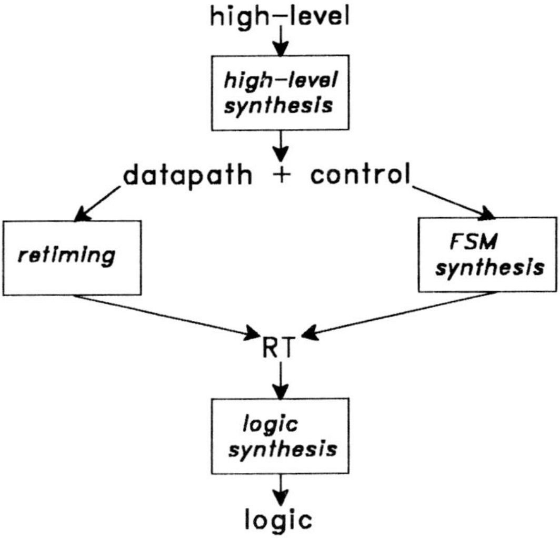

Figure 1. Synthesis above the Logic Level

they must be *partitioned.* Some of the effects of partitioning discussed in [13] are repeated here to show that a good partitioning tool is essential for high-level synthesis to become practical.

Partitioning the combinational logic has, among others, effects on the size, speed and testability of a design. For example, fig. 2 shows how the relative size and delay of 5 designs (d1-d5) is affected by partitioning the combinational logic into 1 to 5 pieces. It can be seen that the effects on both size and speed are rather unpredictable.

Testability may also be affected severely by partitioning combinational logic. With respect to the usual stuck-at fault model, irredundant combinational logic is considered 100% testable. However, in a partitioned design, even if every partition is 100% testable, the complete design may not be 100% testable. In an example of 10000 nets partitioned into 14 partitions, each of them 100% testable, the resulting complete combinational logic was only 80% testable.

So despite of research in this area [2, 8], many aspects of partitioning remain an open problem in high-level synthesis, and further work in this area is necessary.

5. Verification

Verification is another open problem that adds to the difficulty of high-level synthesis use. Verification is highly desirable at all levels, because even though synthesis promises *correctness by construction*, the implementation of the synthesis algorithms may always be faulty. And even if the implementation is correct, the software used

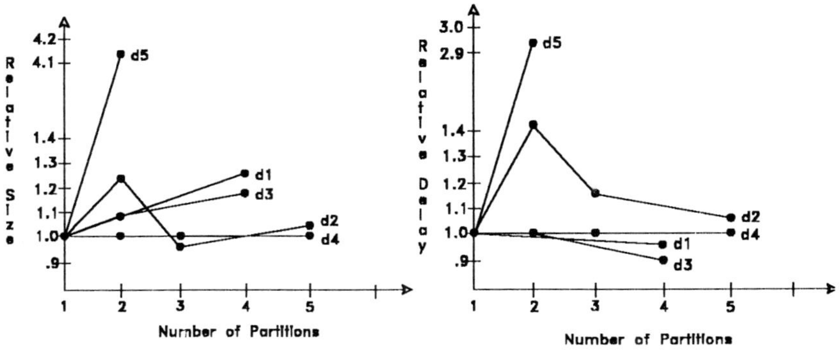

Figure 2. Partitioning effect on size and speed

(compiler, operating system) may be faulty. In large systems, errors seem unavoidable.

Layout is verified by circuit extraction and netlist comparison. Logic synthesis is verified by Boolean comparison. Both techniques are current industrial practice. But how is high-level synthesis (retiming and FSM synthesis) verified? Currently no technique is sufficiently developed to be able to verify really large designs, but there is promising research under way [4, 11]. Practical 'high-level' verification techniques would greatly contribute to the practicality of high-level synthesis.

6. Specification and other Problems

High-level synthesis uses a *behavioral* specification of a design. So the question must be asked, when does a behavioral specification make sense?

Consider a combinational logic design with n inputs and one output. A tabular behavioral specification may require as many as 2^n entries. A structural specification (e.g. netlist) however, is of a size proportional to the implementation, which may be much smaller.

For a finite state machine with n inputs and s latches, a state transition table to specify the behavior may have as many as 2^{n+s} entries. Again, a structural specification is of size proportional to the implementation. In particular, it depends only on s rather than on 2^s.

Thus a behavioral specification may be too large and a structural specification may be more desirable. In high-level synthesis, very rarely a 'pure' behavioral specification can be given ('pure' meaning I/O relationship). Instead, languages which imply certain structures (such as loops) are used. Functional specification of digital systems behavior is not really understood, and analogies to software are often the best that can be achieved. Timing behavior additionally complicates the issue.

High-level synthesis is further complicated by the need to include *asynchronous* and *analog* design parts. Design automation for these kinds of designs is usually not included in digital design automation systems.

Finally, there are also non-technical problems that may potentially prevent the practical use of high-level synthesis. Examples are the education and skills required of a designer, and the current design practice.

7. Conclusions

Although this paper identifies many problems which make the practical use of high-level synthesis difficult, there is reason to believe that high-level synthesis will soon be more practical. First, in the area of digital signal processing, high-level synthesis is already practical. Applying high-level synthesis to restricted problem domains is indeed a powerful paradigm that brings it closer to practicality. Second, the need for high-level synthesis exists and will probably grow as IC markets and digital systems complexity grow. Third, active research is closing existing gaps in higher-level design automation, so that a practical design methodology including high-level synthesis is becoming more feasible. To predict to what extent, and when, high-level synthesis will be actually used in industrial design practice would go beyond the scope of this contribution.

References

[1] Electronics, 1990.

[2] R. Camposano, R.K. Brayton, "Partitioning Before Logic Synthesis," *Proceedings of the ICCAD'87*, pp. 237-246, Santa Clara, Ca, November 1987.

[3] M.C. McFarland, A.C. Parker, R. Camposano, "The High-Level Synthesis of Digital Systems," *Proceedings of the IEEE*, pp. 301-318, February 1990.

[4] A. Cohn, "Correctness Properties of the VIPER Block Model: The Second Level," in G. Birtwistle, P.A. Subramanyam, editor, *Currents Trends in Hardware Verification and Automatic Theorem Proving*, pp. 1-91, New York: Springer-Verlag, 1989.

[5] D.J.McGreivy, "Market Demand," in D.J. McGreivy, K.A. Pickar, editor, *VLSI Technologies through the 80's and beyond*, pp. 3-9, IEEE, 1982.

[6] S. Note, J. VanMeerbergen, F. Catthoor, H. DeMan, "Automated Synthesis of High-Speed Cordic Algorithm with the Cathedral III compilation system," *Proceedings ISCAS'88*, pp. 581-584, June 1988.

[7] S. Devadas, "General Decomposition of Sequential Machines: Relationships to State Assignment," *Proceedings of the 26th Design Automation Conference*, pp. 314-320, Las Vegas, June 1989.

[8] E.M. Dirkes, *Architectural Partitioning for Design Synthesis*, PhD thesis, Electrical and Computer Engineering Department, Carnegie-Mellon University, 1989.

[9] A.E. Casavant et.al., "A Synthesis Environment for Designing DSP Systems," *IEEE Design & Test*, pp. 35-45, April 1989.

[10] IEEE, Standard VHDL Language Reference Manual, New York: The Institute of Electrical and Electronics Engineers, Inc., March 1988.

[11] O. Coudert, C. Berthet, J.C. Madre, "Formal Boolean Manipulations for the Verification of Sequential Machines," *Proceedings EDAC'90*, pp. 57-61, Glasgow, Scotland, March 1990.

[12] S. Malik, E.M. Sentovich, R.K. Brayton, A. Sangiovanni-Vincentelli, "Retiming and Resynthesis: Optimizing Sequential Networks with Combinational Techniques," *Proceedings of the 23rd HICSS*, pp. 397-406, IEEE, January 1990.

[13] R. Camposano, L.H. Trevillyan, "The Integration of Logic Synthesis and High-Level Synthesis," *Proceedings ISCAS'89*, IEEE, 1989.

[14] G. DeMicheli, R. Yip, "Logic Transformations for Synchronous Logic Synthesis," *Proceedings of the 23rd HICSS*, pp. 407-416, IEEE, January 1990.

High Level Synthesis I

Logic and Architecture Synthesis
P. Michel and G. Saucier (Editors)
Elsevier Science Publishers B.V. (North-Holland)

Symbolic Don't Cares and Equivalence in High-Level Synthesis

Bill Lin, Gregory S. Whitcomb, and A. Richard Newton
Department of Electrical Engineering and Computer Sciences
University of California, Berkeley, CA 94720

Abstract

Ongoing work [6, 8] on synthesizing implementable hardware from symbolic high-level description languages is presented. Unlike fully encoded register-transfer level specifications, a data variable in a symbolic description takes values from a set of possibly many symbolic values rather than just 0's and 1's. This permits a higher level of abstraction and a more general specification. Previous approaches make distinction between symbolic and binary level specification. An optimal *symbolic encoding* step [6] is usually invoked to map a multi-level symbolic description into a binary representation suitable for logic synthesis [1]. However, this neglects a significant number of optimizations that can be performed at the symbolic level. Alternatively, one can treat a "symbolic" specification in the same manner as a "Boolean" description with the Boolean case as a special symbolic set $\{0, 1\}$. The synthesis process from a symbolic to the binary level is viewed as a *continuum* rather than as separate stages. In this paper, we take a step in this direction by focusing on a set of optimization involving the use of multi-level *symbolic don't care* and *equivalence* information derived from the symbolic description. Potential savings obtained by exploiting these conditions are difficult to achieve at the Boolean level. We propose optimization strategies for synthesis under these degrees of freedom. We believe a generalized framework for symbolic synthesis offers a powerful paradigm to general purpose high-level synthesis.

1 Introduction

In many application specific designs, it is often more natural to describe high-level behavior in terms of abstract data types rather than at the level of 0's and 1's. A number of hardware description languages (HDLs) including ELLA [7] and VHDL [5] permit the definitions of data types. For example, a variable representing a microprocessor instruction may take values from the symbolic set { ADD, SUB, CMP, LOAD, STORE }. More complex data types can be defined using constructs such as *arrays*, *structures*, and *unions*.

An example of a high-level hardware description is shown in Figure 1. The description specifies the behavior of a process control interface module written in the ELLA language. The ability to represent design at a symbolic level alleviates some of the design decisions to the synthesis process. Hence a higher level of abstraction is achieved. We argue that the data abstraction mechanisms of these languages would be extensively used if effective synthesis strategies can be devised for generating the final logic implementation.

We present ongoing work in synthesizing implementable hardware from symbolic level specifications. Our high-level behavior representation, Hardware Data Flow (HDF), supports data typing of tokens and abstract combinational functionality. In [6, 8], it is indicated how significant reductions in area and delay can be achieved by optimal synthesis of data types. This paper presents further optimization techniques which can be applied to an important subclass of HDF descriptions called *symbolic networks*, an acyclic network consisting exclusively of enumeration, or symbolic, types.

A critical step in our synthesis methodology is the mapping of the symbolic networks into a binary implementable form. Part of this process involves the encoding of symbolic values into the Boolean domain suitable for logic optimization. It was shown in [6] an optimal symbolic encoding step for mapping symbolic values into the that the result of logic synthesis can be very sensitive to the choice of binary codes assigned to symbolic values.

However, optimal encoding alone does not exploit the data abstraction captured at a high-level to the fullest extend. Additional optimizations can be performed before encoding, potentially leading to a reduction in the area and delay of the resulting binary implementation. We view the synthesis process as a continuum rather than as a two-step process of encoding and logic optimization. Techniques developed for logic synthesis can be generalized to a uniform framework for synthesizing both symbolic and binary specifications. In this paper, we take a step in this direction by focusing on a set of optimizations involving the use of *symbolic don't care* and *equivalence* information derived from the symbolic description. These conditions can be used to achieve optimizations which are either impossible or extremely difficult (and expensive) to achieve once the symbolic information is lost.

The leading ideas behind the proposed approach is as follows. For a given internal symbolic function in a multi-level network, there are symbolic input conditions to the function that can never occur due to the dependence on the primary inputs. For these input conditions, the output of the function can be any of the admissible symbolic values. Under certain primary input conditions, the function output can be one of several symbolic values without changing the final values at the primary outputs. Two symbolic values are equivalent if they can be used interchangeably without affecting the final output. Symbolic don't cares and equivalence can be exploited at the symbolic, encoding, or combinational logic level. At the symbolic level, the size of the symbolic data types can be reduced by combining equivalent symbolic values, which may result in the reduction in the number of required encoding bits if appropriate. At the encoding level, the output mapping of an input condition may be one of several possible symbolic output values. Thus, the output binary code for that input condition can be any one of the binary codes assigned to the possible symbolic output values. At the logic level, groups of valid output codes can be merged into a single cube as the output condition.

We have implemented the ideas presented in this paper in a prototype synthesis environment. The symbolic descriptions are generated by compiling an enhanced version of ELLA called EllaX into a high-level representation called Hardware Data Flow. Once in this representation, other transformations, in addition to the ones described here, can be invoked to optimize the design.

The remainder of this paper is organized as follows. In the next section, basic definitions and notation are given. In Section 3, we describe the symbolic don't care and equivalence conditions that can arise in a multi-level symbolic description and give some illustrative examples. In Section 4, optimization strategies are given for exploiting these conditions. These degrees of freedom can be exploited at the symbolic, encoding, or logic level. Some experimental results are given in Section 5 to indicate the possible gains by our optimization strategies.

2 Definitions and Notation

A high-level, combinational logic description can be compiled into a multi-level (cascaded) network of two-level *symbolic functions* called a *symbolic network*. Each symbolic function is a mapping from a set of symbolic-valued inputs to symbolic-valued outputs. A Boolean network is a special case of a symbolic network where all values are from the set $\{0,1\}$. Formal definitions are now given.

A *symbolic variable* v_i has a set of admissible values $S_i = \{s_1, s_2, \ldots, s_k\}$. For functions of n input variables and m output variables, let $S_i{}^I$, $i = 1,2,...,n$, and $S_i{}^O$, $i = 1,2,...,m$, be

```
TYPE bool = NEW(t | f),
    stage = NEW(WAIT | RESET | FETCH),
    insts = NEW(ADD | SUB | CMP).
FN PROCESS_CONTROL =
    (bool: c, flag1, flag2, stage: pfork) →
    (bool, insts, insts, stage):
BEGIN
    FN LATCH = (stage) → stage: DELAY(RESET, 1).
    MAKE LATCH: latch.
    LET internal = CASE pfork
    OF WAIT: CASE NOT flag1 OF
                    t: (f, SUB, CMP, WAIT)
                    f: (t, SUB, CMP, FETCH)
            ESAC,
    OF RESET: CASE c AND flag2 OF
                    t: (t, ADD, CMP, WAIT)
                    f: (f, ADD, CMP, RESET)
            ESAC,
    OF FETCH: CASE flag1 OF
                    t: (t, CMP, SUB, RESET)
                    f: (f, CMP, SUB, FETCH)
            ESAC,
    ESAC.
    JOIN internal[4] → latch.
    OUTPUT (internal[1], internal[2], internal[3], latch)
END.
```

Figure 1: A symbolic high-level description example in ELLA.

the set of admissible values for the corresponding variables ${v_i}^I$ and ${v_i}^O$. The domain of the symbolic function is $S^I = \prod_{i=1}^{n} {S_i}^I$ and the range is $S^O = \mathcal{P}(\prod_{i=1}^{m} {S_i}^O)$, where $\mathcal{P}(A)$ is the powerset of A.

A *symbolic function* of n input variables and m output variables is a *one-to-one* mapping $F : S^I \rightarrow \mathcal{P}(S^O)$. The interpretation is that for every point $s_i \in S^I$ in the input space, there may be several possible symbolic output values. A *completely specified* symbolic function is when every point in the input space maps to a singleton output set. Multiple element outputs designate *symbolic don't care* conditions. A symbolic function can be specified in the truth table form. An example is shown in Figure 2.

A *symbolic network* is a multi-level representation of behavior. It can be specified by a directed acyclic graph (DAG) $G = (N, E)$, where each node $n_i \in N$ is associated with a single-output symbolic function f_i and a symbolic variable v_i. There is a directed edge e_{ij} from n_i to n_j if f_j explicitly depends on the symbolic variable v_i. The symbolic variable v_i is said to be a *fanin* of symbolic function f_j. Correspondingly, variable v_j is said to be a *fanout* of the symbolic function f_i. The *support* of a symbolic function f is the set of variables that f explicitly depends on. A special set of nodes are classified as the *primary inputs* or *primary outputs* of the symbolic network. An example of a symbolic network is shown in Figure 3. Each node in the network is a symbolic function in the tabular form described above.

s_1	s_2	f
A	B	A
A	C	$\{B,C\}$
B	A	$\{A,B\}$
C	$\{A,C\}$	$\{A,C\}$

Figure 2: Symbolic truth table example.

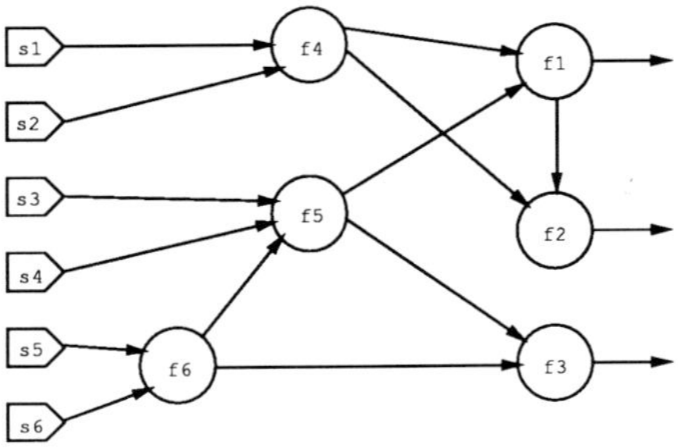

Figure 3: Symbolic network example.

3 Symbolic Don't Cares and Equivalence

One approach to symbolic synthesis is to convert the symbolic variables into binary variables via encoding and optimize each encoded node in the symbolic network with a two-level Boolean minimizer [2] prior to multi-level logic synthesis. However, this ignores the don't care conditions that can arise in a symbolic network.

Like Boolean networks, two sources of don't cares exist when optimizing a symbolic function in the context of a multi-level structure. One is related to input conditions that can never occur due to the structure of the network. A symbolic function can either be considered as a function of its immediate inputs or as a function of all primary inputs and intermediate symbolic variables. The latter is analogous to the satisfiability don't care set, but is more difficult to exploit. A more restricted form is only in terms of only the immediate inputs.

Consider the example shown in Figure 4. The corresponding truth tables are given in Figures 5, 6, and 7. The set shown on each edge corresponds to the set of admissible symbolic values for that variable. The function $F3$ is dependent on the output of $F1$ and $F2$. But since both $F1$ and $F2$ depend on the same inputs, namely $s1$ and $s2$, certain combinations of $F1$ and $F2$ can never occur. For example, the combination $\{c0, d0\}$ and $\{c1, d1\}$ will never appear

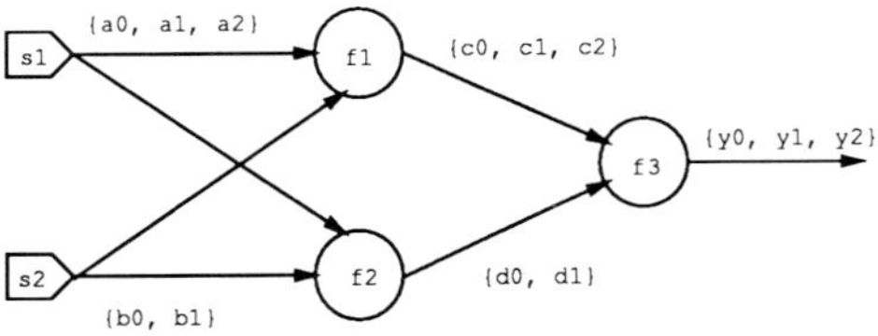

Figure 4: Symbolic don't cares and equivalence.

$a0$	$b0$	$c0$
$a0$	$b1$	$c0$
$a1$	$b0$	$c1$
$a1$	$b1$	$c1$
$a2$	$b0$	$c2$
$a2$	$b1$	$c2$

Figure 5: Truth table for F1.

$a0$	$b0$	$d1$
$a0$	$b1$	$d1$
$a1$	$b0$	$d0$
$a1$	$b1$	$d0$
$a2$	$b0$	$d1$
$a2$	$b1$	$d0$

Figure 6: Truth table for F2.

at the inputs of $F3$. This class of don't cares is related to unsatisfiable input conditions.

Naturally, a class of symbolic don't cares exists for unobservable outputs. Consider the function $F2$. When the input condition at the primary inputs, $s1$ and $s2$, is the combination $\{a2, b0\}$ or $\{a2, b1\}$, the output of $F1$ is $c2$. However, this implies that $F3$ will be $y2$ regardless what appears at $F2$. Hence, for the input conditions $\{a2, b0\}$ or $\{a2, b1\}$, $F2$ can be either $d0$ or $d1$.

Consider instead the function $F1$. Its output is fed into $F3$. But $F3$ will always give the same response whether $F1$ produces $c0$ or $c1$. Thus, $F1$ can safely produce $c0$ instead of $c1$, and vice versa. But since this don't care is not conditional on what appears at the output of $F2$, the symbolic values $c0$ and $c1$ are said to be *equivalent*. An example of a symbolic function with don't cares extracted is shown in Figure 8 for function $F1$.

External don't care conditions can be incorporated in the synthesis process. These don't cares are obtained directly from the high-level specification.

$c0$	$d0$	$y0$
$c0$	$d1$	$y1$
$c1$	$d0$	$y0$
$c1$	$d1$	$y1$
$c2$	$d0$	$y2$
$c2$	$d1$	$y2$

Figure 7: Truth table for F3.

$\{a0, a1\}$	$\{b0, b1\}$	$\{c0, c1\}$
$a2$	$\{b0, b1\}$	$c2$

Figure 8: Truth table for F1 with don't cares.

00	$\{k0, k1\}$
01	$\{k2, k3\}$
10	$k2$
11	$k3$

Figure 9: Symbolic truth table for FF.

4 Optimization Strategies

4.1 Introduction

Viable optimization strategies that can be used to exploit symbolic don't cares and equivalence are presented in this section.

4.2 Symbolic Reduction

The symbolic reduction method attempts to exploit the symbolic equivalence condition. If two symbolic values are equivalent, then they can be combined into one and thus reducing the number of admissible values in the symbolic set. This is loosely analogous to state minimization where the number of internal states in a finite state machine are reduced. Consider again the example shown in Figure 4. The symbolic set for the function $F1$ can be reduced from three elements $\{c0, c1, c2\}$ to just two elements $\{c0, c2\}$ by merging the equivalent symbolic values $c0$ and $c1$ into just $c0$.

4.3 Symbolic Encoding with Don't Cares

Symbolic don't cares can be exploited in the encoding process. The symbolic encoding problem in the usual sense considers the mapping of symbolic values into binary codes. Consider the example FF shown in Figure 9. Assume the inputs are already encoded. If the symbolic set $\{k0, k1, k2, k3\}$ is encoded with the respective binary codes $\{00, 11, 10, 01\}$, then a *Boolean relations* problem [3] would have to be solved to fully exploited the logic don't cares inherited from the symbolic don't cares, as shown in Figure 10. No efficient Boolean relations solver currently exists. However, if the binary codes are chosen carefully, a Boolean relations problem can be avoided (at least as much as possible). Consider instead the the encoding $\{00, 01, 11, 10\}$. After mapping into the binary form, the symbolic don't cares can be mapped directly to logic don't cares *without* solving a Boolean relations problem, as shown in Figure 11. Conventional logic minimizers, both two-level and multi-level, can be used to optimize the encoded logic without modification. Symbolic encoding tools like [6] can be modified to consider this cost function in conjunction into the encoding process.

4.4 Logic Minimization with Don't Cares

After encoding, the symbolic don't cares become logic don't cares. Currently, effective Boolean relations solver is still an open problem. One heuristic strategy is the select the *maximal* subset

00	{00, 11}	
01	{10, 01}	codes = {00, 11, 10, 11}
10	10	
11	01	

Figure 10: Naive encoding for FF.

00	0–	
01	1–	codes = {00, 01, 11, 10}
10	11	
11	10	

Figure 11: Don't care based encoding.

of output choices that can be represented as a single Boolean cube and perform conventional logic minimization.

5 Experimental Results

In this section, experimental results are given for the synthesis techniques described in the previous sections. The techniques have been implemented in a prototype high-level synthesis environment. All the examples presented were written in the EllaX language, our enhanced version of ELLA, and compiled automatically. In our former approach, we did not take advantage of the internal symbolic don't care conditions that may exist in the network. The symbolic network was first encoded into the binary form and then optimized using traditional logic optimization procedures, as shown in Algorithm ENCODE_ONLY in Figure 12.

Our enhanced procedure involves the use of symbolic don't cares. Internal don't care conditions are first extracted from the network. At present, no attempts were made to do this efficiently. Some examples were manually extracted. Once the symbolic don't cares are extracted, they are used in the encoding process. An enhanced version of JEDI has been implemented to account of this degree of freedom. When converting symbolic values into a binary representation, the symbolic don't cares in the symbolic description become Boolean

Algorithm ENCODE_ONLY:

1. Encode each symbolic function in the network using the program JEDI [6], where each symbolic function is essentially a symbolic truth table.
2. Once encoded, each symbolic truth table becomes a multiple output binary PLA. Each encoded PLA is then optimized using the two-level minimizer ESPRESSO [2].
3. Then the multi-level logic synthesis system MIS (version 2.1) [1] is used to optimize the entire encoded network.

Figure 12: Former Synthesis Procedure.

Algorithm ENCODE_DC:

1. Derive internal symbolic don't care information.
2. Encode each symbolic function in the network using an enhanced version of JEDI that considers symbolic don't cares. Each symbolic truth table is essentially a symbolic *relation*.
3. Once encoded, the symbolic truth table is converted into a binary truth table. Mapping the symbolic don't cares into binary form becomes a Boolean relation [4]. We approximate the Boolean relation problem by selecting a maximal subset of output choices that can be represented in a single cube and using it in the form of traditional don't cares. If an efficient Boolean relation minimizer is available, this is not necessary. Minimize the truth table with the don't cares using ESPRESSO.
4. Then the multi-level logic synthesis system MIS (version 2.1) [1] is used to optimize the entire encoded network.

Figure 13: Synthesis Procedure with Symbolic Don't Cares.

relations in the binary form. At the moment, no efficient Boolean relation minimizer is available to fully exploit the output don't care conditions carried down from the symbolic level. For our experimental results, we approximated a Boolean relation by choosing a maximal subset of output choices that can be represented as a single cube, then using a traditional two-level minimizer to exploit these don't cares. Multi-level logic synthesis is then carried out as before. This new procedure is outlined in Algorithm ENCODE_DC in Figure 13.

The results of 9 symbolic network examples are shown in Table 1. Results of the former synthesis process are shown in the first column labeled *former*. The results for the enhanced procedure are shown in the second column labeled *enhanced*. For multi-level logic optimization, MIS was used with the standard command script. In some cases, results obtained with the former approach were improved by running the standard script multiple times on the same network. This was also the case for the results obtained with the enhanced procedure although the gains were not as profound. The best of these runs are presented in the table.

Network	former	enhanced
s17a	54	11
s17b	45	12
snt1	24	14
snt10	122	46
snt7	171	39
snt8	36	16
sym1	55	54
cas2	323	134
cas3	69	38

Table 1: Synthesis Results for Symbolic Networks.

6 Conclusion

Optimization techniques based on the exploitation of symbolic don't care and equivalence conditions in high-level synthesis have been presented. They have been implemented in a prototype synthesis environment. Designs can be generated by compiling an enhanced version of ELLA called EllaX into a high-level representation called Hardware Data Flow. Once in this representation, a number of transformations, including those described here, can be invoked to successively refine the design. Results for several synthesis examples were presented to illustrate the potential gains that can be by incorporating our proposed techniques in the synthesis process.

7 Acknowledgements

This research was supported in part by Defense Advanced Research Projects Agency under contract number N00039-87-C-0182, Semiconductor Research Corporation under Contract 90-DC-008, and Digital Equipment Corporation. They are gratefully acknowledged.

References

[1] R. K. Brayton, R. Rudell, A. Sangiovanni-Vincentelli, and A. R. Wang. Mis: A multiple-level logic optimization system. *IEEE Transactions on Computer-aided design*, CAD-6(6):1062–1081, November 1987.

[2] R.K Brayton, C.T. McMullen, G.D. Hachtel, and A. Sangiovanni-Vincentelli. *Logic Minimization Algorithms for VLSI Synthesis.* Kluwer Academic Publishers, 1984.

[3] R.K. Brayton and F. Somenzi. Boolean relations and the incomplete specification of logic networks. In *Proceedings of IFIP International Conference on Very Large Scale Integration*, August 1989.

[4] R.K. Brayton and F. Somenzi. Minimization of boolean relations. In *Proceedings of the International Conference on Computer-Aided Design*, November 1989.

[5] IEEE Inc., 345 East 47th Street, New York, NY, 10017. IEEE Standard 1076-1987. *IEEE Standard VHDL Language Reference Manual*, March 1982.

[6] B. Lin and A.R. Newton. Synthesis of multiple level logic from symbolic high-level description languages. In *Proceedings of IFIP International Conference on Very Large Scale Integration*, August 1989.

[7] J.D. Morrison. Ella: Hardware description or specification. In *Proceedings of the International Conference on Computer-Aided Design*, November 1984.

[8] G. Whitcomb and A.R. Newton. Abstract data types in high-level synthesis. In *Proceedings of the Design Automation Conference*, June 1990.

Logic and Architecture Synthesis
P. Michel and G. Saucier (Editors)
Elsevier Science Publishers B.V. (North-Holland)

Scheduling of Large Signal Flow Graphs based on Metric Graph Clustering*

F. Depuydt[1], G. Goossens[1], J. van Meerbergen[2], F. Catthoor[1], H. De Man[1]

[1]IMEC Laboratory, Kapeldreef 75, B-3030 Leuven, Belgium

[2]Philips Research Laboratories, P.O. Box 80000, NL-5600 JA Eindhoven, The Netherlands

In this paper, a new strategy for scheduling large signal flow graphs is presented. The main features are the control and the balancing of register usage—by trading off time for registers—and the division of the large schedule problem in smaller schedule problems—yielding smaller design times. The register cost is controlled by constraining the operation mobilities. A hierarchical clustering process defines clusters of operations by means of a metric that is based on min-flow theory.

1 Introduction

The computer-aided synthesis of custom ICs for digital signal processors in the audio and speech application domain (clock-to-sample-rate ratio > 100) using the CATHEDRAL2ND system is based on a target architectural style, with flexible execution units (EXUs) in the data path. Each EXU has a register file at each input [1]. The local and temporary signals of a DSP algorithm are kept in *foreground* memory (register files), whereas the *background* memory (RAM) contains bulk data.

Problems with register file sizes arise when large applications (> 100 operations) are synthesised on this architecture by means of a scheduler that aims at a maximal parallelism (for a minimal execution time). For heuristics like as-soon-as-possible (ASAP) list scheduling, the register usage is quite unbalanced, because early in the schedule there are too many signals alive at the same time. Limiting the operation *mobility* (i.e. the difference between ALAP and ASAP schedule time) can solve this problem. This results in less signal parallelism in the schedule, which means less registers (at the cost of an increase in execution time). In this paper, the operation mobility is controlled by defining clusters of operations in the signal flow graph, which are scheduled locally. A *new partitioning problem* for the identification of these clusters can be formulated: finding an optimal partitioning of the signal flow graph in order to meet *a prescribed* register cost such that the register usage is *balanced.* The clusters are found by means of a hierarchical clustering algorithm, based on a fast min-flow evaluation. As macronodes, they are scheduled *sequentially*, basically trading off time for less registers.

Another problem of large signal flow graphs is the non-feasibility of CPU-intensive scheduling techniques like integer linear programming. Partitioning the large graph in smaller graphs makes these techniques feasible again for the scheduling of the subgraphs.

*This research is sponsored by the ESPRIT 2260 ("SPRITE") project of the EC.

In Section 2, an overview of some alternative approaches in the literature is given. The global scheduling strategy is outlined in Section 3. The following subtasks of the scheduling strategy are described: metric clustering (Section 4), distance calculation (Section 5), internal and macronode scheduling (Section 6). In Section 7, the results of some experiments are given. Conclusions are drawn in Section 8.

2 Alternative approaches in literature

There are several approaches to the register minimization problem in the literature. In general, most of the existing systems can be classified in two groups : systems with a *global* (HAL, HYPER, ALPS) and systems with a *local* register minimization paradigm (SPAID, CATHEDRAL-II, EASY, REAL). A global minimization system integrates somehow a register cost in the hardware cost function steering the scheduler. In HAL [12], a storage distribution graph contributes in the force calculations. The HYPER system [2] uses global hardware utilization ratios, including registers. The scheduling process is organized as a min-max search through the design space. The ALPS system [10] uses integer linear programming scheduling, where the sum of the lifetimes of the signals contributes to the minimizing function.

On the contrary, systems like SPAID [6], CATHEDRAL-II [11], EASY [13] and REAL [9] deal with register minimization locally (e.g. for each register file) in a *post-scheduling* phase, in an attempt to reduce the complexity of the algorithms. Mainly graph colouring or clique partitioning methods are used here. In these systems however, there are no means of imposing an upper bound on register cost during scheduling.

Since for architectures with local memory in the vicinity of the functional units, register minimization is important, and a global optimization during scheduling for the whole graph seems less attractive, we adopted the strategy of a *pre-scheduling clustering phase*, as the best of both worlds.

3 The global scheduling strategy

The global scheduling strategy will be outlined first (fig. 1). Basically, a scheduling iteration consists of three phases: clustering as a preprocessing phase, internal scheduling of each cluster and macronode scheduling (the clusters being the macronodes). For the internal scheduling, different schedulers can be hooked up onto the system (e.g. list scheduler, force directed scheduler, ILP scheduler). The results are evaluated by the user, who can interact with the clustering and with the internal scheduling by using another scheduler or by changing the scheduling constraints. A schedule can for instance be improved by initially taking the number of cycles from the list schedule as cycle count for the ILP scheduler, and iteratively increasing the cycle count.

Initially, all macronodes are scheduled sequentially. In this way, basically, time is traded off against registers. In a post-processing step however, small overlaps between macronodes are introduced to shorten the schedule without affecting the register cost.

4 Metric graph clustering

The goal of the clustering step is the definition of *tightly connected groups of operations*, limiting the operation mobilities and the corresponding signal lifetimes. Also, the global register usage should be balanced over the different clusters.

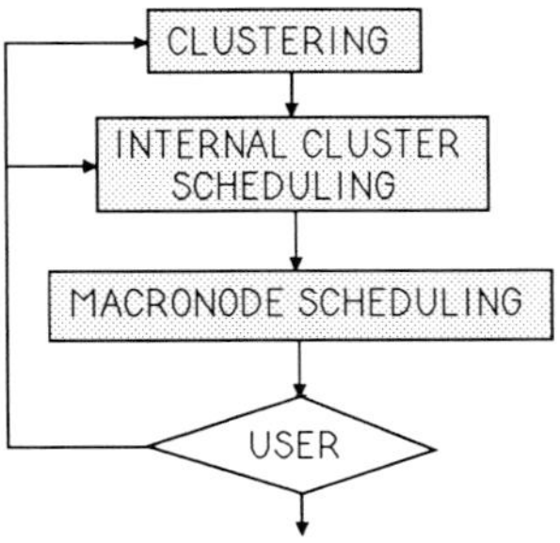

Figure 1: The global scheduling strategy

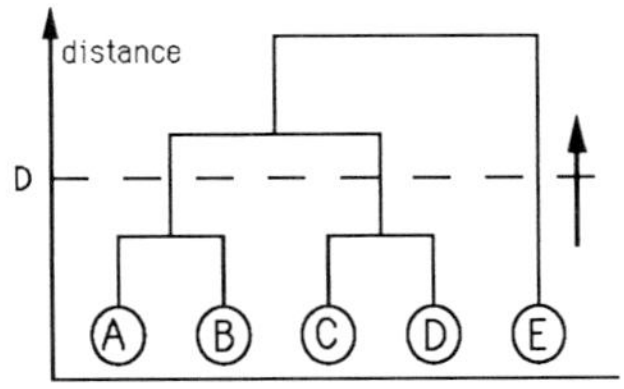

Figure 2: Hierarchical clustering scheme

4.1 Basics

Clusters in the signal flow graph are obtained in a constructive way by making use of a *hierarchical clustering algorithm* [8]. Based on a metric (distance) between adjacent nodes or clusters, smaller clusters can be merged into larger ones. In fig. 2, initially there are five nodes — A to E. Assume that a distance can be computed between the adjacent nodes, say A and B, and C and D. The value of the distance is shown as the height of the joins A-B and C-D. The distance between the non-adjacent nodes (e.g. D and E) is assumed to be infinite. At each step of the process, the nodes (clusters) at the smallest distance are merged. For the resulting new cluster configuration, the distances are recalculated. For instance, during the first step in fig. 2, nodes A and B and nodes C and D are merged. During the next step, clusters A-B and C-D are merged.

In Section 4.3 a control strategy will be introduced to decide on the level up to which the hierarchical clustering algorithm is applied.

4.2 Definition of the metric

If the distance between two clusters (or nodes) is defined as follows:

Definition 1 (Distance) *The* distance *between two adjacent clusters in a signal flow graph is the maximal global foreground register cost for the union of these two clusters.*

then there is a direct and clear relation between the metric and register cost. This distance equals the sum of all registers to store the local signals in the union of both clusters plus the registers to store the global signals that are alive at the same time. Since we want to apply the clustering algorithm without having to schedule the graph, we will rely on an approximative distance calculation, which will be given in Section 5.2.

4.3 Motivation and consequences

Since it is assumed that the clusters are scheduled sequentially, the operations in a cluster have a limited mobility and the global signal parallelism is reduced. Gradually merging clusters together increases the potential signal parallelism. Two important motivations for Definition 1 are given.

In the first place, the *minimum* distance at a certain level of the hierarchical clustering equals—in case of an exact distance calculation—the global register cost at the next level of the clustering (i.e. after merging the clusters at that minimum distance).[1] Therefore, if at a certain moment during the clustering process the minimum distance exceeds a certain distance corresponding to a user-defined upper bound on register cost (e.g. D in fig. 2), the clustering is stopped. In other words, the global register cost is controlled during clustering.

Second, because a new cluster configuration is obtained by merging clusters at the same distance from each other, the register usage in the global schedule will be more or less *balanced* over all clusters.

4.4 Initialization of the clustering

The first clusters in the hierarchical process appear where maximal *global* signal parallelism occurs. The *minimal flow* through the graph (Section 5) identifies a maximal cut[2]. All cut edges represent signals that *can* be alive at the same time. The producing and consuming operations of each such edge are grouped in a cluster. Since clusters are scheduled sequentially, this yields a maximal degradation of signal parallelism. Therefore, in the first step of the clustering algorithm the distances between two adjacent nodes is set to the flow between these nodes. In this way the clustering is initiated at the places of global minimal flow.

4.5 An example

In fig. 3, some steps of the clustering of the signal flow graph of a syndrome generator algorithm are shown. The algorithmic description in the Silage input language [7] is given below:

```
func main(x1,x2,x3:num)syn1,syn2,syn3:num=
begin
   syn1 = x1 + x2 + x3;
   syn2 = beta*x1 + beta*x2 + beta*x3;
   syn3 = gamma*x1 + gamma*x2 + gamma*x3;
end;
```

The maximal register costs are more or less the same for each cluster at each level, as was intended. Moreover, the clusters at level 9 coincide with the computational kernels of the algorithm. These kernels could be scheduled in parallel or in an interleaved way, but now they are identified and executed sequentially. Note that not all steps of the clustering are shown in fig. 3.

[1]It can be shown that this is true if the clustering is started where the potential register cost is largest [3].

[2]Calculating the minimal flow is sufficient, since the maximal cut and the minimal flow are dual.

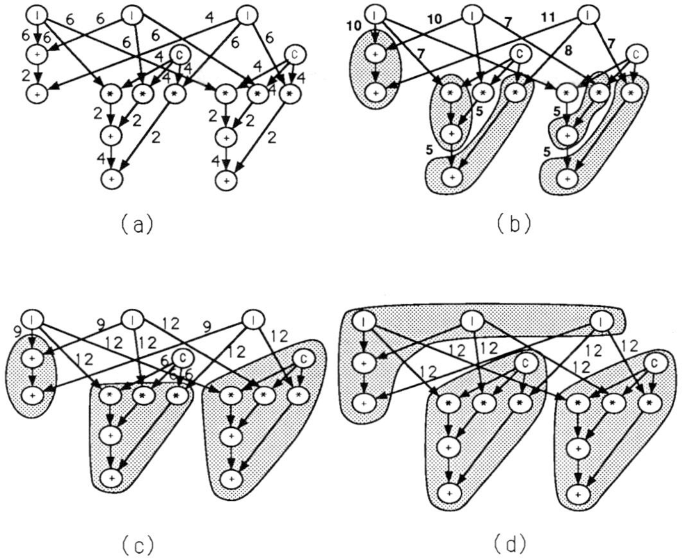

Figure 3: Clustering of syndrome generator. The numbers along the edges are distances. (a) Initial distances. (b) After clustering at $d = 2$. (c) After clustering at $d = 5$. (d) After clustering at $d = 9$.

5 Min-flow and distance calculation

The basis for all register cost and distance calculations is the *minimal flow* principle, which is explained in subsection 5.1. The results are used to compute the distances for the hierarchical clustering process, as explained in subsection 5.2.

5.1 The maximal register cost based on min-flow

A signal flow graph corresponding to a DSP algorithm can be translated in an acyclic network by adding a source node S and a sink node T and discarding state signals (fig. 4(a)). Only unconditional and non-hierarchic graphs are considered for the time being. Nodes without input edges are connected to S through an artificial edge. All nodes without output edges are connected to T. Artificial edges do not have a signal equivalent. According to network theory [4], a *flow* through a network is a set of integer values (flow values) attached to the edges of the graph such that the sum of the incoming flow values equals the sum of the outgoing flow values in each node[3]. The *total flow value* is the value of the flow leaving the source (or entering the sink).

Definition 2 *A* minimal *flow is a flow with minimal total flow value such that the flow value for each edge is at least* 1.

If the following general architectural assumptions are made: parallel signals (i.e. signals for which the corresponding edges do not lie on the same $S \longrightarrow T$ path) require separate

[3]In considering flows through graphs, a current law—similar to Kirchoff's Current Law—is always obeyed.

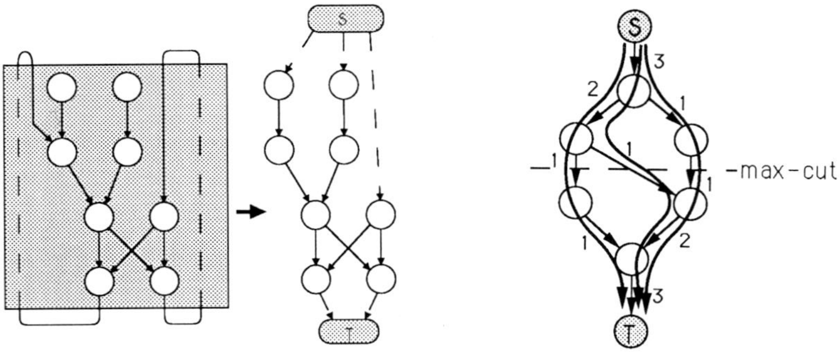

Figure 4: (a) Projection of a signal flow graph onto a network. (b) Minimal flow. The numbers along the edges are flows.

registers, and consecutive signals(i.e. signals lying on the same $S \longrightarrow T$ path) can be stored in the same register. Then the following theorem holds:

Theorem 1 (Min-flow max-cut) *If in a network $N(E,V)$, corresponding to a signal flow graph, a minimal flow F is constructed, then:*

- *The maximum number of registers NR_{max} that any possible schedule of N can require is:*
$$NR_{max} = |F| \tag{1}$$
where $|F|$ represents the total flow value of F.
- *A cut through $|F|$ edges of E identifies a set of signals that can be alive simultaneously.*

The proof of this theorem can be found in [3]. In fig. 4(b), the minimal flow F with total flow value $|F| = 3$ is decomposed in three elementary flows. The flow values are indicated along the edges. Since F is minimal, F the elementary flows are disjoint in at least one edge. These edges correspond to concurrent signals and may need physically separate storage. As stated in the theorem, a cut through 3 edges (e.g. max-cut in fig. 4(b)) indicates maximal signal parallelism, and a schedule of this graph can require at most 3 registers. The same cut identifies three initial clusters (sharing some common nodes in this case) in the initialization phase of the clustering process.

Once the signal flow graph is converted in a network, its minimal flow can be constructed by augmenting the total flow through the network through adding elementary flows until saturation occurs, i.e. until all edges have a flow value $f(e) \geq 1$. To assure the minimality of the flow, all *reducible paths* have to be traced and reduced.

Definition 3 (Reducible paths) *A* reducible path *ρ is a directed $S \rightarrow T$ path in a network $N(E,V)$ such that :*
$$\forall e \in \rho : f(e) > 1 \tag{2}$$
where $f(e)$ is the flow value of an edge of the network.

A reducible path can be reduced by substracting an amount of flow ΔF from the flow in ρ:
$$\Delta F = \min_{e \in \rho} f(e) - 1 \tag{3}$$

As a result, at least one edge on the path has $f(e) = 1$. The reducible path in fig. 5 can be reduced with one unit of flow. The algorithm to construct the minimal flow through a network is given below:

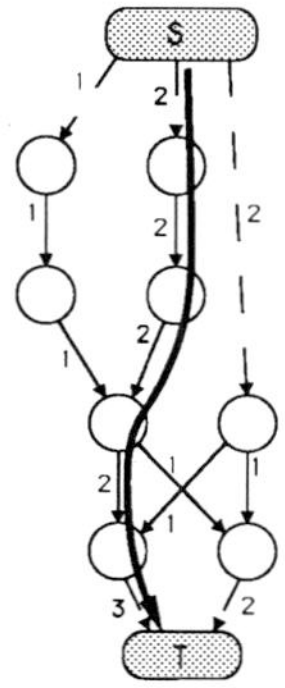

Figure 5: A reducible path. The numbers along the edges are flows.

Algorithm 1 (Min-flow)

1. *$\mathcal{E}$ = set of all edges.*
2. *While $\mathcal{E}$ is not empty {*
   (a) *Take an edge $e \in \mathcal{E}$.*
   (b) *Trace a path $S \rightarrow e$, remove each encountered edge from set $\mathcal{E}$ and increase the flow through that edge.*
   (c) *Trace a path $e \rightarrow T$, and perform the same tasks as in (b).*

   }
3. *Look for reducible paths and reduce them.*
4. *$|F|$ is the total flow leaving the source S.*

Because only acyclic networks are considered here, the Dinic algorithm from T to S [4] to reduce the reducible paths (which is $O(|V|^3)$) can be left out of consideration. Clearly, Algorithm 1 has a $O(|V|.|E|)$ complexity.

An upper bound on register cost is given in Theorem 1. However, a *lower bound* on register cost can also be computed by counting simultaneous overlaps of ASAP and ALAP lifetimes of signals, as indicated by Paulin [12]. If the number of register files at the inputs of functional units is larger, then the lower bound is put to this number.

The previous theory can be extended for architectures where the assumptions of Theorem 1 do not hold.

5.2 Distance calculation

Fig. 6 shows a generalized configuration of two clusters, C_i and C_j, with a number of edges e_{ij} between them. e_{ik} $(1 \leq k \leq n)$ is the number of ingoing edges into a destination cluster of an outgoing edge of C_i. Similarly, e_{jk} $(1 \leq k \leq m)$ is the number of outgoing edges of a source cluster of an ingoing edge of C_j. As macronodes, clusters are scheduled sequentially.

According to Definition 1, the distance d between two clusters has to reflect the global register cost for their union. In this Section an approximative distance function is introduced which can be computed without scheduling the graph. It contains contributions of the individual minimal flows of the two clusters (maximal register cost, see Theorem 1), the edges between the clusters and edges corresponding to signals produced

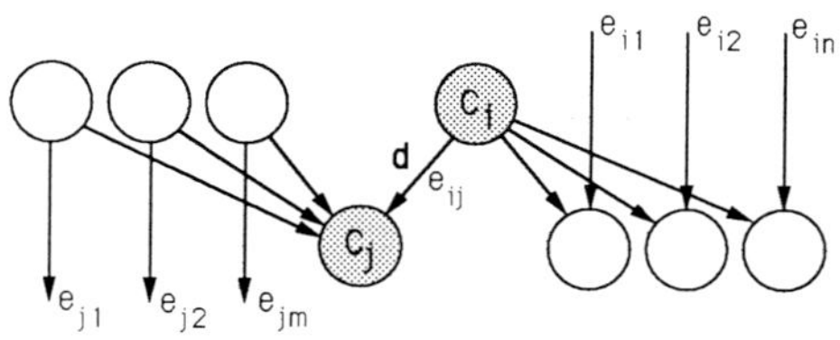

Figure 6: Definition of distance between two clusters

by operations with a precedence to one of the clusters and consumed by operations with a precedence from one of the clusters (bridging edges), and is approximated by:

$$d = \begin{cases} F(e) \text{ (at first clustering level)} \\ |F_i| + |F_j| - e_{ij} + \sum_{k=1}^{n} e_{ik} + \sum_{k=1}^{m} e_{jk} \end{cases} \tag{4}$$

In fact, this formula is a sum of two factors, more or less outweighing each other: a maximum register cost $|F_i| + |F_j| - e_{ij}$ and a minimum cost for the extra effects of bridging edges. Since tracing all bridging edges involves complex graph searching, only a minimal contribution is traced with a depth first search. As shown in the examples (Section 7), approximation (4) is satisfactory.

6 Internal and macronode scheduling

After reducing the parallelism by clustering the signal flow graph, the obtained clusters are separately scheduled. If a cluster is small enough, and if proven to be worth while, globally optimal scheduling of the cluster can be done with *integer linear programming* (ILP) techniques [10]. In other cases, heuristic schedulers [5][12] can be used.

After substituting the scheduled clusters by macronodes in the graph, the macronode graph has to be scheduled on the same hardware as well. Since the number of macronodes is small, this task is not a complex one. However, as a post-scheduling phase, the separate schedules can be partially overlapped, to reduce the total number of cycles without changing the individual schedules. In this way a better functional unit occupation is obtained while the register cost is maintained. In general, N clusters have to be sorted such that the sum of all intercluster overlaps is maximized and the register cost stays the same.

Because the problem of scheduling a large graph is transformed in the problem of scheduling several smaller graphs, a substantial decrease in CPU time can be expected. This favors *user interaction.* On the other hand, the user might want to perform *optimal* scheduling (e.g. using ILP techniques) for the separate clusters, since these techniques are not feasible for the total graph.

7 Examples

As a first example, the previously described syndrome generator was scheduled with the ATOMICS list scheduler [5] on one adder and one multiplier. Constants are in ROM and inputs are in RAM. Without clustering the schedule requires 8 cycles and 8 registers. With clustering up to e.g. a distance $D = 6$, aiming at a global register cost of 6, and without schedule overlapping, the schedule requires 14 cycles and 6 registers. The number of cycles is reduced to 10 by schedule overlapping (fig. 7).

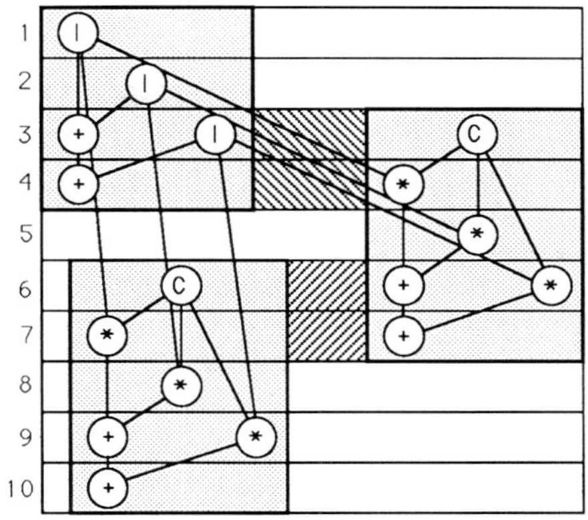

Figure 7: Schedule of syndrome generator. The shaded areas are overlaps.

QMF1	D=6	D=10	D=12	D=15	D=17	ATOMICS
clusters	14	7	5	4	3	1
cycles	52	43	37	37	33	-
jamming	15	12	12	9	6	-
cycles	37	31	25	28	27	23
registers	11	10	10	13	13	19

QMF2	D=13	D=16	ATOMICS
clusters	7	5	1
cycles	47	41	-
jamming	9	7	-
cycles	39	34	31
registers	14	14	21

Figure 8: Results of clustering and scheduling.

The second and third example are 2 resp. 3 times unrolled loop body of a quadrature mirror filter (QMF). These examples contain 62 and 105 operations respectively. The rightmost column of the tables below give the ATOMICS list scheduling results. These results are compared with the results of a clustering up to different distances, aiming at different register costs. Also, the number of clusters, the number of cycles without overlapping and the number of cycles gained by overlapping are given. The clusters were scheduled manually (list scheduling) or scheduled with ILP, taking the cycle count of the list scheduling as a parameter. The results are summarized in fig. 8. These results show that a register cost vs. time trade-off is possible. In the QMF1 example, a register cost reduction of almost 50% can be obtained at the expense of only 10% cycle time increase. For example QMF2, 30% of the registers can be traded off for 10% more cycles. Because the register cost is upper and lower bounded, the lowest and highest distance values deviate strongly from the actual register cost. The CPU-times to do a complete clustering, i.e. up to one cluster—the original graph—were measured on a DECstation 3100: 17 sec. for QMF1 and 32 sec. for QMF2. The scheduler is being implemented in C++ and is called TRON.

8 Conclusion

A new scheduling strategy with emphasis on register cost control is proposed. It consists of three steps: clustering, internal scheduling of clusters, and macronode scheduling. Tightly connected groups of operations with a balanced register usage are identified and scheduled in the first and second steps, and they are scheduled as macronodes in the third step. Examples point out that the register cost can indeed be controlled, at the expense of time. The metric clustering uses global information on lifetime parallelism, but is fast as well. This makes it suitable for interaction with the designer. As a consequence of the clustering, a faster scheduling of more but smaller clusters can be done. Future work will be done on hierarchic and conditional signal flow graphs.

Acknowledgements

The authors are grateful to all colleagues at IMEC's VSDM division for their contributions to the presented work. In particular, significant inputs are recognized from Gert Goossens and Jef van Meerbergen.

References

[1] F. Catthoor, J. Rabaey, G. Goossens, J. L. V. Meerbergen, R. Jain, H. D. Man, and J. Vandewalle. Architectural strategies for an application-specific synchronous multiprocessor environment. *IEEE Transactions on Acoustics, Speech and Signal Processing*, 36(2):265–283, February 1988.

[2] C. Chu, M. Potkonjak, M. Thaler, and J. Rabaey. Hyper: an interactive synthesis environment for high performance real time applications. In *Proceedings of ICCD*, pages 432–435, 1989.

[3] F. Depuydt. *Scheduling of Large Signal Flow Graphs based on Metric Graph Clustering.* SPRITE Report, IMEC Laboratory, Leuven, Belgium, 1989.

[4] S. Even. *Graph Algorithms.* Computer Science Press, 1979.

[5] G. Goossens. *Optimization techniques for automated synthesis of application-specific signal processing architectures.* PhD thesis, Katholieke Universiteit Leuven, June 1989.

[6] B. S. Haroun and M. I. Elmasry. Automatic synthesis of a multi-bus architecture for dsp. In *Proceedings of the ICCAD88*, pages 44–47, November 1988.

[7] P. N. Hilfinger. *Silage: A Language for Signal Processing.* Technical Report, University of California, Berkeley, August 1985.

[8] S. C. Johnson. Hierarchical clustering schemes. *Psychometrika*, 32(3):241–254, September 1967.

[9] F. J. Kurdahi. *Area Estimation of VLSI Circuits.* PhD thesis, UCLA, August 1987.

[10] J. Lee, Y. Hsu, and Y. Lin. A new integer linear programming formulation for the scheduling problem in data path synthesis. In *Proceedings of Papers of the ICCAD89*, pages 20–23, Dept. Computer Science, Tsing Hua University, November 1989.

[11] H. D. Man, J. Rabaey, J. Vanhoof, G. Goossens, P. Six, and L. Claesen. Cathedral-ii: a computer-aided synthesis system for digital signal processing vlsi systems. *Computer-Aided Engineering Journal*, 55–66, April 1988.

[12] P. G. Paulin and J. P. Knight. Force-directed scheduling for the behavioral synthesis of asics. *IEEE Transactions on Computer-Aided Design*, 8(6):661–679, June 1989.

[13] L. Stok and R. van den Born. Easy: multiprocessor architecture optimisation. In *Proc. Int. Workshop on Logic and Architecture Synthesis for Silicon Compilers, Grenoble*, 1988.

Logic and Architecture Synthesis
P. Michel and G. Saucier (Editors)
Elsevier Science Publishers B.V. (North-Holland)

Profit-Loss-Gain Algorithm for Data-Path Synthesis *

Vites Techangam †

Computer Engineering Department
Chulalongkorn University
Bangkok 10330, THAILAND.

Anucha Pitaksanonkul

Intergraph Corporation
One Madison Industrial Park
Huntsville, AL 35807-4201, U.S.A.

Chidchanok Lursinsap

The Center for Advanced Computer Studies
The University of Southwestern Louisiana
Lafayette, LA 70504, U.S.A.

Abstract

One of the major problems in the data path synthesis is the resource allocation problem. Tseng and Siewiorek have formalized this problem into clique partitioning problem. They proposed a GCP algorithm to partition nodes in the weighted compatible graph into cliques with maximum profit gained. In this paper, we propose a cluster partitioning algorithm, PLG. The PLG algorithm maintains only a single compatible graph instead of multiple subgraphs maintained by the GCP algorithm. Since PLG algorithm considers both the profit of the edge and the profit of the neighbor edges during the process, the PLG designs have area 5-10 percent less than the GCP deigns, as shown in the experiments. We also show how this algorithm can be applied to the problem of functional unit allocation.

1 Introduction

During the past decade, significant effort has been devoted to the development of the automated data-path synthesis from high-level description [1]- [10]. One of the major tasks in the automated data-path synthesis is the resource allocation which deals

*This works was partially supported by a grant from the Board of Regents, State of Louisiana, LEQ SF (1987-1989)-RD-A-29

†Supported by a grant from the Ananda Mahidol Foundation, Bangkok, THAILAND

with the allocation of registers, functional units and interconnection units. A particularly powerful technique for the resource allocation has been proposed by Tseng and Siewiorek [9, 10] by unifying the resource allocation problems into the clique partitioning problem. They introduced a *Clique Partitioning* (CP) algorithm to partition the nodes of a *lifetime compatible graph* into a minimum number of disjoint cliques to obain the minimal hardware cost.

Frequently, the profit of grouping some set of nodes into a cluster is greater than the profit of grouping another set of nodes into a cluster. Thus, Tseng and Siewiorek also proposed the *General Clique Partitioning* (GCP) algorithm based on two notions: the notion of divide-and-conquer and the notion of the transitive property. The GCP algorithm classifies the edges in the *lifetime compatible* graph into several categories according to the profit measure of grouping each pair of connected nodes. Then the GCP algorithm partitions the compatible graph into several subgraphs, whose edges belong to the same category. The algorithm reduces the subgraph with the best profit before it successively reduces the lesser profit subgraphs.

However, by maintaining a separate graph for each category, the GCP cannot practically handle general profit measure of grouping each pair of connected nodes in the compatible graph. Furthermore, the notion of divide-and-conquer used in the GCP algorithm also limits the algorithm from obtaining the global optimal design.

In this paper, we propose a more general algorithm, *Profit-Loss-Gain* (PLG) algorithm, which can handle the general profit measure of grouping each connected node pair. Furthermore, the notion of divide-and-conquer is eliminated from the PLG algorithm. Thus, our PLG algorithm yields the same or better results than the GCP algorithm. We will illustrate the technique by focusing on the problem of functional unit allocation. However, the PLG can also be applied to the register and interconnection unit allocation.

2 Profit-Loss-Gain Algorithm

For the functional unit allocation, an important issue is to group an appropriate set of functional units to form an ALU. Two functional units can be grouped together if they are not used at the same time, which is called *lifetime compatible*. Thus, as long as there are lifetime compatible functional units, the profit can be gained by merging those functional units into an ALU. However, the benefit of grouping some set of functional units are greater then the benefit of grouping another set of functional units. Unfortunately, after two functional units are grouped into an ALU, some set of functional units are excluded from the later grouping. Hence, it is important to select those groupings that will result in the most profit in the final design.

Before describing the PLG algorithm, we must define the weighted graph G as follows:

Definition 1 *G is a* **weighted graph** *$\langle V, E\rangle$, where V is a set of vertices and E is set of weighted edges, $E = \{(v_1, v_2, w) | v_1, v_2 \in V$ and w is a weight $\}$.*

From a scheduled code sequence, a weighted graph G can be built. Let the nodes in the graph represent the functional units. There is an edge connected nodes i and j if

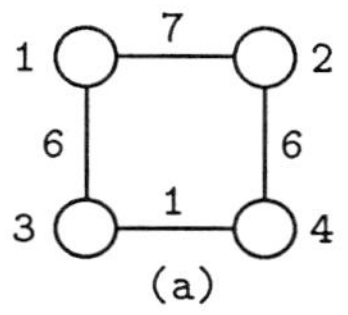

Edge	Net Gain
(1,2)	-5
(1,3)	-2
(2,4)	-2
(3,4)	-11

(b)

Figure 1: The simple compatible graph and the edge net gain.

functional units i and j are not used at the same time. The edge (i, j) is also assigned with a weight w_{ij} when grouping functional unit i and j would save w_{ij} units of area. We call the weighted graph, G, *weighted compatible* graph.

To further describe the PLG algorithm, we must define some additional terms:

Definition 2 *A node k is called a* `neighbor` *of an edge (i, j), if k is connected to nodes i or j or both in the weighted compatible graph G.*

Definition 3 *A node k is called a* `common neighbor` *of an edge (i, j), if k is connected to both nodes i and j in the weighted compatible graph G.*

Definition 4 *When nodes i and j are combined to form a composite node $\langle i, j\rangle$, a* `deleted edge` *of a composite node $\langle i, j\rangle$ occurs in either of the following conditions:*

- *The edge (i, j) itself must be deleted.*
- *Let Node k be a neighbor of edge (i, j). Once node i and j are discarded, the edge (i, k) or (j, k) or both have to be deleted.*

Definition 5 *When nodes i and j are combined to form a composite node $\langle i, j\rangle$, an* `added edge` *$(k, \langle i, j\rangle)$ of a composite node $\langle i, j\rangle$ is added to graph G if a node k is a common neighbor of edge (i, j).*

From the compatible graph shown in Figure 1(a), it is obvious that reducing the subgraph with best profit first would not generate optimal design. The result partition would be $\{(1,2),(3,4)\}$ which saves 8 units of area $(7+1)$ instead of $\{(1,3),(2,4)\}$ which saves 12 units of area $(6+6)$. Furthermore, using the number of common neighbors and the number of deleted edge as the secondary priority is not enough to obtain the optimal design since each edge also has different weight. Thus, we also define the following terms:

Definition 6 *The* `potential weight loss`, *l_{ij}, of an edge (i, j) is the sum of the weights of the deleted edges of a composite node $\langle i, j\rangle$ except the weight of the edge (i, j) itself.*

Definition 7 *The* `potential weight gain`, *g_{ij}, of an edge (i, j) is the sum of the weights of the added edges of a composite node $\langle i, j\rangle$.*

$v_3 = v_1 +_1 v_2$
$v_5 = v_4 -_2 v_3; \quad v_7 = v_3 +_3 v_6$
$v_9 = v_5 *_4 v_8; \quad v_{10} = v_7 +_5 v_5$

Table 1: Example code sequence

Function Unit	Area
Adder	10
Subtractor	10
ALU (+,-)	14
Multiplier	60

Table 2: Function unit relative size

Definition 8 *The* **net gain**, G_{ij}, *of an edge* (i, j) *is*

$$G_{ij} = w_{ij} + g_{ij} - l_{ij}.$$

Given a weighted compatible graph G, we apply the following algorithm to partition the graph G into clusters.

The PLG Algorithm
While there is an edge in the graph G **do**

- For each edge (i, j), compute the net gain G_{ij}.
- Select the edge (m, n) which has the *most net gain*.
- Combine the nodes m and n into a composite node $\langle m, n \rangle$.
- Delete all the deleted edges of the composite node $\langle m, n \rangle$.
- Add all the added edges of the composite node $\langle m, n \rangle$ and also assign weight to these added edges.

end while.

The PLG algorithm partitions the compatible graph in Figure 1 into 2 clusters: $(1, 3)$ and $(2, 4)$.

3 An Illustrative Example

To illustrate the PLG algorithm, we will use a simple example as shown in Table 1. Table 2 shows the relative size of the function units used in the code sequence.

From the given code sequence, we obtain the compatible graph as shown in Figure 2(a). Let edge (i, j) be labeled with (w_{ij}, G_{ij}). Since there is no function unit which can perform both addition and multiplication or both subtraction and multiplication, the edges $(1, 4), (2, 4)$, and $(3, 4)$ are discarded. From Figure 2(a), there are three edges which have the most net-gain: $(1, 3), (1, 5)$, and $(3, 6)$. The algorithm arbitrarily chooses edge $(1, 3)$ and combines the node 1 and 3 into a composite node $\langle 1, 3 \rangle$. Consequently, we would obtain the reduced compatible graph with new (w_{ij}, G_{ij}) shown in Figure 2(b). Then, the algorithm chooses the edge $(\langle 1, 3 \rangle, 5)$ and combines the node $\langle 1, 3 \rangle$ and 5 into a composite node $\langle 1, 3, 5 \rangle$. Hence, the algorithm partitions the compatible graph into 3 clusters: $\{(1, 3, 5), (2), (4)\}$. The code sequence is allocated with one adder, one subtractor, and one multiplier.

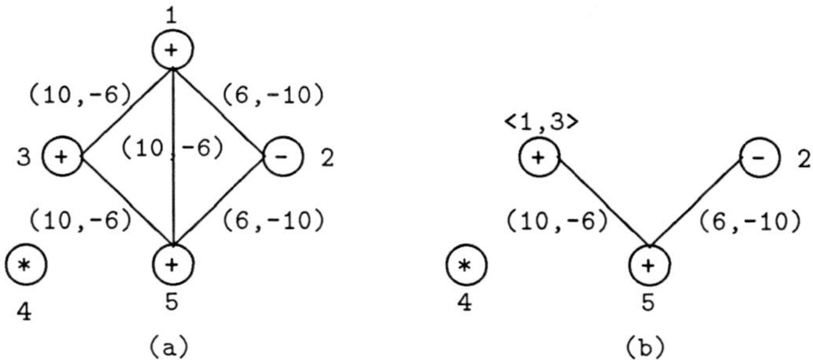

Figure 2: The compatible graph(a) and its reduced graph(b).

4 Experiments

The GCP and PLG algorithms were applied to several randomly generated examples on VAX 11/750. These examples have different number of nodes and edges. Some graphs have weights ranging from 1 to 30 and some have weights ranging from 1 to 15.

Table 3 shows the number of clusters and area saved for each examples. From this table, PLG produces the same or a fewer number of clusters than GCP. In most cases, PLG saved more area than GCP. GCP saves a little more area than PLG in only two cases. Therefore, PLG trends to produce better results than GCP.

# of Nodes	*# of edges*	*GCP*		*PLG*	
		# of clusters	*area saved*	*# of clusters*	*area saved*
10†	30	5	130	4	144
10‡	30	4	74	3	78
10†	40	2	189	2	196
10‡	40	3	85	2	91
20†	60	10	203	8	215
20‡	60	9	118	8	117
30†	120	14	355	11	383
30‡	120	14	188	12	184

Table 3: The results obtained from GCP and PLG algorithms

†weights ranging from 1 to 30
‡weights ranging from 1 to 15

5 Conclusion

We have presented the cluster partitioning algorithm, PLG, which is more general than the previously proposed GCP algorithm. The PLG algorithm maintains only one compatible graph instead of multiple subgraphs maintained by the GCP algorithm. Since the PLG algorithm considers both the weight of the edge and weight of the neighbor edges in the selection process, it trends to obtain the better result than the GCP algorithm. Even though, we only apply the PLG algorithm for functional unit allocation, it is applicable to general resource allocation in automatic data path synthesis.

References

[1] R. L. Blackburn, D. E. Thomas, and P. M. Koenig. CORAL II: Linking Behavior and Structure in an IC Design System. In *Proceeding of the 25th IEEE/ACM Design Automation Conference*, pages 529–35, 1988.

[2] E. F. Girczyc and J. P. Knight. An ADA to Standard Cell Hardware Compiler Based on Graph Grammars and Scheduling. In *Proceedings of the IEEE International Conference on Computer Design*, pages 726–31, October 1984.

[3] T. J. Kowalski, D. J. Geiger, W. H. Wolf, and W. Fichtner. The VLSI Design Automation Assistant: from Algorithms to Silicon. *IEEE Design and Test*, August 1985.

[4] G. D. Micheli and D. C. Ku. HERCULES – A System for High-level Synthesis. In *Proceeding of the 25th IEEE/ACM Design Automation Conference*, pages 483–8, 1988.

[5] B. M. Pangrle and D. D. Gajski. Slicer: A state synthesis for intelligent silicon compilation. In *Proceedings of the IEEE International Conference on Computer Design*, pages 42–5, October 1987.

[6] A. C. Parker, G. T. Pizarro, and M. Mlinar. MAHA: A Program for Datapath Synthesis. In *Proceeding of the 23th IEEE/ACM Design Automation Conference*, pages 461–66, July 1986.

[7] P. G. Paulin and J. P. Knight. Force-Directed Scheduling for the Behavioral Synthesis of ASIC's. *IEEE Transactions on Computer-Aided Design of Integrated Circuits*, CAD-8(6):661–79, June 1989.

[8] H. Trickey. Flamel: A high-level hardware compiler. *IEEE Transactions on Computer-Aided Design of Integrated Circuits*, CAD-6(2):259–69, March 1987.

[9] C. J. Tseng and D. P. Siewiorek. Facet: A Procedure for the Automated of Digital Systems. In *Proceeding of the 20th IEEE/ACM Design Automation Conference*, pages 490–6, 1983.

[10] C. J. Tseng and D. P. Siewiorek. Automated Synthesis of Data Paths in Digital Systems. *IEEE Transactions on Computer-Aided Design of Integrated Circuits*, CAD-5(3):379–95, July 1986.

Logic and Architecture Synthesis
P. Michel and G. Saucier (Editors)
Elsevier Science Publishers B.V. (North-Holland)

A Novel Scheduling/Allocation Approach for Datapath Synthesis based on Genetic Paradigms [1]

N.Wehn M.Held M.Glesner

Technical University of Darmstadt
Institute for Microelectronic Systems
Schlossgartenstrasse 8
D-6100 Darmstadt

Abstract

In this paper we present a new scheduling/allocation approach for the structural synthesis of synchronous datapaths. Compared to existing techniques the new approach performs a global simultaneous design space exploration of time and operation parallelism. Beside the minimization of time steps and multicycle (pipelined) functional units, the interconnection complexity is modelled in an appropriate way. To deal with the completeness of the corresponding optimization problem genetic paradigms are used. Benchmarks from the literature were chosen to illustrate the advantage of our approach and to compare them with existing techniques.

1 Introduction

Due to the increasing complexity of VLSI circuits and IC-designer shift away from device-level considerations towards to architectural ones (meet in the middle strategy), high-level synthesis becomes more and more popular. On the field of physical design a lot of good systems competitive to designers results are available which transform a structural description into silicon, whereas high-level synthesis is still an unsolved research topic.

High level-synthesis [1] consists of a number of steps, which transform successively a behavioural circuit description down to a structural (register-transfer) one. These transformations can be divided into two tasks: *behavioural synthesis* performs optimizations similar to that of language compilers like loop unrolling, meta-variable expansion, flattern of block hierarchy, variable and constant unfolding etc. Output is an intermediate representation like dataflow graphs - abbreviated in the following as DFG - or parse trees which contain the informations on used variables, operations and data dependencies; *structural synthesis* transforms the intermediate form into a structural representation performing design trade-offs between speed,

[1]This work was partly sponsored by the ESPRIT Basic Research Action 3281 (ASCIS) of the E.C.

area and power. Subtasks to be solved while structural synthesis are scheduling, allocation, module binding and partitioning.

In this paper we will present a new approach that solves one of the main problems in structural synthesis, namely the scheduling/allocation problem [1].

2 The Scheduling/Allocation Problem

In the context of high-level synthesis, *scheduling* (*time* domain in the design space) is the assignment of operations to time slots (control steps) taking into account data dependencies of a *DFG*. It is obvious that the assignment of operations to control steps fixes the operation parallelism. Since all operations within one time slot must be executed in parallel - excepted they are mutually exclusive - at least one functional unit must be provided for each operation in this control step. As a consequence, scheduling interrelates strongly with *allocation* (*space* domain in the design space), which maps operations and their variables onto hardware resources like functional units, registers, busses and multiplexers. *Module binding* decides how each component of the datapath is to be implemented. This paper is emphasized on the scheduling/allocation problem.

To reduce the problem complexity due to the interrelation of the *time* and *space domain* during scheduling and allocation, many systems restrict one domain while optimizing the other one. Typical examples are ASAP [2,3,4], LIST [5,6,7] and FDLS [8,9] scheduling techniques where hardware constraints are specified and the algorithm attempts to minimize the total execution time taking into account special priorities when resource conflicts occur. On the other hand the well known force directed scheduling algorithm FDS [10,11] uses a global time constraint to evaluate time frames. Based on these time frames the algorithm attempts to minimize the resources required to meet exactly these time constraints. Recently a scheduling algorithm based on neural networks was published providing good results [12]. Although there are some iterative strategies known, which are based on stepwise refinements with alternating scheduling and allocation [13,14], they lack in their global view.

3 The New Approach

Compared to existing techniques the novel approach performs a simultaneous design space exploration in time *and* space. Starting point for our considerations is a DFG which consists of a set of operation nodes $V = \{o_1, .., o_o\}$ and a set of directed edges E which describe the data dependencies. These edges represent variables. We can abstract the scheduling/allocation problem by using some conventions: let us denote $F = \{f_1, ..f_f\}$ a set of functional units, where each functional unit f_i is able to perform a subset of operations, $M = \{m_1, ..m_m\}$ a set of storage elements on which the variables are mapped, $X = \{x_1, ..., x_x\}$ a set of multiplexers and IC the interconnections complexity. The number of time steps needed for execution is represented by a set $S = \{s_1, ..., s_s\}$. With these conventions the scheduling/allocation problem can be formulated as a quadratic assignment problem QA:

$$QA : (V, E) \Rightarrow F \times S$$

In the following paragraphs it is shown that for a given assignment α, the sets F, M, X and S are fixed in an unique way.

3.1 The Costfunction

The function C to be minimized by QA takes into account execution time and hardware costs. An assignment α fixes the sets S and F, in the following abbreviated by $S(\alpha)$ and $F(\alpha)$. Variables needed in different timeslots have to be stored in memory elements. The corresponding set $M(\alpha)$ can be evaluated by lifetime analysis and the left-edge algorithm [15,16]. In addition α fixes the inputs and outputs of a functional unit and also the number of different sources and sinks within a time slot. As shown later these values permit a quite good estimation on the set $X(\alpha)$ and the interconnection complexity. The complete costfunction $C(\alpha)$ is defined as a sum of different weighted components:

$$C(\alpha) = w_1 * |S| + w_2 * area(F, M, X) + w_3 * IC$$

w_i are weighting factors. The function *area* provides abstracted area costs for the functional units, memory elements and multiplexers, and IC represents the interconnection costs.

3.2 Evaluation of Interconnection Costs

In [17] it is shown that the interconnect hardware strongly influences decisions in the design space and must be modelled in an adequate way while scheduling and allocation. We denote $\pi(var)$ as the minimum number of registers needed to store a set var of variables taking into account the lifetime table for all variables [15]. In our approach each functional unit is assumed to have 2 inputs and 1 output. The set of variables belonging to a certain input or output of a functional unit due to an assignment α is denoted as $i_1(f_i)$, $i_2(f_i)$ and $o(f_i)$ respectively. We provide a multiplexer with $\pi(i_1(f_i))$ inputs at the first input of a functional unit f_i, if this number is greater than 1. For example a set of variables $var = \{a, b, c, d\}$ with life time intervals $a : [1,5], b : [2,6], c : [7,8], d : [9,10]$ and a set $i_1(f_i) = \{a, b, d\}$ would provide a $\pi(i_1(f_i)) = 2$ input multiplexer. Note that this strategy provides an exact lower bound on the number of multiplexer inputs. The same procedure can be used to evaluate the multiplexer for the second input. The interconnection complexity is evaluated as:

$$IC(\alpha) = \sum_{f_i \in F} (\pi(i_1(f_i)) + \pi(i_2(f_i)) + \pi(o(f_i))) + \; w_b * \max_{s_i \in S}(dist\_sources/sinks(s_i))$$

The second term estimates the number of busses with w_b as a weighting factor. Examples have shown that this stragegy provides a good estimation on the interconnection complexity. The evaluation of the sets $M(\alpha), X(\alpha)$ and $C(\alpha)$ is dominated by the left edge algorithm and can be done with a computation complexity of $O(n^2)$.

To illustrate the cost function a small example is presented. Given is the following assignment:

time	$f_1(+)$	$f_2(*)$
1	$v_1 = i_1 + i_2^{(1)}$	
2	$v_3 = i_4 + i_5^{(2)}$	
3	$v_2 = v_1 + i_3^{(3)}$	
4		$v_4 = v_2 * v_3^{(4)}$

For clearness the operations are numbered. The corresponding numbers are specified in brackets. We assume that the input variales $i_1..i_5$ are stored in registers respectively. The evaluation of the multiplexer and interconnection costs provides: $i_1(f_1) = \{i_1, i_4, v_1\}$. Since these three variables all have overlapping time intervals a 3-input multiplexer is necessary at the input1 of f_1. Similiar $i_2(f_1) = \{i_2, i_5, i_3\}$ provides a 3-input multiplexer too at the second input of f_1. The output of f_1 must be routed to at least 2 different registers since $o(f_1) = \{v_1, v_3, v_2\}$. The maximum number of different sources/sinks is 2.

3.3 Pipelining

Up to now it was assumed that the functional units are nonpipelined multicycle circuits. With a small modification we can find for a given assignment the maximum possible number of pipelined functional units that can be used. For an assignment α a graph $G_{ty(f)}(V, E)$ is build up for each type $ty(f)$ of functional unit. The nodes correspond to the operations wich are realized on this unit type. There is an edge between two nodes if the corresponding operations can be realized in a pipelined version. We denote t_s as the first time slot of a operation executed on a certain functional unit type and $t_a(o)$ as the actual time slot of a operation realized on the same unit type. Two nodes o_i, o_j are connected by an edge if:

$$(((t_a(o_i)-t_s) \bmod latency) = 0) \wedge (((t_a(o_j)-t_s) \bmod latency) = 0) \wedge (t_a(o_i) \neq t_a(o_j))$$

The graph $G_{ty(f)}(V, E)$ is an interval graph. A maximum clique cover for this graph gives the maximum possible number of pipelined functional units for this type. All operations that are in the same clique can be realized in the same pipelined unit. Similar to the register evaluation problem the clique covering for an interval graph can be calculated in polynomial time with the left-edge algorithm.

3.4 Genetic Paradigms

In the previous sections the scheduling/allocation problem was formulated as a quadratic assignment problem. Based on this representation a cost model was derived which represents the underlying VLSI structure in an appropriate way. However the global minimization of the cost function which represents a NP-complete problem was not addressed.

There are different solutions to solve this problem e.g. by simulated annealing which has been proven to be a general optimization technique. This technique has been successfully applied to various VLSI problems [18,19]. However the large number of states generated while annealing restricts the complexity of the cost function. On the other hand gradient based optimization techniques explore locally the configuration space, the configuration space must be restricted e.g. fixing some parameters like the number of functional units or time steps.

To overcome this problem our approach uses a general optimization technique known as genetic algorithms [20,21,22,23]. Genetic algorithms are search algorithms based on the mechanics of natural selection and natural genetics. They combine survival of the fittest among string structures with a structured yet randomized information exchange to form a search algorithm with some of the innovative flair of human search. In every generation, a new set of artificial creatures (strings) is created using pieces (building blocks) of the fittest of the old: an occasional part is tried for good measure. Though randomized, genetic algorithms are no simple random walk. They efficiently exploit historical information to speculate on new search points with expected improved performance. The outline of the genetic algorithm is shown below:

procedure **genetic_algorithm**

```
pop:=initial_population; P:=| pop |;
for i:=1 to max_generation do
    newp:={}; k:=1;
    while k≤c*P do
        mate1:=select(pop); mate2:=select(pop);
        newp:=newp ∪ crossover(p_c,mate1,mate2);
        k:=k+2;
    end_while;
    for each string ∈ newp do with p_m mutate(string);
    pop:=newp; i:=i+1;
end_for;
solution:=best_string(pop);
```

The function **select** selects a string propabilistically according to its *fitnessvalue* (costfunction). Our approach is based on an *expected value model* instead of the simple *stochastic sampling* technique [22]. The function **crossover** generates two new strings out of four substrings of two strings. The function **mutate** changes randomly one bitposition of a string. These functions are responsible for the overall performance of the genetic algorithm. p_c and p_m are probabilities according to them the corresponding functions are executed. The theory of genetic algorithms is based on the *building block hypothesis*, which is explained e.g. in [22]. A linear dynamic scaling of the fitness values is used to avoid entropy loss and premature convergence due to extraordinary individuals in the initial population.

3.5 Genetic Paradigms vs. Simulated Annealing

It is of interest to compare genetic algorithms to simulated annealing since both techniques are general optimization strategies. The main difference is that simulated annealing improves iteratively *one solution* whereas genetic algorithms develop simultaneously *several solutions* comparable to islands in the design space until the best one survives. Since at every time during the optimization process a complete set of solutions is visible the user can study efficiently the influence of the weighing factors in the cost function. Although the number of strings in a population is quite small i.e. 50-100, a large entropy is guaranted i.e. having n strings in a population $O(n^3)$ schematas are processed [22]. Furthermore genetic algorithms can be efficiently parallelized without modifying the algorithm itself. On the other hand genetic paradigms involve more computation overhead due to the storage of a complete population (no local updates) and this technique is also very parameter sensitive (problem specific tuning). In general the number of states examined by genetic algorithms are less than the number examined by simulated annealing. Therefore the cost function can be more complex.

3.6 Coding

In our case an assignment α has to be coded as a string st. So every population represents a set of quadratic assignments which can have different sizes in both dimensions. We need a clever coding (e.g. no enumeration based coding) to permit simple crossover for legal configurations. The importance of using a small alphabet for coding is pointet out in [22,23], so *small* integer values are used as string components (=*genes*) in our approach.

Two independently processed strings st_1 and st_2 are used to code legal solutions. The space dimension of α is represented by a set of genes: for each type of functional unit one value v_{ty} is provided in st_1 indicating the number of available units of type ty. The gene which represents v_{ty} can vary from 1 to $vmax_{ty}$. $vmax_{ty}$ can be specified by the user or evaluated by an ASAP scheduling without limiting the available functional units.

The coding of the time dimension isn't as simple as the coding of the space dimension, because the data dependencies have to be taken into account during decoding. Additionally we have to make sure that crossover and mutation always produce valid assignments. To fulfill these requirements we use a modified ASAP scheduling technique as a decoding procedure for α. Every operation o_i is scheduled by applying the following procedure. It is assumed that operation o_i is scheduled if all other operations o_j with a precedence relation $pred(o_j, o_i)$ to o_i were already scheduled (simple graph algorithm running $O(|E|)$):

1. Determine the first possible time slot for operation o_i
2. Deferre this operation by δ_i time steps. The value δ_i is represented by a gene in st_1
3. If there is no functional unit available to perform operation o_i in the so calculated time slot, then deferre o_i again to the next possible time slot

Thus a $\delta_i > 0$ results in a "gap". The string st_1 consists of two parts st_{1f} and st_{1t}: genes (v_{ty}) representing the space domain (functional units) are located in st_{1f}; genes (δ_i) to evaluate the corresponding time slots are located in st_{1t}. Thus the length of st_1 is $|V|$ (length of st_{1t}) plus the number of different functional unit types (length of st_{1f}). The second string st_2 which is processed independently of the first one is responsible for the distribution of operations within one time step i.e. it fixes operations on functional units: assuming we have two adders as functional units coded in st and two add-operations in a time slot, then there are two possibilities to map the two additions on the two functional units. Genes of st_2 determine which addition is performed on which adder. We also code the order of inputs of symetric operations in this string. The assignment of our small example can be represented by the following string:

op_1	op_3	op_2	op_4	+	*
0	1	0	0	1	1

The last two genes represent the number of adders and multipliers which are 1 respectively. There is one gap coded for op_3. Assuming that the operations are scheduled according to the specified gene order (see next chapter) this gap is located between op_1 and op_3. op_2 is deffered into this gap since there is no adder (only 1 coded) available in the corresponding ASAP time slot. The adder is already used by op_1 in the first time slot.

3.7 Gene Ordering

As already stated crossover needs short building blocks (i.e. small substrings) which result in an above average fitness value. Therefore the order of string elements is very important for the successful application of genetic algorithms. On the other hand the operations have to be scheduled in such a way that the precedence relations are fullfilled. In our case we search for longest subpathes in the data flow graph by depth-first order in a way that no data precedence relations are violated. According to this traversal the genes are concatenated in st_{1f}. In our small example the gene ordering of the operations in st_{1t} is: (1)(3)(2)(4). The operations are scheduled according this order.

3.8 Initial Population

As pointed out in the genetic algorithm, a first population has to be set up. It is important to have a large entropy in the initial population. Heuristic initializations often result in premature convergence. In our approach we sensefully restrict the state space and randomly generate an initial population by ensuring large variety in the gene values. The upper bound of the genes in st_{1f} was already discussed in the coding chapter. The genes δ_i associated with critical operations should be smaller than those for less critical ones. To find upper bounds ($=\Delta_i$) for these genes an ASAP scheduling without limiting the hardware resources is performed to determine the minimum number of time steps needed. We specify a maximum number of time steps and carry out an ALAP scheduling and calculate a $\Delta_i = ALAP_i - ASAP_i$ for each operation. It is obvious that this strategy provides small Δ_i for genes of the critical paths. A further improvement can be gained: search for operations which form a path p in the DFG with the same Δ_i for each operation on p, generate a random Δ_{rnd} with $0 \leq \Delta_{rnd} \leq \Delta_i$ and randomly distribute δ_i along p, so that $\sum_{o_i \in p} \delta_i = \Delta_{rnd}$ e.g. having a critical path of 17 tsteps and an upper bound of 20 tsteps, then the sum over all gaps (δ_i) on this path must be ≤ 3. However note that due to the actual coded number of functional units, assignments with more than 20 tsteps can be generated due to the deferring which is performed in step 3 of the decoding procedure. This processing drastically improves the convergence behaviour of the genetic algorithm, since meaningful gene values are generated in the initial population. Although we introduce some restrictions e.g. we specify a maximum number of time steps, these values are not fixed since they are modified while performing i.e. crossover and mutation. They only serve to set up good gene values for the initial population.

3.9 Crossover and Mutation

The described coding allows the use of simple crossover and mutation operators. There is no need to introduce complex operators like PMX [23]. For crossover, two strings out of the population are randomly selected, two crossing sites are determined and the string positions between them are exchanged.

4 Experimental Results

The described algorithm is implemented in the program ADaPaS. The algorithm was tested on different known examples, e.g. the differential equation example from

HAL, the bar code example [24] and the well known fifth order elliptic filter [25] from which there are different versions known in the literature. In the following we present some results of the filter example and compare them with other techniques. Multipliers ($*$) are assumed to require two time steps for execution and adders ($+$) only one.

The first row of the table shown below points out the number of adder and multiplier allocations for different *timing constraints* as obtained from the FDS algorithm. The second row was obtained by taking the FDS allocation of 17, 19 and 21 time steps, setting these as a maximum *FU limit* and running the FDLS algorithm to obtain the shortest execution time. The optimal 18 time step solution was also obtained by the EASY-system [11] and the CADDY-system [26]. The former use a slightly modified version of the FDS algorithm. The CADDY-system uses time frames and distribution graphs as in FDS with a slightly different force function. The third row represents the schedule obtained in [27] using ASAP scheduling with conditional deferment. The fourth row represents the results obtained by the SAW system [28] using a list scheduling. The fifth row shows the results obtained by HAL using two stage pipelined multipliers ($*^p$). The last row shows the results obtained with ADaPaS. Permitting pipelined multipliers the pipelined versions were favoured. As already mentioned there are different versions of the filter example known in the literature e.g. in [19] a less complex version is used. The results obtained with simulated annealing in [19] were 17 tsteps, 3 adders and 2 multipliers which was obtained with ADAPAS too.

It is obvious that the results are at least as good as the published ones, however no strong restrictions in time or space are necessary. All results were obtained with the same initial population only by changing the weight factors in the cost function. 4 adders and 4 multipliers were specified as upper bounds, which corresponds to the hardware needed for an ASAP schedule. An upper time limit of 25 tsteps was fixed to distribute the δ_i as described before. Both restrictions are not hard and don't require a detailed knowledge of the underlying problem.

Fifth Order Elliptic Filter				
Alg./Time	17 tsteps	18 tsteps	19 tsteps	21 tsteps
FDS [10]	$3+, 3*$	$3+, 2*$	$2+, 2*$	$2+, 1*$
FDLS [8]	$3+, 3*$	$2+, 2*$		$2+, 1*$
ASAP [27]	$4+, 4*$			
LS [28]			$2+, 2*$	
FDS,FDLS	$3+, 2*^p$	$3+, 1*^p$	$2+, 1*^p$	
ADaPaS	$3+, 3*$ $3+, 2*^p$	$2+, 2*$ $3+, 1*^p$	$2+, 1*^p$	$2+, 1*$

In the following some strings of the initial population and the 11. population are shown. The first number in a string specifies the necessary time steps, the second and third number the adders and multipliers respectively. 'r' is the estimated number of registers and 'i' the interconnection costs. The last number is the actual cost value according the described cost function:

```
1.Generation:
 (22t,2add,2mul,11r,56i 370) (21t,2add,3mul,11r,60i 373)
 (22t,3add,1mul,11r,58i 360) (27t,3add,1mul,10r,53i 425)
 (19t,2add,2mul,10r,54i 328) (18t,3add,3mul,12r,58i 339)
```

```
11.Generation:
 (18t,3add,2mul,11r,54i 321) (18t,2add,2mul,10r,54i 314)
 (17t,3add,3mul,11r,59i 325) (21t,2add,1mul,10r,49i 336)
 (19t,2add,2mul,11r,56i 329) (20t,2add,2mul,11r,60i 345)
```

The final value after running 25 generations was (18t,2add,2mul,9r,47i 310). Note that already in the 11. generation all known solutions are present. Due to the selection of the weighting factors in the cost function, the 18 time step solution survived. The parameters of the genetic algorithms were chosen as follows: $pop\_size = 100$, $p_c = 0.4$ and $max\_generation = 25$, which was sufficient in all cases to obtain convergence. The CPU time measured on a VAXstation 3100 was about 210 seconds.

5 Conclusion

In this paper we have presented a novel scheduling/allocation approach for data-path synthesis based on genetic paradigms. The new approach performs a global simultaneous design space exploration of time and operation parallelism. Beside the minimization of time steps and multicycle (pipelined) functional units, the interconnection complexity is modelled in an appropriate way. Genetic paradigms were used to solve the corresponding optimization problem. The performance of the new approach were demonstrated on the well known fifth order elliptic filter benchmark. Future work is emphasized on the adaption of this system to our silicon compiler ALGIC [29], the exploration of the parallelism of genetic algorithms and further extensions towards behavioural synthesis, e.g. loop folding and operation chaining.

References

[1] McFarland, M.C., Parker, A.C., Camposano, R. Tutorial on High-Level Synthesis. *Proceedings of the 25th Design Automation Conference,* ACM and IEEE, March, 1988, pp. 330–336.

[2] Tseng, C. and Siewiorek, D.P. Automated Synthesis of Data Paths in Digital Systems. *IEEE Transactions on Computer-Aided Design of Integrated Circuits and Systems CAD-5,* 3 (July, 1986), 379–395.

[3] Hitchcock, C.Y., Thomas, D.E. A Method of Automatic Data Path Synthesis. *Proc. of the 20th Design Automation Conference*, June 1983, pp.484–489

[4] Trickey, H. Flamel: A High-Level Hardware Compiler. *IEEE Trans. on CAD* CAD-6, 2, 1987, pp.259–269

[5] Davidson, S. et al. Some Experiments in Local Microcode Compaction for Horizontal Machines. *IEEE Trans. on Computers*, C-30, 7, Jul.1981, pp.460–477

[6] McFarland, M.C. Using Bottom-Up Design Techniques in the Synthesis of Digital Hardware from Abstract Behavioral Descriptions. *Proceedings of the 23rd Design Automation Conference,* IEEE and ACM, June, 1986.

[7] Gooesens, G. et al. Optimization-based Synthesis of multiprocessor chips for digital signal processing with Cathedral II, *Logic and Architecture Synthesis for Silicon Compilation*, North Holland 1989, pp. 283–297

[8] Paulin, G.P., Knight, J.P. Force-Directed Scheduling for the Behavioral Synthesis of ASICs. *IEEE Trans. on CAD*, VOL.8, NO.6, June 1989, pp.661–679

[9] Paulin, G.P., Knight, J.P. Scheduling and Binding Algorithms for High-Level Synthesis *Proc. of 26th Design Automation Conference*, June 1989, pp.1–6

[10] Paulin, P.G. and Knight, J.P. Force Directed Scheduling in Automatic Data Path Synthesis. *Proceedings of the 24th Design Automation Conference,* ACM and IEEE, June, 1987, pp. 195–202.

[11] Stok, L., van den Born, R. EASY: Multiprocessor Architecture Optimisation, *Logic and Architecture Synthesis for Silicon Compilation,* North Holland 1989, pp. 331–328

[12] Hermani, A., Postula, A. NISCHE: a neural net inspired Scheduling algorithm. *Proc. of the 4th. International Workshop on High-Level Synthesis*, Kennebuntport, Oct.89

[13] Parker, A.C., Pizarro, J., and Mlinar, M. MAHA: A Program for Datapath Synthesis. *Proceedings of the 23rd Design Automation Conference*, ACM and IEEE, June, 1986, pp. 461–466.

[14] Brayton, R.K., Camposano, R., DeMicheli, G., Otten, R.H.J.M., and vanEijndhoven, J. The Yorktown Silicon Compiler. *Silicon Compilation*, D.D. Gajski, Ed. Addison-Wesley, Reading, MA, 1988, pp. 204–311.

[15] Kurdahi, F.J., Parker, A.C. REAL: A Programm for REgister ALlocation. *Proc. of the 24th Design Automation Conference*, June 1987, pp.210–215

[16] Hashimoto A., Stevens J. Wire Rouring by Optimizing Channel Assignment Within Large Apertures. *Proc. 8th Design Automation Workshop*, 1971, pp.155–169

[17] McFarland, M.C. Reevaluating the Design Space for Register-Transfer Hardware Synthesis. *Proc. of IC-CAD 1987*, pp. 262–265

[18] Kirkpatrick, S., et al. Optimization by simulated annealing. *Sience*, Vol.220, No.4598, 1983, pp.671–680

[19] Devadas, S., Newton, R. Algorithms for Hardware Allocation in Data Path Synthesis. *IEEE Transactions on Computer-Aided Design CAD-8*, 7 (July, 1989), 768–781.

[20] Cohoon, J.P., et al. Genetic Placement. *Proc. of ICCAD* 1986, pp.422–425

[21] Cohoon, J.P., et al. Floorplan Design Using Distributed Genetic Algorithms. *Proc. of ICCAD* 1988, pp.452–455

[22] Goldberg, D.E. *Genetic Algorithms in Search, Optimization and Machine Learning*. Addison Wesley Publishing Company, 1989

[23] Davis, L. *Genetic Algorithms and Simulated Annealing*. Pitman Publishing, 1987

[24] Duzy, P., Krämer, H., Neher, M., Pilsl, M., Rosenstiel, W., Wecker, T. CALLAS - Conversion of Algorithms to Library Adaptable Structures. *VLSI 89, Proc. of the IFIP TC 10/WG 10.5 Int. Conf. on VLSI*, G. Musgrave and U. Lauther (Editors), pp. 197–208

[25] Detjiens, E., Boriello, G. *Workshop on High Level Synthesis*. Orcas Island, Jan.1988

[26] Kramer, H. et al. DataPath and Control Synthesis in the CADDY system, *Logic and Architecture Synthesis for Silicon Compilation*, North Holland 1989, pp. 229–242

[27] Kung, S.Y., et al. *VLSI and Modern Signal Processing*. Prentice Hall, 1985, pp.258–264

[28] Thomas, D.E., et al. The Systems Architect's Benchmark. *Proc. of 25th Design Automation Conference*, July 1988, pp.337–343

[29] Schuck, J., Wehn, N., Glesner, M., Kamp, G. The ALGIC Silicon Compiler System: Implementation, Design Experience, Results. *Proc. of the 24th Design Automation Conference* June 1987, pp.370–375

High Level Synthesis II

Logic and Architecture Synthesis
P. Michel and G. Saucier (Editors)
Elsevier Science Publishers B.V. (North-Holland)

PORT ASSIGNMENT IN MULTIPORT MEMORIES FOR INTERCONNECTION MINIMIZATION IN DATAPATH SYNTHESIS

Thomas C. Wilson, Dilip K. Banerji, Anupam Basu,
Jay C. Majithia, and Arun K. Majumdar*

VLSI-CAD Group
Department of Computing & Information Science
University of Guelph
Guelph, Ontario
CANADA N1G 2W1

* Department of Computer Science & Engineering
Indian Institute of Technology, Kharagpur, INDIA 721 302

1. INTRODUCTION

A major objective in VLSI design is to minimize interconnections among the various circuit components, especially the number of inter-module connections and multiplexers. Interconnections can be reduced by allocating registers to multiport memories (register files). This permits registers in the same memory to share the same physical paths between the memory and various operator units. The use of single port memories for register files was first suggested in [1,2]. Methods of placing registers in multiport memories during datapath synthesis have been reported in [3-5]. These methods start with a register-transfer level (RTL) description of circuit behavior and place the registers so that the number of registers accessed in a control step from a memory never exceeds the number of ports in the memory. This assures the existence of some feasible access pattern to each memory. However, placing the registers in memories is not the whole story. The connections to implement required memory accesses still need to be determined, and that problem is the subject of this paper.

We assume that operator scheduling and binding to operator units have already occurred. We also assume that register identification and assignment of registers to memories have taken place. At this point, the memory contents and the operator terminals that must connect to each memory are known, but the details of the interconnection are not yet known. What concerns us here is the judicious assignment of interconnections between operator terminals and specific memory ports so that concurrent accesses to multiple registers in a memory can always occur through distinct ports. The problem is to do this assignment using the minimum number of interconnections. This "port assignment problem" applies to each multiport memory separately. Although the interactions between a memory and all operation unit terminals must be considered, we assume no direct connections between memories, so that other memories are not an issue here.

Balakrishnan et al [3] concentrate on assigning a memory's resident registers to its ports on individual control steps. The solution to this problem implies which interconnections are required and thereby addresses the interconnection problem, albeit indirectly. Their approach involves solving a 0-1 Linear Programming problem for each control step, which only provides a locally optimum solution. Instead of assigning registers to ports, we concentrate directly on minimizing connections between memory ports and operator terminals. Once the paths have been established, the association of registers to ports becomes simple. Our approach deals with a much smaller set of objects (ports and terminals vs. registers and control steps). This allows more extensive searches for good assignments and very fast processing. Our technique surpasses the L.P. approach in speed, typically by a factor of 2 to 5.

2. A CLOSER LOOK AT THE PORT ASSIGNMENT PROBLEM

We distinguish between *ports*, which are access points of memories, and *terminals*, which are access points of operation units. We define a **link** to be an association between a register and a particular operator terminal on an individual control step. A link is a logical connection that reflects one instance of a register and a terminal communicating. A register may have many links to the same terminal, as well as links to many terminals. A **path** is an actual connection between a specific port and a particular terminal. Each link must ultimately follow an actual physical path, and the number of such paths is to be minimized. Ideally, we hope to map all links between a memory and a specific terminal to a single path. If this is not possible, then there will be *multiple paths* between (different ports of) the memory and the terminal. In certain contexts we refer to *assigning* a terminal to one or more ports. This is equivalent to providing a path between the terminal and the port(s).

Any two terminals that access a memory in the same control step are said to **conflict.** If two terminals conflict, they must access the memory through different ports in the conflicting steps. The complexity of the conflict pattern is what may require multiple links to the same memory. The global conflict pattern for any given memory is summarized in a **conflict graph** G, where each terminal is represented by a node, and two nodes are joined by an edge if they conflict in at least one control step. Figure 1 shows a 2-port memory containing registers r1, r2, and r3. The RTL description causes each pair of involved terminals to conflict. We can provide a single path for any two of the terminals, but the third terminal will always need paths to both ports, in order to provide access through separate ports on all control steps. Figure 1 shows one of these four-path configurations and the conflict graph. Although all conflicts in this example are between pairs of terminals (as they must be in a two-port memory), this fact is not evident from the graph. G would have appeared the same if all three terminals had required concurrent access (to a 3-port memory).

3. A GRAPH APPROACH TO SINGLE PATH ASSIGNMENT

Although the port assignment problem has been formulated using 0-1 Integer Programming [6], the number of constraints for each memory is $O(P*T^2)$ where P and T are the number of ports in a memory and the number of operator terminals linked to the memory, respectively. The same problem can be solved directly and much faster by analysing the conflict graph.

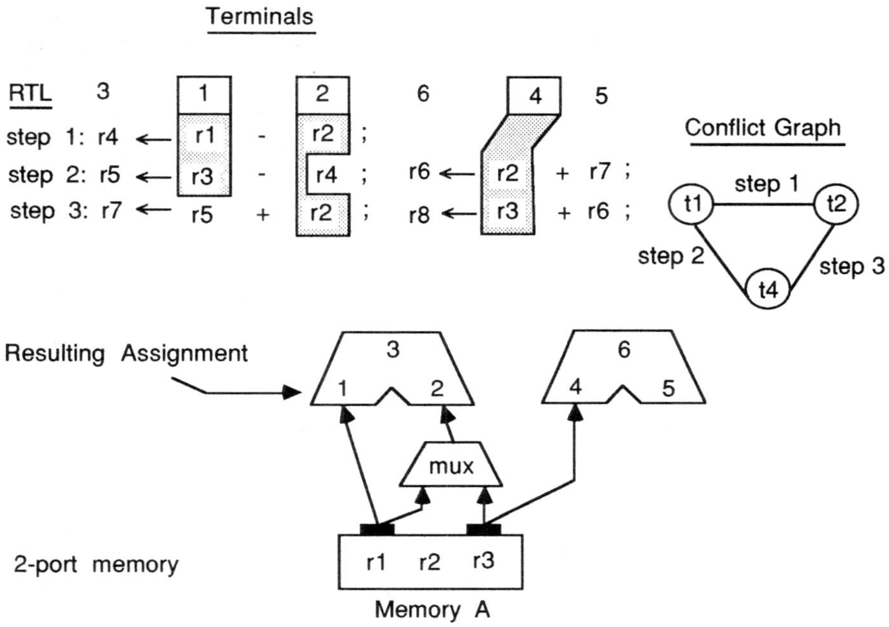

Figure 1: Assignment with Two Paths to One Terminal

By **single path assignments** we mean port assignments in which every terminal is connected to the memory by at most one path and these are unmistakably optimum. The assignment in Fig.1, though optimum, is not a single path assignment. The conflict graph does not have to be trivial in order to yield a single path assignment. Consider the situation in Fig. 2. A tree-structured graph such as this always produces a simple, single path assignment to two ports. By designating any node as the root, the notion of being an odd or even distance from this root has a consistent interpretation for each node. By assigning even nodes to one port and odd nodes to the other port, no two conflicting terminals will ever enter through the same port. In fact, this holds true for any bipartite graph, of which a tree is a special case. In general, for a memory having P ports, a single path assignment exists if the conflict graph is P-partite. Unfortunately, testing a graph in general for the P-partite property is not so simple when P>2.

There is, however, a simple procedure that often succeeds in partitioning the nodes of G into P sets with no edge having both ends in the same set -- especially when G is not too richly connected. This procedure is based on the following observation, where V_i is the valency of node t_i, that is, the number of edges that meet t_i :

> *When assigning node t_i to a set, we must choose a set which does not contain any nodes connected by an edge to t_i. If $V_i < P$, and if the V_i nodes with which t_i conflicts are all assigned to separate sets (a worst case situation), there remain $P-V_i$ sets to which t_i can be assigned without any of its incident edges having both ends in the same set.*

RTL step	Active terminals
1	1,5
2	3
3	2,4
4	1
5	2,3
6	1,5
7	1,2

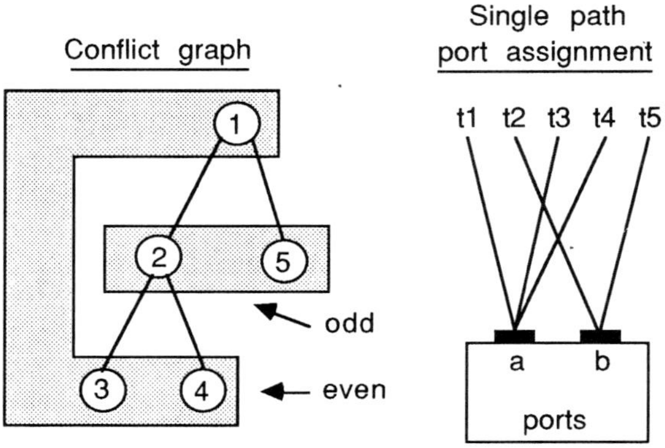

Figure 2: Single Path Port Assignment

Equating node sets with ports, this rule says that any terminal t_i having $V_i<P$ can be assigned to any of the $P-V_i$ ports that do not have paths to the terminals with which t_i conflicts. Note that the terminals with which t_i conflicts may have paths to fewer than V_i ports, because these terminals may not conflict with each other. This would provide even more than $P-V_i$ choices of ports for terminal t_i. For this rule to be useful, we assume that all conflicting terminals have only one path to a memory. However, a terminal can have paths to more than one port in reality; this situation is dealt with later. The preceding observation suggests (temporarily) ignoring any node t_i with $V_i<P$. Once the "other" nodes are assigned to (single) ports, t_i can be assigned safely to any of the remaining ports. Ignoring t_i is equivalent to removing it and its incident edges from G, as if t_i had never been involved in potential conflict. With t_i removed from G, some other node, t_j, might also have its $V_j<P$ and be similarly removable from G. If we continue to remove nodes having valency < P, we may eventually remove all the nodes from G. If this happens, a single path assignment is, in fact, within reach. All we must do is assign nodes to ports in the order opposite to their order of removal. Accordingly, the single path assignment algorithm consists of two parts: (1) Decomposition of G (Phase 1); (2) Assignment of terminals to ports in the reverse order of node removal (Phase 2). A stack is used to convey the results of Phase 1 to Phase 2. This is summarized in the following procedures P1 and P2:

P1: {decompose G...}
while the remainder of G contains node t_i with $V_i<P$ **do**
remove t_i from G, and push t_i onto the stack;

P2: {assign ports...}
while the stack is not empty **do**
pop node t_i from top of stack;
{ t_i conflicts with fewer than P already assigned nodes }
assign t_i to any port that has not had a conflicting node assigned to it.

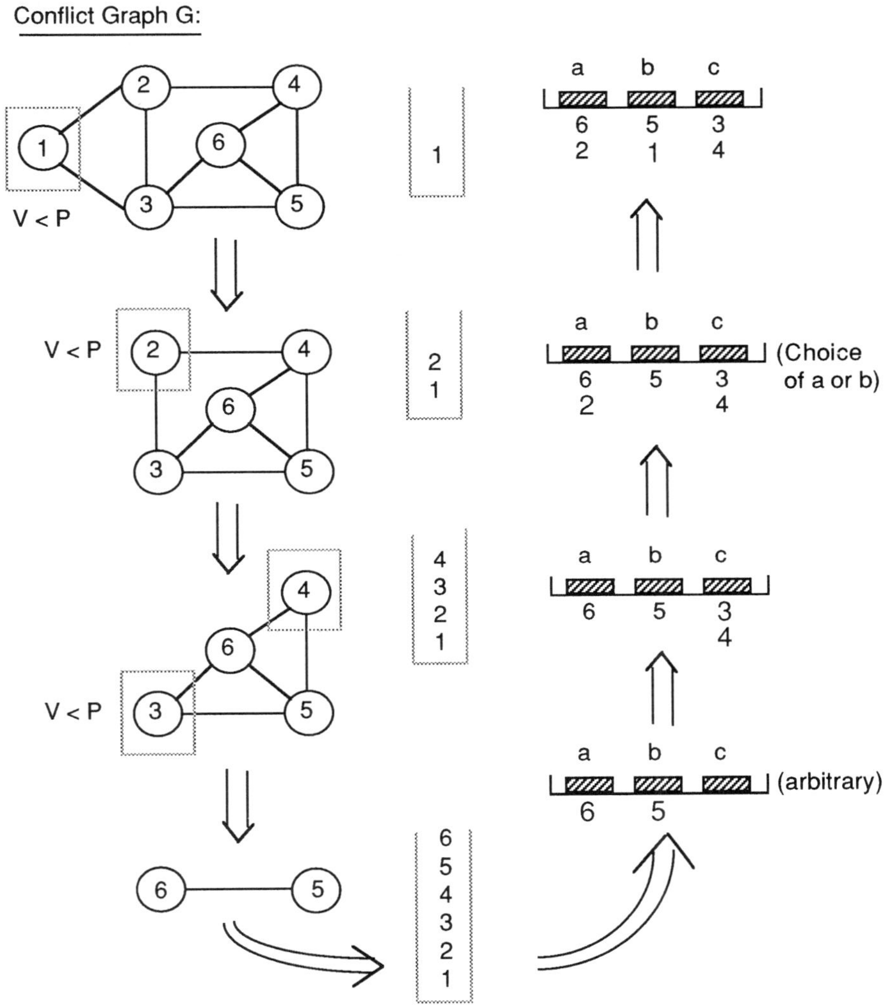

Figure 3: Completely Decomposable Conflict Graph

The conflict graph shown in Fig. 3 is fairly complex but still leads to a single path assignment for a memory with 3 ports. The figure demonstrates how the above procedures decompose this conflict graph and assign terminals to ports. The resulting paths of the single path assignment are also shown.

4. THE GENERAL DECOMPOSITION PROCESS (Multiple Path Assignments)

Unfortunately, not all P-partite graphs can be decomposed using procedure P1. The bipartite test provides easy recognition of single path assignments when P = 2, but it does not extend conveniently to $P>2$.

Furthermore, many conflict graphs are not P-partite at all, and do not yield a single path assignment. We may still employ the decomposition procedure to remove nodes but we can no longer predict how the decomposition and assignment will turn out. At some point in the decomposition, we may find that no node remaining in G has $V_i<P$. It is always worth testing what remains of G for the bipartite property. If G is bipartite, the quest for a single path assignment is successful.

If the decomposition cannot proceed and G is not bipartite, we remove any node, expecting that the corresponding terminal will be linked to more than one port. Two heuristic rules apply in this case: (1) Removing a node having largest valency is most likely to leave the graph in a condition allowing further reduction; and (2) Removing a node that did not conflict with any previously removed nodes, will not limit the availability of free ports when these previously removed nodes are assigned.

Because this removed node may lead to multiple links, a special warning symbol (X) is also placed on the stack. (Since this choice may preclude optimality, a practical implementation could try alternative nodes for removal, seeking a truly optimum solution, as our implementation does. However, for this presentation, we simply remove an arbitrary node and accept the outcome.)

Even when $V_i>P$, it may happen that the nodes with which t_i conflicts can be assigned to fewer than P ports. For example, if the nodes remaining in G form a bipartite graph, they can be assigned to only two ports. If P > 2, the node removed by arbitrary choice can be assigned to any of the P - 2 remaining ports. A more complete decomposition procedure P3 incorporating these features follows.

P3: General Decomposition Procedure (without backtracking)
while not all nodes of G have been removed **do**
  begin
  while the remainder of G contains node t_i with $V_i<P$ **do**
    remove node t_i from G, and push t_i onto the stack;

  if not all nodes of G have been removed **then**
    select some node t_j , preferably with largest V_j and
    not conflicting with any previously removed node;
    remove node t_j from G and push t_j onto the stack;
    if G is bipartite **and** P > 2 **then**
      place both node SETS onto the stack, exhausting G
    else place "X" warning on the stack
  end {decomposition loop}

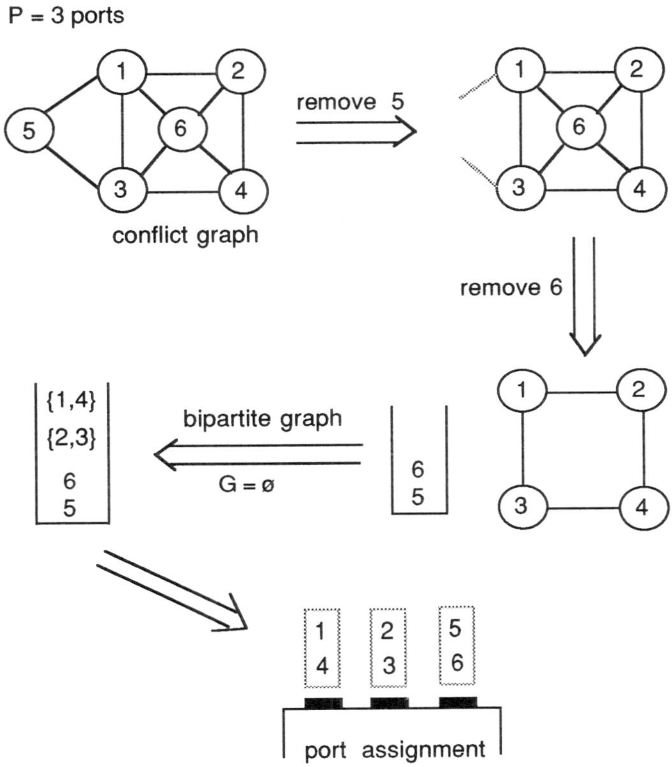

Figure 4: Decomposition Giving Single Path Assignment

Figure 4 shows an example of its operation. The node with largest valency is removed to break the impasse, leaving a bipartite graph. Fig. 5 wil show a situation where a single port assignment is not possible.

5. THE GENERAL PORT ASSIGNMENT PROCESS

In Fig. 4 we were lucky; we did not have to really deal with the impasse. However, these situations often require multiple paths from the same terminal, and multiple paths invalidate the assumption on which the decomposition is based. Therefore, the port assignment procedure must be expanded accordingly.

If the first two entries on the stack are node sets, constituting a bipartite subgraph of G, they are assigned to two available ports. Other nodes popped from the stack are assigned as before -- unless an "X" warning is popped from the stack. If "X" is encountered, the next terminal will likely be linked to more than one port. To

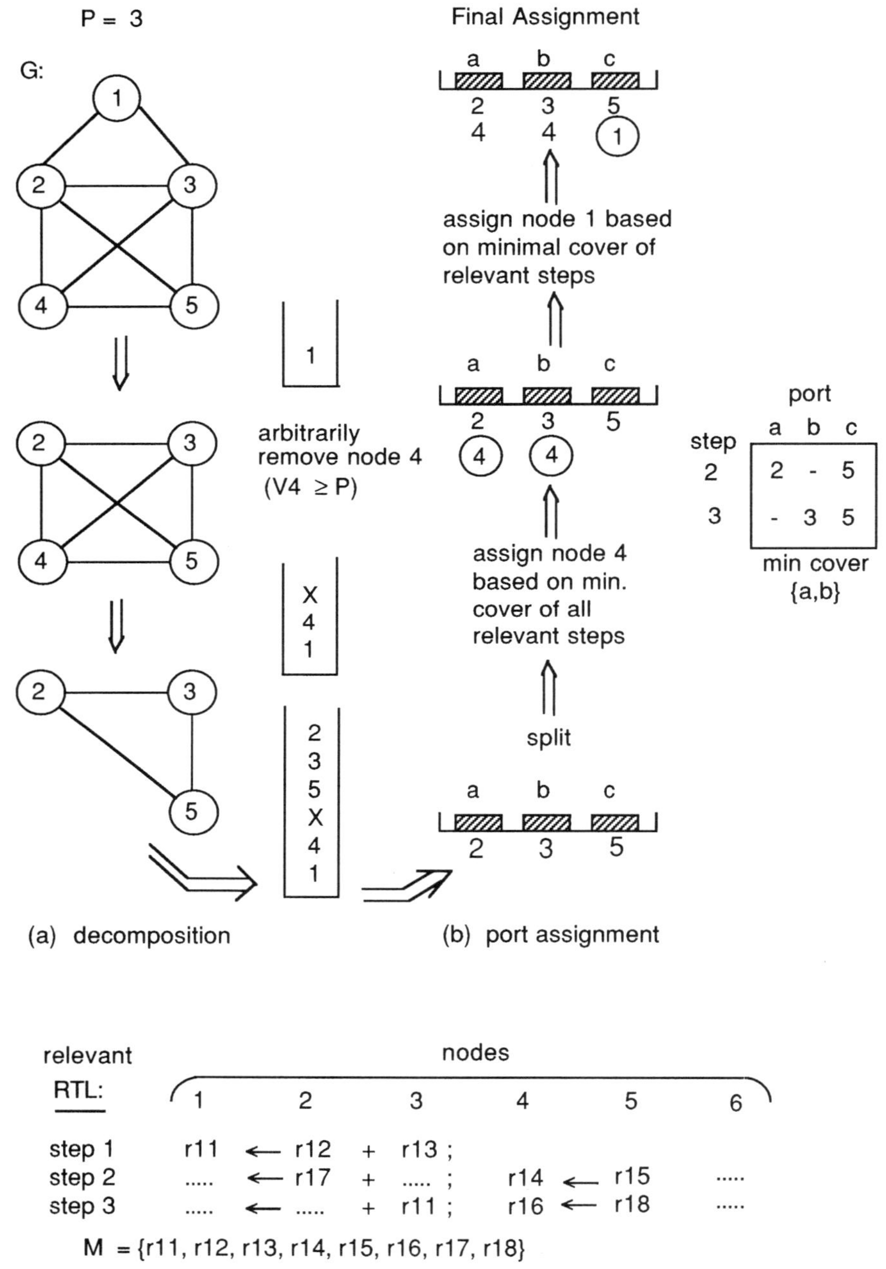

Figure 5: Multiple Port Assignment

determine which ports, we inspect the original RTL description to consider individual steps where that terminal has links. Each step will have certain available ports, not used by previously assigned nodes. A minimum set of ports is chosen that includes an "available" port for each relevant step -- a standard minimum cover problem. This process is denoted by *port_cover_relevant_steps*(t_i) in the port assignment algorithm, whose general form is shown in procedure P4.

Once a terminal is actually assigned to more than one port, all subsequent terminals should be assigned with similar care, in case the effect of multiply-linked ports cascades. Of course, terminals that conflict only with the terminals assigned to single ports, can be assigned exactly as before. However, to simplify the presentation, we assume that the first occurance of "X" on the stack triggers the use of *port_cover_relevant_steps* for all subsequent assignments and that this procedure decides whether inspecting the RTL and performing minimum cover are warranted.

P4: **while** stack not empty **and** not(top(stack) = "X") **do**
    pop node t_i from top of stack;
    {t_i conflicts with fewer than P already assigned nodes,
      and t_i may be a node set}
    assign t_i to any port that has not had a conflicting node assigned to it.
  while stack not empty **do**
    pop t_i from stack;
    port_cover_relevant_steps(t_i);

Figure 5 shows a 5-node graph and associated RTL that does result in a multiple path assignment. The presence of "X" on top of the stack causes "careful" assignment of the remaining nodes. Because this memory links to terminal 4 in steps 2 and 3, only these steps are involved in the minimum cover. Because terminal 1 does not conflict with terminal 4, it can be assigned as before, without bothering to find a minimum cover: assign it to any port not having a path to terminals 2 or 3.

CONCLUSION

We have examined the problem of finding the minimum number of paths between a multiport memory and operator (functional unit) terminals with which the resident registers of the memory must communicate. A graph-oriented algorithm has been presented that almost always yields an optimum solution and that operates sufficiently fast to allow real-time exploration of the design space. These algorithms have been implemented and tested in both C and PASCAL.

The major contribution of this work is to model clearly and directly the terminal to port interconnection problem. Compared to the earlier integer LP approach [3], which solves the problem in a step-by-step manner, thereby yielding a locally optimum solution, our method considers all the control steps at a time in order to approach a globally optimum solution. The worst case time complexity of the proposed algorithm has been estimated as $O(T^2*S)$ with no backtracking, where T is the number of operator terminals and S is the number of control steps. Experimental results from several test cases show a 2 to 5 times speed enhancement over the integer LP approach, with the result always as good as LP and often better, typically by 10 percent.

REFERENCES

[1] Tseng, C.J., "Automated Synthesis of Datapaths in Digital Systems", **Ph.D Dissertation**, Department of Electrical and Computer Engineering, Carnegie Mellon University, 1984.

[2] Tseng, C.J. and D.P. Siewiorek, "Automated Synthesis of Data Paths in Digital Systems," **IEEE Trans. on CAD**, vol. 5, no. 3, July 1986, pp.379-395.

[3] Balakrishnan, M., A.K. Majumdar, D.K. Banerji, J. Linders, and J.C. Majithia, "Allocation of Multiport Memories in Data Path Synthesis," **IEEE Trans. on CAD**, vol. 7, no. 4, April, 1988, pp.536-540.

[4] Wilson, T.C., D.K. Banerji, J.C. Majithia, and A.K. Majumdar, "Optimal Allocation of Multiport Memories in Datapath Synthesis," **Proc. 32nd Midwest Symposium on Circuits and Systems**, Urbana Ill., Aug. 1989, pp.1070-1073.

[5] Stok, L., "Interconnect Optimization during Datapath Allocation", **Proc. European Design Automation Conference**, Glasgow, Scotland, March 1990.

[6] Wilson, T.C., D.K. Banerji, Anupam Basu, J.C. Majithia, and A.K. Majumdar, "On the Port Assignment of Multiport Memories for Interconnection Minimization", **Tech. Report #CIS-90-D3**, Dept. of Computing and Info. Science, Univ. of Guelph, Feb. 1990.

Logic and Architecture Synthesis
P. Michel and G. Saucier (Editors)
Elsevier Science Publishers B.V. (North-Holland)

Automatic Generation of Single Chip Multi Processor Systems

Heinrich J. Krämer, Wolfgang Rosenstiel

Computer Science Research Centre at the University of Karlsruhe (FZI)
Department Automation of Circuit Design (Prof. D. Schmid)
Haid-und-Neu-Straße 10-14, 7500 Karlsruhe, FRG

In this paper the environment and the algorithms for the synthesis of a multiprocessor implementation are presented. The starting point of the synthesis is a behavioural description of the desired circuit. The described procedure transforms a scheduled data flow graph into a set of subtasks each implemented by a single processor. The system automatically generates the global control and the communication network.
Finally the construction of the data path for the single processors is described. The used algorithms are based on graph colouring

1. Introduction

The recent improvements in the IC fabrication technology make it possible to implement whole systems as one integrated circuit. One major task in the design of a complete system is the selection of a suitable architecture for a given algorithm.
To support the designer in exploring the possible design space, synthesis systems should be able to generate different architectures on certain demands from the designer. Finally the designer can choose the most cost effective solution for his specific problem.
This paper presents a synthesis system, which is able to generate a RT-structure from an algorithmic specification. The RT-structure may consist of a various number of processors. The system is integrated into an already existing synthesis framework.

2. Previous Work

Today two different approaches in high level synthesis can be distinguished. The first are systems like the CATHEDRAL systems [CCDG88] or the PARSIVAL system [HaCo88] that are specially designed to meet the demands in DSP. This systems are able to generate solutions with multiple processors. Also pipelining is widely used to increase the throughput rate necessary for real time DSP applications. But due to the low flexibility in the generated architecture, these systems are more suitable for the implementation of DSP algorithms, such as filters, by an integrated circuit.
The other type of systems generate a single microprocessor-like solution from any kind of algorithms. These systems are often referred as high level synthesis systems. As stated in [McPC88] this task can be divided into two major steps the scheduling problem and allocation problem. An overview on currently used scheduling and allocation techniques can be found in [Paul89].
The system presented here tries to combine the two approaches and generates multiprocessor solutions but with a more flexible architecture of the generated data path.

3. The Synthesis Environment

The system is embedded into the already existing CADDY synthesis tool [DKNP89]. The behaviour of a desired circuit may be specified either in DSDL[DSDL88] or DSL [CaWe84].

An example in the input language DSL is shown in the appendix. It is a tiny micro processor inspired by the MIPS architecture.
CADDY generates the *data flow graph* (DFG) and an initial *control flow graph* (CFG). Also included is the possibility of behavioural transformations such as the detection of data independent loops, loop unrolling, constant expression calculation and dead code elimination. After this transformations the DFG is generated using data flow analysis techniques.
The next step included is a scheduling phase described in [KrRo90]. The result of these steps is a scheduled DFG. This may then be mapped onto a single processor solution by performing register assignment module allocation described in chapter 6. The resulting structure may further be optimised using local transformations.
The generated symbolic control automaton may be mapped onto a boolean description using state assignment described in [ANRR89]. This whole description, which consists now of a data path, additional glue logic and a hardware implementation of the controller is then mapped onto a layout using the VENUS® macro cell system.
In this paper we want to describe the extensions needed for the generation of multi processor systems and the generation of the data path of a single processor.

4. The Target Architecture

The target architecture consists of a tightly coupled multi processor system (fig 4.1). The communication is done via global shared memories. Each processor consists of its own data path that performs a part of the specified algorithm. This part of algorithm is a calculation sequence, that is controlled by a local controller. The synchronization of the different processors is done by a global controller. This global controller is also responsible for the correct I/O sequence. If the data path of one single processor is simple or the calculation sequence consists of one or few steps and further the global memories are simple registers or latches each processor may be denoted as a pipeline stage.

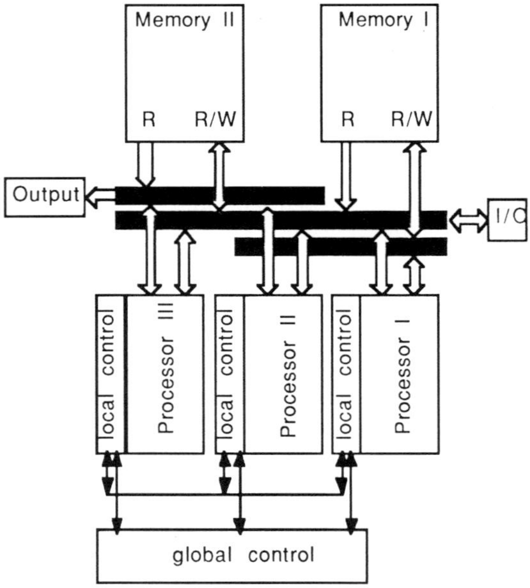

Figure 4.1 The overall target architecture of the synthesis system.

® VENUS is a SIEMENS trade mark

The architecture of a single processor is shown in figure 4.2. Each data path has its own controller. The operational part may consist of an unlimited number of buses, operation units, registers and bus connectors that connect an input (BC_i) or an output (BC_o) of an hardware unit to a bus. This bus connectors serve as some sort of multiplexor/demultiplexor units.

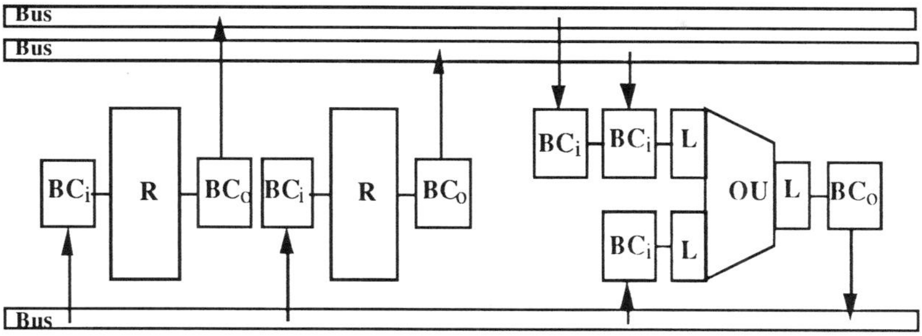

Figure 4.2 The architecture of a single processor.
The data path is clocked with a non overlapping two phase clock. The operation units have latches at the inputs and at the outputs. Two different clocking schemes are possible:

- The operands are transferred in the same cycle as the results are written back to the registers. This is the fastest possible type of scheme but may increase the number of buses. On the other hand it may decrease the number of registers since a result may directly be used as an input operand in the case of accumulator type of operations.

- The operands are read in in one cycle and in the next cycle the results are written back to the registers. This operating scheme is slower than the previously presented one. But the number of needed buses decreases, since the results may be transferred using the same buses as the input operands. In this scheme the accumulator operations are no longer possible and the intermediate values must be stored in a separate register.

5. Procedure for the Automatic Generating of the Multi Processor Systems

5.1 Generation of the Control
The starting point for the construction is a scheduled data flow graph. This is then partitioned by the user into subtasks that will be performed by one processor. Two different problems have to be solved in this step. First the preservation of the I/O sequence, second the integrity of the used data.
The order in which the I/O operations have to be performed is determined by the input specification and the scheduling phase. This order is then directly mapped to a global controller. This controller enables a new I/O operation in the specific local controller only if the previous one is successfully executed. This preserves the sequence even in a pipelined system. The state of the global controller may directly be used as condition for the activation of the I/O operation in the local controller.
The problem for the integrity of the data is to prevent the overwriting of still needed values and the use of outdated values. The approach taken in this system is based on the analysis of the algorithm. This algorithm can be divided into sequences that run unconditionally. This can be regarded as basic blocks of the algorithm. In the previous compilation steps the variables are transformed to single assignment variables. So each variable has only one defining operation and at least one consuming operation. Variables that are transferred between different processors are called global variables, the others are called local variables. We generate a graph DTG = (V,E) with V = { b_i | with b_i is a basic block } and E = { (bi,bj) | with bi,bj $\in$ V, b_j consumes a global variable defined in b_i }. This graph combined with the I/O sequence is the base for the global control. Each basic block is represented by a flag

register. This is set to one if the basic block finished execution. This concurrent state register is read by the the local controller. It may only enter a new basic block if all predecessors in the DTG have finished.

5.2 Generating the Interconnect Network

The different processors work in a pipeline fashion. So each processor must be supplied with the data produced in the correct iteration step here called a *time zone*. This is usually done by shift registers. Each time the processors start a new execution cycle the date is shifted one place. If one value is used by several processors in different time zones the shift register must be taped at the appropriate places.

To determine the length and the taps of the shift register the DTG is transformed to the processor stage graph PSG. In this graph each node represents a processor or pipeline stage and the arcs the data transfers. First all arcs in the DTG that refer to feedback variables must be removed. These are the global variable with a definition time after the first use. Then we introduce super nodes $P_i \in \mathcal{P}$ which includes all basic blocks, that are assigned to one processor. Multiple arcs from one super node to another are replaced by one. The result is a acyclic reducible graph. This can be sorted using the TOPSORT algorithm. The result of the sorting is a function $T : \mathcal{P} \rightarrow N$ that assigns each processor the time zone. Due to the preprocessing each variable has only one defining statement and at least one consuming statement. If v is a global variable then there exist only one processor P_p which produces v and a set $\mathcal{P}_c$ of processors P_i $i = 1..c_v$ of consuming processors. The length of the shift register is than determined by $T(P_p) - \max_{P_i \in \mathcal{P}_c} (T(P_i)) + 1$ The tabs are determined accordingly.

The global data transfers are performed by bidirectional buses. The assignment can be done analogously to the later bus assignment in the single processor data path. But an additional condition for no overlap is that the basic blocks in which the data transfer is performed are in a predecessor relation or are assigned to the same processor.

5.3 Optimising the Global Interconnect Network

If there are no data dependencies in the algorithm then for each processor additional optimisation are possible to reduce control complexity, wiring are or memory area. First no synchronization is necessary because the longest execution time determines the latency and the restart of each processor. Second since all processors have a fixed but perhaps shorter than the maximum execution time, the restart of a processor may be delayed to minimize the global bus requests. Third the shift registers may be replaced by global memories with additional addressing logic. This leads to a smaller implementation than the shift registers.

5.4. A small Example

The DSL description of this example can be found in the appendix. It is a part of the MIPS instruction set described in DSL. The intention is to demonstrate the system generation steps with a more data dependent algorithm than commonly used in the literature.

The "processor" is able to store a word of 32 bits in each of its 4 internal registers (The small number of registers is due to presentation purpose) and to write it back to main memory. It can perform an ADD or SUB operation on the data stored in the registers, a relative jump operation and a return by a address stored in register 3. After the preprocessing and the scheduling the scheduled data flow graph is shown in figure 5.1. A memory access is assumed to take two cycles. In the first cycle the address is written in the second the data is ready.

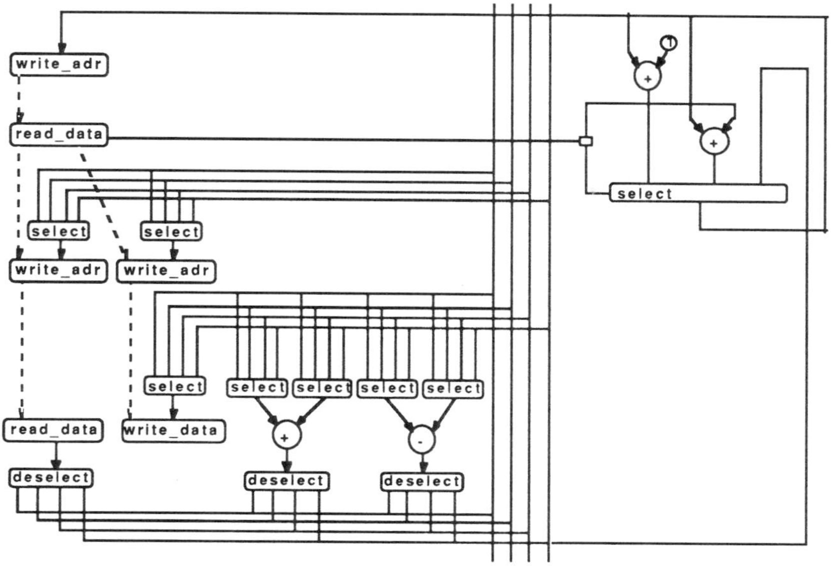

Figure 5.1 The scheduled data flow graph for the tiny microprocessor

The structure shown in figure 5.2 of the data part for the tiny micro processor example is automatically generated by the synthesis tools.

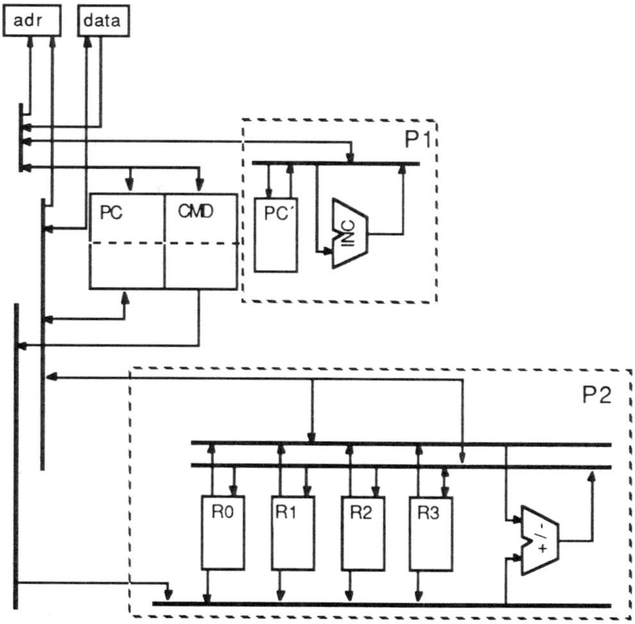

Figure 5.2 data part of the tiny microprocessor example.

6. Construction of the Single Processors

This step consists of three different subtasks. First the assignment of local variable values to dedicated data path registers. Here variables with non overlapping life times should be assigned to the same registers. Second the assignment of operations to specific operation units. Please note that only the incarnation of a module type has to be assigned in this system, because the type is already determined by the scheduling algorithm. This is a much easier problem than the assignment of type and incarnation of operation units. Third the assignment of data transfers to buses. The sequence is interchangeable but has an influence to the result. The HAL [Paul89] system first assigns the operations to hardware modules then the registers are assigned. Because only unidirectional data links are considered, this determines the interconnect network. In the system described in [Stok89] the bus assignment is done first and the variable values are then distributed onto several register files each connected to one bus. This may lead to a larger register request but tries to minimize the number of buses.
In this system the register assignment is done first, then the operation unit assignment is done, finally the bus assignment is performed. These steps determine the needed bus connectors. Each step can be done optimal, or nearly optimal in the case of register assignment, separately. But the optimal assignment of hardware units in each case may lead to a bad solution in terms of bus connectors. So the whole algorithm is designed to minimize the number of bus connectors.
Therefore first the variable values are assigned to registers. This has the largest freedom, since the operations are already preassigned to operation unit types. Then the operations are assigned to operation units. This has an influence on the bus assignment because of the accumulator type operations. Finally the buses are assigned.

6.1 Register Assignment

In [CACC80] and [Pfah87] the register assignment problem was formulated as a graph colouring problem. We will extend this formulation on all three steps since the problem of assigning hardware resources to behavioural constructs is similar.
We first construct a constraint graph CG = (V,E) with V = {v | v is a variable} and with $E = \{(v_i,v_j) \mid$ with $v_i,v_j \in V$ and v_i,v_j have overlapping life times}. In the case of loops with feedback values this leads to a circular-arc graph. In this case the left edge algorithm proposed by [KuPa88] is no longer valid. Secondly in the REAL system data dependent branches are handled by heuristics, which is not necessary in the case where feedback values are not regarded.
In this system the node colouring is based on a subclass of perfect graphs, the triangulated graphs proposed by [Golu80]. For this subclass of graphs polynomial algorithms for commonly referred problems are known. Also the triangulated property can be tested in polynomial time. If the graph is triangulated then the optimal algorithm for this kind of graph from [Golu80] is used. In the general case a heuristic algorithm has to be used.
This leads to an optimal or nearly optimal register assignment. But also the number of bus connectors have to be minimized. So the colouring is controlled by preferences. This preferences are introduced by a second graph $PG = (V,E^-)$ the preference graph. Preference arcs are weighted arcs between vertices without a constraint arc. Preferences for example are:

- Two values are are result of the same type of operation unit
  The operations may be assigned to the same unit, then only one bus connection is needed for both operation unit and register.
- Two variables are the source for the same type of operation unit
- Two values are sink for a global memory access
  The global bus driver is only needed once.

The algorithm for colouring a graph with constraint arcs and preference arcs is presented in fig 6.1.

```
COLOR GENERAL GRAPH
{ The algorithm returns the number of used_colours and a coloured graph }

  The function COLOURABLE(vertex,colour,preference_weight) returns
  true, if the vertex and all its preferred vertices with
  preference ≥ preference_weight can be coloured with colour.
  The procedure COLOUR(vertex,colour,preference_weight) colours the
  vertex and all its preferred vertices with preference ≥ preference_weight
  with colour.

  Sort vertices with non increasing degree;
  used_colors := 0;
  for all vertices in order of list do
    while list of vertices is not empty do
      act_colour := 1;
      if colourable(vertex,act_colour,preference) then
        colour(vertex,act_colour,preference_weight);
        remove vertex from list;
      else
        act_colour := act_colour + 1;
        used_colors := max(act_color,used_colors)
      fi;
    od;
    next vertex;
  od;
```

Figure 6.1 Algorithm for colouring of general graphs and for graphs with preferences.

This algorithm normally leads, due the preferences, to a node colouring with a larger number of colours than possible, this means additional hardware resources. This increment in area has to be controlled by the algorithm. So first the minimal possible number of colours is calculated. Then the preferences are regarded. The preferences are removed until the growth in area is acceptable. The colouring algorithm for the assignment problem is shown in fig 6.2.

```
COLOR GRAPH
complete graph is the graph with constraints and preferences

if triangulated (constraint graph) then
     determine chromatic number
else
     COLOR GENERAL GRAPH(constraint graph)
fi;
preference_weight := 0;
repeat
     COLOR GENERAL GRAPH(complete graph);
     preference_weigth := preference_weight + increment;
until used colours ≤ chromatic number + threshold

If failed then use optimal algorithm
```

Figure 6.2 Colouring algorithm for graphs with weighted preferences

6.2. Operation Unit Assignment
The operations are assumed not to overlap over the end of a loop. This assumption prevents optimisations like loop rolling which on the other hand can be simulated by an explicit loop rolling at the behavioural level. Combined with preassignment of the operation unit type this leads to a triangulated graphs for the operation unit assignment. In this case an optimal assignment is possible.
Not regarded in this assignment is the minimization of the number of bus connectors. This is here also introduced by preferences. The colouring is then done using the algorithm proposed for the register assignment problem.
In this case the preferences for example are:

- Sources of two operations are the same register
- Destinations of two operations are the same register

6.3. Bus Allocation and Assignment
The bus allocation always leads to a triangulated graph, if only single cycle data transfers are considered. This usually the case in word-oriented architectures. This may also be coloured using the standard algorithms for triangulated graph. The preferences are introduced using the algorithm depicted in figure 6.2. Preferences are

- Two data transfers to or from the same operation unit
- Two data transfers to or from the same register

7. Results and Conclusions

The different results of a representative set of examples is reported in tables 7.1 to 7.3. The first example is the description of a kalman filter which was proposed as a benchmark for the 4th Int. Workshop on High Level Synthesis 1989. This example has a large amount of operations and data transfers. Also one conditional branch is included. The second example is the here presented tiny micro processor as an example for a heavily data dependent execution path.
The fifth order digital wave filter (5th order DWF) is from [Paul89]. This is an example with no data dependencies. The last example is a 16 point FIR filter presented in [PaPa86]. The first table shows the scheduling results obtained with our scheduling algorithm. We have used a library which contains a pipelined multiplier (p-mult) with two cycles execution time and one cycle latency, a normal multiplier (mult) with two cycles execution time and an ALU that is able to perform arithmetic operation except multiplication in one control step and an ADD/SUB unit with one cycle execution time and a comparator.

Example	resources			control steps	exec. time
	p-mult 2 cycles	mult 2 cycles	add/sub/comp 1 cycle		all / main
Kalman filter	1		1	27	2442/1362
		1	1	28	2458/1378
Simple μ-processor			1	4	3.33
5th order DWF	1		2	19	19 / 19
		2	2	18	18 / 18
16 p. FIR filter		2	2	19 (8 IO)	19 / 11
	1		2	19 (8 IO)	19 / 11

Tab 7.1 Scheduling results for the designs

For the first two examples we have chosen a multi processor implementation. The speed up in terms of latency in the main execution in the case of the kalman filter is 25% but with the penalty of over double hardware request. The numbers in brackets are the resources specified to run in parallel in one logical control step. In the case of the processor the speed up is about 100% but with an area penalty of less then 100% since the registers are mostly separated over the two processors.

design example	operations	op. units		calculation time
	mult add sub comp	p-mult add/sub comp	processors global memories	complete latency
Kalman filter	4	2 (1)	3	1953
	4	2 (1)		
	10		1 (dual port)	1056
	12	2 (1)		
Simple μ-processor	-	-	2	4
(Part of MIPS)	5	2 (1)		
	1		1 (4 registers)	~2 (one bus cycle)
	-	-		

Tab 7.2 Results for multiprocessor implementations

For the last two examples we have chosen a single processor implementation. The numbers in brackets show the number of bus connectors if the colouring is done without preferences. In each case a significant decrease in the number of bus connectors was obtained.

Example		#registers	#buses	#BCi	#BCo
5th order DWF	1 p-mult, 2 add	10	5	21 (41)	20 (37)
	2 mult, 2 add	10	5	29 (32)	26 (29)
16 p. FIR filter	2 mult, 2 add	24	8	39 (53)	36 (47)
	1 p-mult, 2 add	24	8	36 (51)	33 (42)
	shift-registers	22	7	34 (49)	30 (38)

Tab 7.3 Results for single processor implementations

The system is currently implemented in PASCAL on an APOLLO workstation. The response times for the examples on a DN4500 are less than 10 minutes.

8. References

[ANRR89] R. Amann, M.Neher, G. Rietsche, W. Rosenstiel
CASTOR: FSM Synthesis in a Digital Circuit Synthesis System
Proc. ESSCIRC, 1989

[CACC80] G.J. Chaitin, M.A.Auslander, A.K.Chandra, J. Cocke, Martin E.Hopkins, P. W.Markstein
Register Allocation via Coloring,
IBM Research Report RC 8395, 1980

[CaWe84] R. Camposano, R. Weber
DSL-Eine Sprache zur Spezifikation digitaler Schaltungen
Internal Report, FZI/University of Karlsruhe Nr 24/84, 1984

[CCDG88] L. Claesen, F. Catthor, D. Lanneer, G. Goossens, S. Note, J. Van Meerbergen, H. De Man
Automatic Synthesis of Signal Processing Benchmark using the CATHEDRAL Silicon Compilers
Proc. IEEE CICC, 1988

[DKNP89] P.Duzy, H.Krämer, M.Neher, M.Pilsl, W. Rosenstiel, T.Wecker
CALLAS -Conversion of Algorithms to Library Adaptable Structures
Proc. VLSI, 1989

[DSDL88] DSDL Sprachbeschreibung
Siemens AG, ZTI KOM 212

[Golu80] M.C. Golumbic
Algorithmic Graph Theory and Perfect Graphs
Academic Press, New York, 1980

[HaCo88] R.I. Hartley, P.F. Corbett
A Digital-Serial Silicon Compiler
Proc. 25th Design Automation Conference, 1988

[KrRo90] H.Krämer, R. Rosenstiel
System Synthesis Using Behavioural Descriptions,
Proc EDAC 90, 1990

[McPC88] M.C. McFarland, A.C. Parker, R. Camposano
Tutorial on High-level Synthesis
Proc. of 25th Design Automation Conference,1988

[PaPa86] N. Park, A. Parker
Sehwa: A Program for Synthesis of Pipelines
Proc. of 23rd Design Automation Conference, 1986

[Paul89] P. Paulin
Scheduling and Binding Algorithms for High-Level Synthesis
Proc. 26th Design Automation Conference, 1989

[Pfah87] Peter Pfahler
Automated Datapath Synthesis: A Compilation Approach
Proc. of EUROMICRO, 1987

[Stok89] L. Stok
Interconnect Optimisation during Data Path Allocation
Proc. 1st EDAC, 1990

DSL-Specification of a Tiny Micro Processor.

```
CIRCUIT  uproc
CONST
   LW      = H'0005';
   SW      = H'000A';
   ARIT    =.H'0010';
   JUMP    = H'0012';
   RET     = H'0013';
INTERFACE
   VCC                  : 5V;
   ground       : GND;
   clk                  : CLOCK;

   adr(31..0)           : OUTPUT;
   data(31..0)   : BIDIRECTIONAL;

VOLTAGE 5V;

VAR
  reg                   : ARRAY[3..0] OF LOGICAL(31..0);
   pc                   : LOGICAL(31..0);
   rs,rt,rd             : LOGICAL(4..0);
   funct                : LOGICAL(5..0);
   op                   : LOGICAL(15..0);

CLOCKBASE clk;

SEQUENTIAL ucycle
   adr := pc;
   cmd := data;
   op := cmd[31..16]
   CASE op OF
          LW      : (* load word *)
                  pc := pc + 1;
                  rs := cmd[25..21]; rt := cmd[20..16];
                  adr := reg[rs];
                  reg[rt] := data;
                  :
          SW      : (* store word *)
                  pc := pc + 1;
                  rs := cmd[25..21]; rt := cmd[20..16];
                  adr := reg[rs];
                  data := reg[rt];
                  :
          ARIT    : (* arithmetic *)
                  pc := pc + 1;
                  rs := cmd[25..21]; rt := cmd[20..16];
                  rd := cmd[15..11]; funct := cmd[5..0];
                  CASE funct OF
                         ADD     : reg[rd] := reg[rt] + reg[rs]; :
                         SUB     : reg[rd] := reg[rt] - reg[rs]; :
                  FO;
                  :
          JUMP    : (* jump *)
                  target := cmd[25..0];
                  pc := pc + target;
                  :
          RET     : (* return *)
                  pc := reg[3];
                  :
   FO;
OD;
```

Logic and Architecture Synthesis
P. Michel and G. Saucier (Editors)
Elsevier Science Publishers B.V. (North-Holland)

A HIGH-LEVEL SYNTHESIS TOOL FOR EXPLOITING PIPELINES IN SPECIAL-PURPOSE SYSTEMS*

Björn FJELLBORG

Department of Computer and Information Science
Linköping University
S-581 83 Linköping, Sweden
E-mail: bfj@ida.liu.se

PiX operates as part of a high-level synthesis system to identify possibilities for pipelining. Unlike previous work, it does not rely exclusively on exploitation of loops to achieve pipelining, but can handle arbitrary designs without apparent pipelining properties. Loops are, however, handled as an important special case. PiX aims at reducing interconnect cost by achieving a close correspondence between required data flow and interconnect structure. Results show that PiX can locate pipelines in arbitrary designs, and also improve the possibility for cost-efficient interconnect for designs that have apparent pipelining opportunities (loops).

1. INTRODUCTION

An important concern in high-level synthesis is the cost-efficient exploitation of concurrency. This can be aided by tools for automatic exploration of different architectures and by considering at an early stage factors like the required amount of multiplexers and interconnect. We present in this paper a tool for exploring the possibilities for *pipelining* in a design, in a way that reduces interconnect by a close correspondence between required data flow and interconnect structure. Pipelining [3] achieves a high degree of concurrency and structural regularity through exploitation of locality in communication. Concurrency is achieved by partitioning of a computation into a sequence of operations, such that subsequent computations overlap. A pipelined processor provides a specific processing element for each operation, and allows a new computation to initiate its first operation as soon as the previous computation has finished executing its first operation. A typical target for pipelining is *loops*, where different iterations can overlap. However, pipelining need not be restricted to this case, since sometimes overlap can be achieved between individual (non-loop) computations, which, if similar enough, can be mapped onto a common pipeline. Such a generalized approach also improves the possibilities to apply pipelining to *parts* of a design and not only to the whole design as contained in an outer loop.

Also if the main speed-up is attained from loop concurrency, pipelining of non-loop computations can serve to further improve speed and regularity of the design (e.g., by using free cycles in a pipeline primarily used by a loop). We believe that this approach is especially suitable for designing special-purpose systems, like embedded systems with tight

* This work is supported by the Swedish Board for Technical Development (STU).

constraints on speed and area, and specialized applications that cannot employ standard solutions with already well-known pipelining properties.

This paper describes the pipeline extractor *PiX* and its underlying strategy. PiX is built as a preprocessor to the *CAMAD* design system [7], see Fig. 1. There is a translator from a Pascal-like algorithmic design description language (ADDL), to ETPN (Extended Timed Petri Nets), the control/data flow representation used by CAMAD. PiX analyzes the ETPN and points out any parts feasible for pipelining by proposing pipeline structures and preliminary schedules for these. CAMAD then performs further high-level synthesis transformations to produce a register transfer data path and, optionally, a finite state machine controller.

PiX is implemented in C on Sun Workstation under SunOS Unix ®.

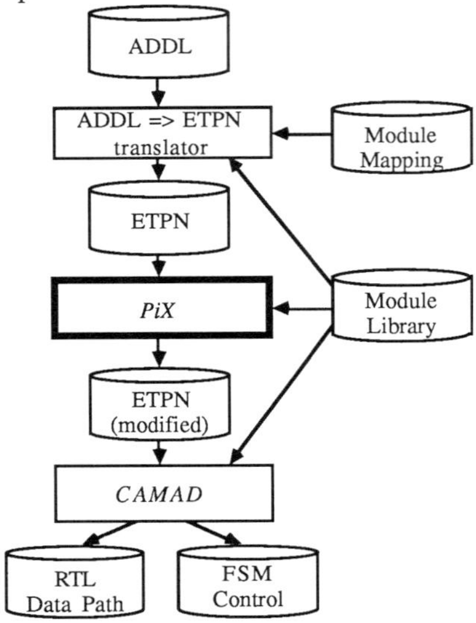

Fig. 1. System overview.

2. PIPELINE EXTRACTION

Pipeline synthesis generates a pipelined implementation of a design, given a high-level specification. *Pipeline extraction* can be defined as a preprocessing phase, that (automatically) identifies, from a high-level specification, those parts that are feasible for pipelined implementation. Full synthesis also includes *generation* of a data path and a control scheme, for both pipelined and non-pipelined parts.

In general, a computation is *pipelinable* if it can be partitioned into a staged pipeline structure, and has instances frequent enough to supply overlapping tasks. Two strategies are useful in collecting computation instances. Either localize a specific computation that is repeatedly executed, or collect distinct but similar computations that each are executed once. Loops are typical targets for the former strategy, and this is used by current pipeline synthesis systems to provide the necessary supply of tasks. Though this is an efficient method to find pipelines that will provide significant performance enhancement in many cases, it also narrows the range of designs to be considered for pipelining. PiX instead exploits the possibility of constructing efficient pipeline structures by regarding a design as a collection of distinct computations which, if having enough similarity and temporal proximity, can be mapped onto a common pipeline. The necessary supply of tasks is thereby ensured from the amount of such computations, and not primarily by the repetition of a specific computation. However, "loop pipelines" form an important special case, and deserve special treatment to exploit their explicit regularity. Fig. 2 shows a data flow graph, with a pipeline used by a loop and another pipeline used by three non-loop computations.

Grouping of computations into pipelines and selection of an optimal set of pipelines are problems of exponential complexity. By considering only linear, i.e., non-branching, computations, and employing a heuristics-guided iterative search strategy, PiX attains a practical pipeline extraction procedure of polynomial complexity.

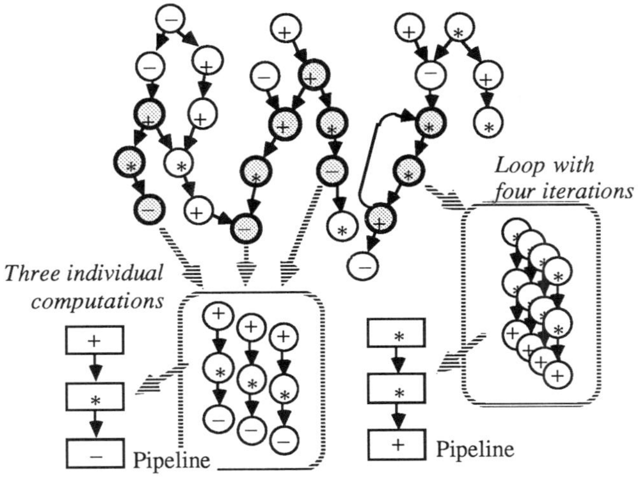

Fig. 2. Pipelining of loop and non-loop computations.

3. RELATED WORK

Several systems for *pipeline synthesis* have been proposed. *Sehwa* [5] is a dedicated pipeline synthesis tool, that allocates function modules for a pipelined data path for operations enclosed by an outer loop with constant latency. No interconnect is synthesized, but its cost is estimated. *HAL* [6] and Devadas and Newton [1] are general-purpose high-level synthesis tools, with pipelining as an option. Both assume an outer loop for the whole design, and pipelining takes place for the different iterations of this loop. The cost for modules, registers, and interconnect is optimized by global algorithms. The use of *pipelined function units*, as in *Slicer* [4] and *HAL* [6], allows overlap also between non-loop computations. However, it is restricted to short pipelines for primitive functions, and achieves a very restricted amount of pipelining unless relying on a loop. *Loop winding* transformations [2] is a preprocessing step to achieve a higher amount of overlap in ordinary data path synthesis.

These systems provide explorative tools for evaluating parts of a larger design. However, since they concentrate on loops, they cannot exploit any potential for pipelining of computations outside loops. This also means that the overall design cannot be improved by automatically identifying *parts* in it that may benefit from pipelining. Loop winding [2] does allow transformations to be applied to part of the design, but these decisions are made manually. PiX, on the other hand, employs a general approach that automatically locates opportunities for pipelining *both* in loops and outside loops.

Furthermore, those systems that allocate interconnect (*HAL* [6] and Devadas and Newton [1]) do so without special consideration of the regularity found in pipelined data paths. PiX makes this regularity explicit in its output.

However, when comparing PiX with the above presented systems, it should be kept in mind that PiX is a *preprocessing* tool, not a complete synthesis tool. PiX should therefore be regarded as a complement rather than as an alternative to these systems.

4. PiX – A PIPELINE EXTRACTION TOOL

We now outline an algorithm for extraction of pipeline structures from a data flow graph, under a constraint (T_{MAX}) for the design's execution time. To reduce complexity, the search is restricted to linear, non-branching computations.

The internal design representation ETPN [7] uses an abstract data path to describe data movements and a Petri net to describe control flow, i.e., precedence constraints between statements in the input design description, conditional execution, and loops. A module library gives cost and delay for the operations. PiX operates on a data flow abstraction of the ETPN, where nodes represent primitive operations in the input description, and arcs represent actual data movements. Fig. 3 shows a simple example.

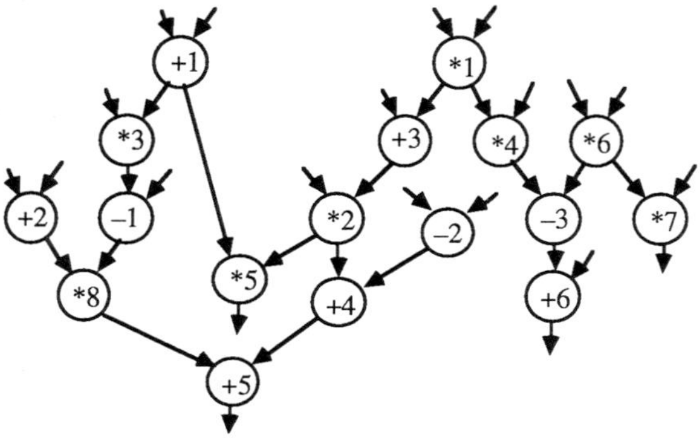

Fig. 3. Example data flow graph.

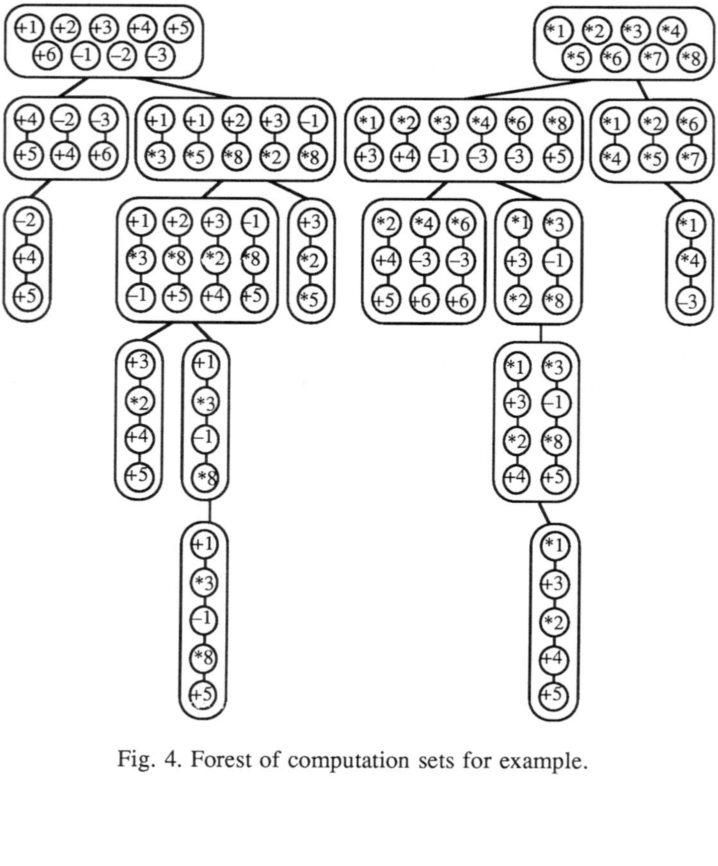

Fig. 4. Forest of computation sets for example.

A connected group of operations in the data flow constitue a *computation*. *Compatible computations* can execute on the same hardware. *Sets of compatible computations* can be arranged into a forest of trees, rooted in the shortest computation sets, and with each branch representing a potential linear pipeline. Fig. 4 shows the complete forest for the example of Fig. 3. (It is assumed that there are two module types, one for + and – and one for *.)

PiX identifies those sets (branches) that are beneficially realized as pipelines. First, only the roots are identified. The branches are then developed selectively, through an iterative extension/search:

1. Select a computation set.
2. Extend the corresponding branch, i.e, identify the computations in the branch; repeat 1-2 (as far as T_{MAX} allows).
3. Evaluate whether the identified computations would be cost-efficient if pipelined.
4. If feasible for pipelining, bind the operations to a pipeline and remove them from further consideration.

This repeats until no more pipelines are found. Locating a feasible pipeline reduces the remaining forest, since certain branches then contain bound operations. This avoids having to actually identify all the potential computations.

The algorithm is presented in somewhat more detail below.

Algorithm

Input: A data flow graph and a time constraint T_{MAX}.
Output: A set of pipelined hardware structures with ordered lists of computations, indicating a scheduling order.

1. Determine ASAP and ALAP schedules, and from these, the minimal execution time T_{min}.
2. If $T_{min} > T_{MAX}$, stop and return error.
3. Identify all sets of computations of length 2.
4. Set $i = 2$.
5. Select a pipelinable computation set of length i.
6. If no computation set was found, then if $i \leq 2$ terminate otherwise go to 9.
7. Mark the computation set as selected and extend it to (i.e., identify the) computation sets of length $i+1$.
8. Set $i = i + 1$. Go to 5.
9. Evaluate whether the selected (and pipelinable) computation sets form a feasible pipeline.

10a. If the computations were feasible:
   i. Add the corresponding pipeline to the output.
   ii. Mark the computations as bound.
   iii. Mark all computation sets as unselected.

10b. If the computations were not feasible:
   i. Mark the longest selected computation set as non-pipelinable.
   ii. Set $i = i - 1$.

11. Go to 5.

The specific procedures for selection, identification, and evaluation can employ various heuristics to guide the search in a way that can be tailored to specific goals and constraints. PiX uses a greedy extension strategy that aims at finding the most useful pipelines first (e.g., loops). For evaluation, PiX checks whether the computations can be equally or better implemented by sequential (non-overlapping) execution on the corresponding non-pipelined hardware; if the amount of overlap is small, pipeline staging may result in a longer total execution time. In general, this step involves selecting a subset of the evaluated computations, which again can be formulated to primarily pipeline any heavy loops; non-used cycles could be used by smaller loops or non-loop computations.

Information concerning potential for overlap, temporal mobility, and criticality is derived from *as soon as possible* (ASAP) and *as late as possible* (ALAP) schedules. These are incrementally modified (using methods similar to [4]) as binding decisions are made. Fig. 5 shows the ASAP schedule for the example data flow graph in Fig. 3.

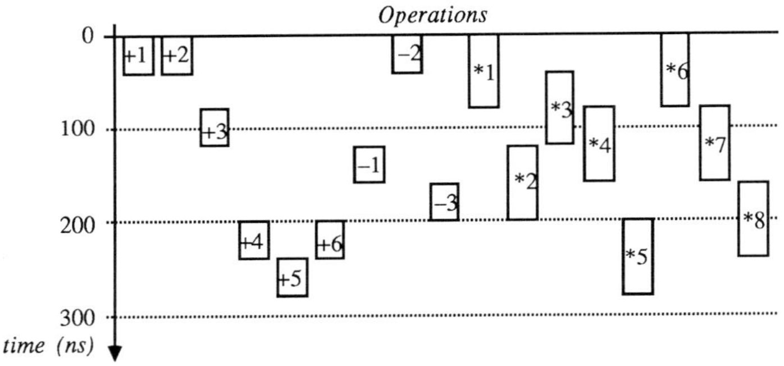

Fig. 5. ASAP schedule for example.

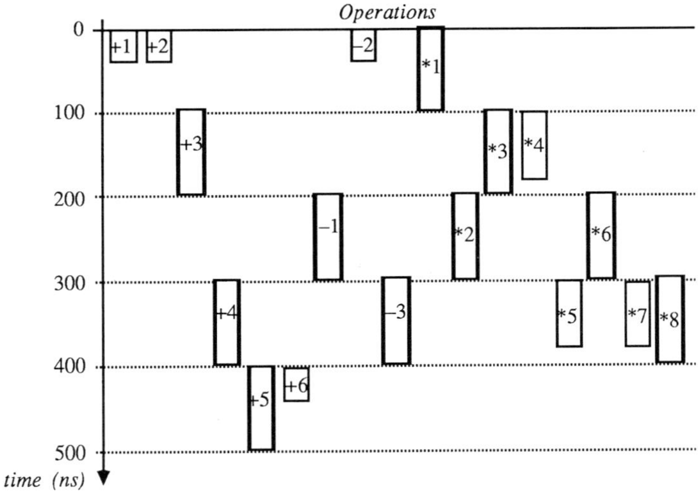

Fig. 6. ASAP schedule for example, with pipeline.

We illustrate the algorithm by the example data flow graph in Fig. 3. We assume a module library with add (having delay 40 ns), multiply (80 ns), and stage latch (20 ns). Figs. 7a-e show the sequence of events, for T_{MAX} = 500 ns. Shown parts of the forest correspond to identified computation sets.

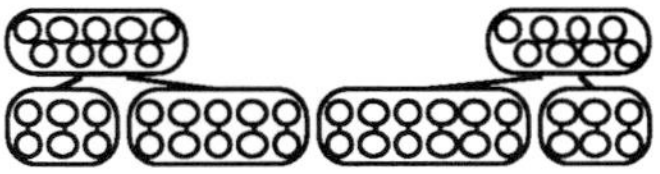

Fig. 7a. Initial state: potential pipelines of length 2 identified.

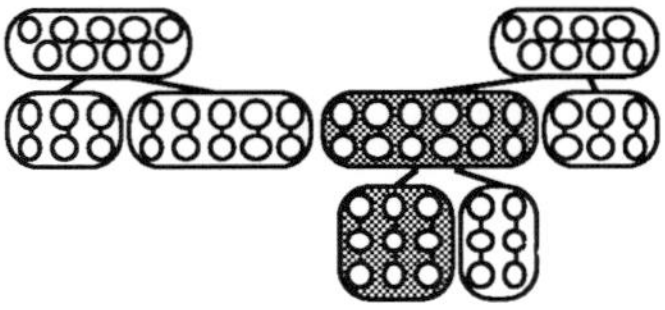

Fig. 7b. Selection of one branch, followed by extension. Selected sets (shaded) evaluated: not feasible.

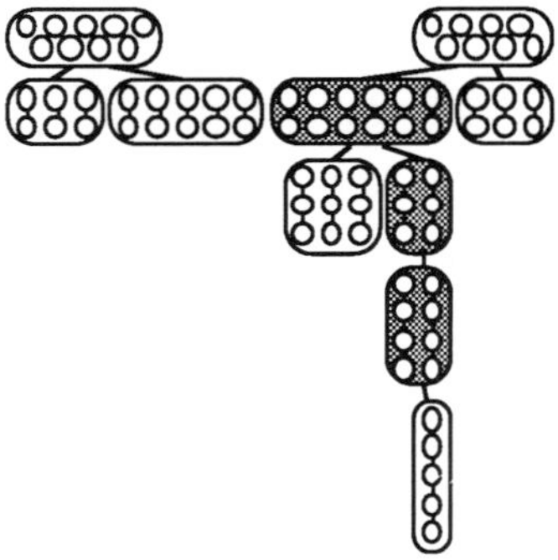

Fig. 7c. Selection and extension of new branch. Evaluation: feasible!

The result is one four-stage pipeline used by three computations, as shown in Fig. 8. The resulting ASAP schedule (with pipelined computations marked) is shown in Fig 6.

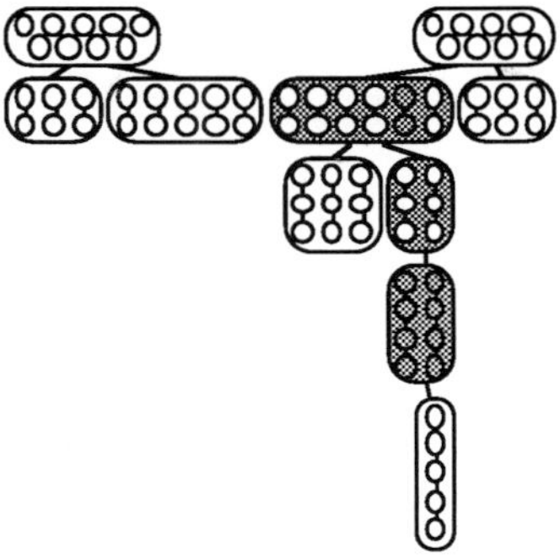

Fig. 7d. Bind computations (shaded) to pipeline.

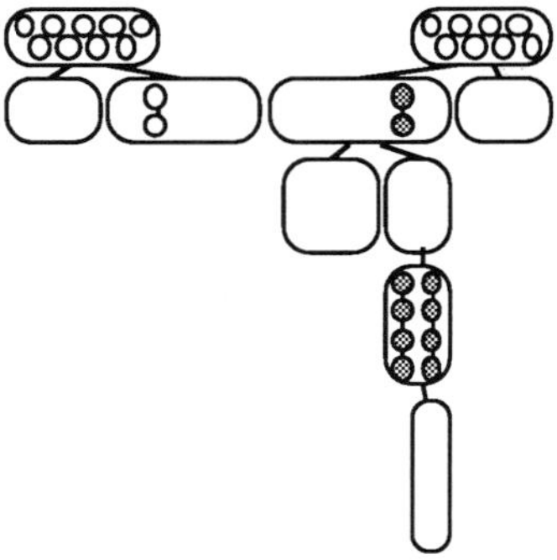

Fig. 7e. Reduce forest by excluding colliding computations. No usable sets left; terminate.

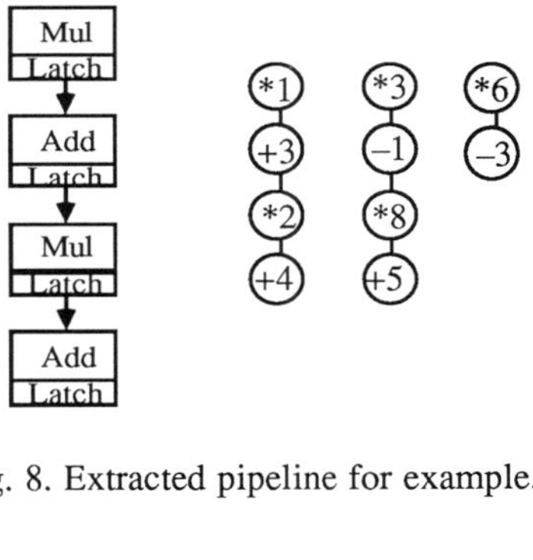

Fig. 8. Extracted pipeline for example.

Conditions affect pipeline extraction in that computations become mutually exclusive and in that their execution probability influences a potential pipeline's feasibility. PiX assumes conditions to have probabilities assigned to the outcomes of the test. These are used as *weights* when determining what computation set to extend. Also, mutual exclusivity is exploited by sharing of hardware.

Loops could be encompassed by the general approach by complete unrolling. However, PiX exploits a loop's regularity by regarding its iterations as a single computation wherever possible. This computation is attributed with an iteration count, which serves as a *weight* (as for conditions). For loops with loop-carried dependencies, a minimal inter-iteration latency is calculated. This sets a limit for the amount of pipelining that can be attained for the loop. In the ASAP and ALAP schedules, only the first/last instance is represented explicitly.

The handling of conditions and loops thus requires information about expected branch probabilities respectively number of iterations. This is a problem common to all parallelism extraction methods, and is not considered in this paper.

For designs with maximized pipeline depth and outdegree of operations, the worst-case complexity of the algorithm is $O(n^3)$, where n is the number of operations in the design.

5. RESULTS

Fig. 9 shows the data flow graph for a 16-point digital FIR filter (from [5]). The data flow graph is contained within an outer loop, which continuously provides new sets of data.

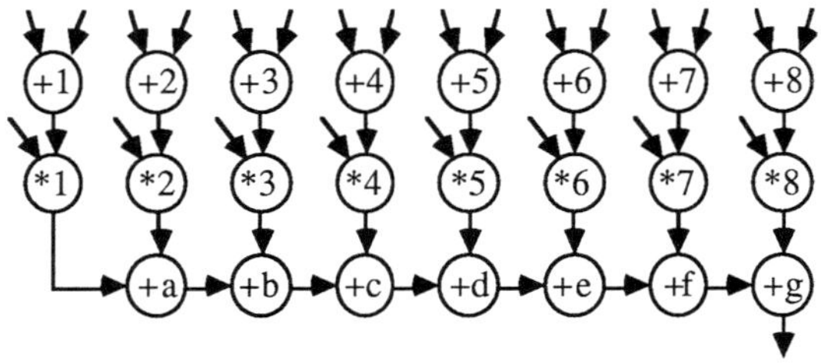

Fig. 9. 16-point digital FIR filter.

We use a module library with a 40 ns adder and an 80 ns multiplier. New data sets arrive every 300 ns, which is modeled as an inter-iteration latency for the surrounding loop. To handle the case of a loop with, in principle, an infinite number of iterations, a throughput constraint is more appropriate than a constraint on the maximum execution time. However, it is possible to model this by calculating the time needed to complete N iterations as $T_{MAX} = 900 + (N - 1) \cdot 300$, in this case. Using $N = 10$, we have $T_{MAX} = 3600$ ns. From this, PiX produces two pipelines, as shown in Fig. 10a (the dashed parts show a possible implementation for the remaining, non-pipelined, computations, which are *not* produced by PiX). Fig. 10b shows a straightforward interconnect solution for the module allocation by Sehwa's feasible scheduling algorithm [5] (Sehwa does not allocate interconnect itself), under the same assumptions. The number of modules is the same, but PiX clearly indicates the opportunities for regular interconnect. This is a result of PiX being designed around the concept of identifying pipeline *paths* in the design. The CPU time for this example is a few seconds on a Sparc workstation.

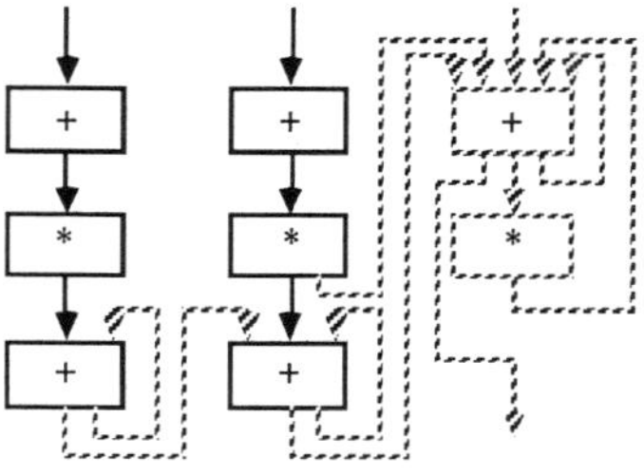

Fig. 10a. PiX allocation for filter.

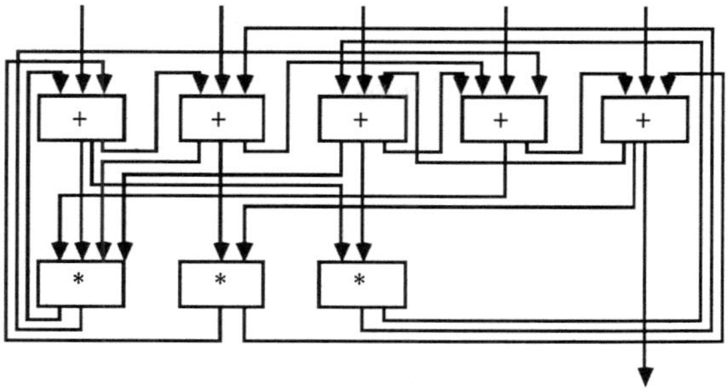

Fig. 10b. Sehwa allocation for filter.

Since the pipeline extraction strategy in PiX is not based primarily on location of loops, it is interesting to consider a single iteration of the filter, which itself contains some potential for pipelining. Doing so, with T_{MAX} = 900 ns, yields a single pipeline, as shown in Fig. 11.

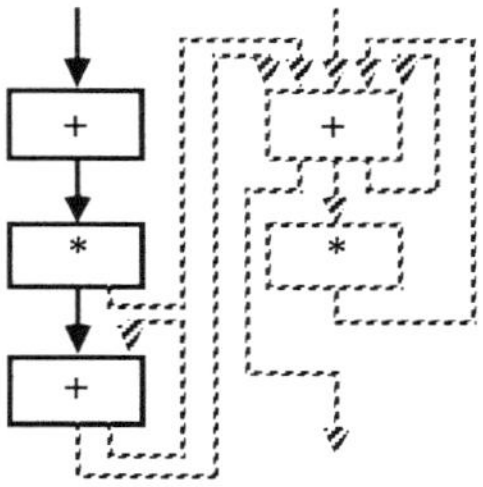

Fig. 11. PiX allocation for single filter iteration.

We have also applied PiX to other designs without apparent pipelining opportunities, such as a single calculation of a numerical square root algorithm. Depending on the given constraints, PiX can pipeline 40-80% of this design.

To test PiX's ability to handle a larger design, an algorithm for the Data Encryption Standard (DES), containing about 2000 operations, was coded. In this design, PiX found from 1 to 74 pipelines, covering 48-87% of the operations. The median CPU time for this design was about 5 minutes.

6. CONCLUSION AND FUTURE WORK

We have presented a new high-level synthesis tool, PiX, which can improve the quality of designs by pipelining of selected parts. PiX employs a new methodology for extracting pipelines, that do not rely exclusively on loops, but considers overlap between individual computations. Still, the special case of loops deserves special treatment, and can be efficiently encompassed within the approach. Results from a prototype of PiX show that this approach can locate pipelines in designs without apparent pipelining opportunities. Also for designs that have apparent pipelining opportunities (loops), PiX can improve the possibility to create cost-efficient interconnect.

An interesting question is how much *extra* pipelining one can find by using the methodology presented here, as compared to relying exclusively on loops. This depends totally on the specific problem formulation. The method presented here will not reveal large amounts of pipelining from non-loop computations only. As with all parallelism extraction methods, the result depends on how much intrinsic parallelism there is in the algorithm formulation.

The amount of "extra" parallelism in the form of pipelining also depends on how successful the algorithm designer has been in trying to make any parallelism explicit by using loops. An alternative view on the approach presented here, is that it can find computations that perhaps *could* have been formulated as loops, but for reasons of, e.g., clarity of design formulation, have not been so.

An algorithm based on this strategy may use various techniques for solving the subproblems involved, concerning, e.g., design space search, evaluation of potential pipelines, and stage partitioning. The algorithm is open for improvement by inclusion of a wider range of heuristics to guide these phases. There are also various extensions to consider, such as multi-function pipelines (which can rather straightforwardly be implemented by considering several tree branches instead of one), multi-level pipelining, and other design constraints. Also, some performance-impacting factors can be taken into account, such as the amount of hazards and interrupts [3].

REFERENCES

[1] S. Devadas and A. R. Newton: Algorithms for Hardware Allocation in Data Path Synthesis, *IEEE Trans. Computer-Aided Design* 8 (1989) 768-781.

[2] E. F. Girczyc: Loop Winding – A Data Flow Approach to Functional Pipelining, *Proc. IEEE Int. Symp. Circuits and Systems* Vol. 2 (1987) 382-385.

[3] P. M. Kogge: *The Architecture of Pipelined Computers* (McGraw-Hill, 1981).

[4] B. M. Pangrle and D. D. Gajski: Slicer: A State Synthesizer for Intelligent Silicon Compilation, *Proc. IEEE Int. Conf. Computer Design* (1987) 42-45.

[5] N. Park and A. C. Parker: Sehwa: A Software Package for Synthesis of Pipelines from Behavioral Specifications, *IEEE Trans. Computer-Aided Design* CAD-7 (1988) 356-370.

[6] P. G. Paulin and J. P. Knight: Force-Directed Scheduling for the Behavioral Synthesis of ASIC's, *IEEE Trans. Computer-Aided Design* 8 (1989) 661-679.

[7] Z. Peng: Synthesis of VLSI Systems with the CAMAD Design Aid, *Proc. 23rd Design Automation Conf.* (1986) 278-284.

Logic and Architecture Synthesis
P. Michel and G. Saucier (Editors)
Elsevier Science Publishers B.V. (North-Holland)
IFIP, 1991

A High-Level Synthesis Algorithm Based On Area Oriented Design Transformations

Wolfgang Schenk*
University of Dortmund
FB Informatik, LS XII
D-4600 Dortmund 50, W. Germany

April 5, 1990

1 Introduction

High-level synthesis of VLSI processor structures from behavioural specification is measured in terms of chip area and execution speed. Problems with current CAD systems arise from the lack of knowledge about low level details in early design decisions. With the integration of floorplanning into high-level synthesis area efficient designs are generated. Each area oriented design decision is based on the floorplanners estimation.

The design specification consists of constraints on the structure to be generated. In order to meet the users ideas about the resulting structure, we allow for a rich set of constraints in the specification. The complexity of the synthesis process depends on the size of the design space. The more constraints are given to the synthesiser, the smaller is the design space to be explored. A powerful feature of our synthesis system is the handling of complex predesigned substructures. They are present in the library of available modules, and either instantiated by the choice of the synthesis system, or by the users advice.

In this paper we describe the new SuccAss synthesis algorithm. SuccAss stands for *succ*essive *a*proximation *s*ynthesis *s*ystem. The algorithm synthesises synchronous, non-pipelined, microprogrammed processor structures with a single phase clock. It is integrated into the MIMOLA design environment [Bro89]. SuccAss features the usage of complex partial structures, and the integration of low level details into the design process.

1.1 The Phase Coupling Problem

In high level synthesis the phase coupling problem is caused by the decomposition of the design problem into subproblems (scheduling, allocation, binding of operations, generation of control), which are solved in a fixed order. We believe that scheduling is of secondary importance for the design problem. If the synthesis system starts with scheduling, the minimum number of operators in the final design is determined without any knowledge about their costs. If the schedule of operations and the set of modules are fixed, the binding of operations to modules implies the interconnections between them.

1.2 Integration of Module Selection, Allocation, Binding, and Scheduling

The synthesis algorithm proceeds by applying area decreasing transformations to an initially generated structure until an upper limit in chip area is achieved. The main design decisions are the selection of modules from the library, and the generation of interconnections. The integration of module selection, scheduling, and binding of operations into the transformations circumvents the classical phase coupling

*Tel.: ++49 (231) 755–5101 Fax: ++49 (231) 755–2047 E-mail: schenk@ls12.informatik.uni-dortmund.dbp.de

problem. The binding of operations to modules has to assert, that no module is used in different ways at the same time. Nevertheless a legal operation to time step binding has to be set up. Given the modules and their interconnections, the scheduling reduces to the microcode compaction problem.

The selection and application of the transformations are based on estimation of the resulting chip area. In order to do so a simple floorplanner is integrated into the system. Thereby, multiplexer and interconnection costs are taken into account. The floorplanner does the placement based on a mincut algorithm. Wiring area is considered by the generation of channels for global routing. The floorplanner returns the total occupied chip area, the locations of the parts, and the length of each interconnection. The synthesiser makes use of the information in order to recognise parts of the structure to be optimised.

1.3 Design Constraints

Since most designers do not start from scratch, they almost always have a pretty good idea of some parts of the final design. We think the designer should be allowed to put his knowledge into the specification of the design problem [Bro89a]. Therefore the MIMOLA language has been extended by constructs that allow the description of additional constraints and lower level details of the design. The SuccAss algorithm makes use of the designers knowledge, which leads to better designs and shorter turn around time. The library of available modules, the maximum occupied chip area, the size of the control storage, the maximum cycle time, the delay of operations, and the maximum number of temporary variables are considered as constraints.

Furthermore, the presence of a component may be required by the designer. Thus, the designer is allowed to specify key components, that are incorporated into the final design by the synthesis system. In the behavioural system specification the designer may refer to the key components by means of bound operations and variables [Mar89]. Storages are considered as key components. The available memories and registers serve as constraints for the maximum allowed number of parallel reads and writes.

1.4 Partial Structures

Partial structures are specified either in the module library, or declared by the designer as key components. Key components are included into the design unconditionally. In either case their behavioural view is described.

The problem arises, whether or not a module is useful to implement some part of the system specification. Thus we have developed a mechanism that treats the equivalence problem between the specification and the behavioural description of the module.

2 Representation of the Design Problem

In this section the design problem, as treated by the SuccAss algorithm, is described. Especially the representation of software and hardware structures influences heavily the efficiency of the synthesis system.

2.1 Representation of the Behaviour

The external representation of the behaviour is a MIMOLA program [Jöh89]. The front end of the system is a compiler that translates the specification into an intermediate code. The intermediate code is expanded to a RTL-program, which essentially is a set of basic blocks. A basic block is a sequence of parallel executable blocks. Each parallel block is a set of statements, which is an assignment or an activation statement. All variable occurrences are replaced by storage accesses.

The destination (left hand side) of an assignment is a register or memory cell. The right hand side of an assignment is an expression made up from constants, storage-accesses, register-reads and operations among them: $e ::= const \mid mem[e_{addr}] \mid reg \mid \circ(e_1, \ldots, e_n)$, where $\circ \in O = \{\circ_1, \ldots, \circ_o\}$.

An activation $M(s_1 \leftarrow e_1, \ldots, s_n \leftarrow e_n)$ of a module M is the specification of the values e_i for inputs s_i of M. A bound activation refers to a component of a structure rather than to a library entry.

2.2 Structure Representation

A structure $Struct = (Parts, Conn)$ consists of a set of parts and their interconnections. Parts are incarnations of modules from a library. With the exception of multiplexers and tri-state drivers, the synthesis algorithm instantiates modules from the given library only.

2.2.1 Partial Structures

The designer declares a set of modules the synthesis system is allowed to use. The library entries could be generated for example by module generation tools. Even AMD-type bitslices can be used by the synthesiser. The synthesis task is technology dependent only as far as the library is.

2.2.2 Module Specifications

A module is an entry in the library $Lib = \{M_1, \ldots, M_m\}$. The modules are described in their behavioural view.

A module $M \in Lib$ is characterised by

- $In(M)$ set of inputs
- $Out(M)$ set of outputs
- $State$ set of internal states (e.g.: registers or memories) of M
- $Fct(M)$ set of guarded assignments $/guard/dst \leftarrow f$, where
  - $guard$ is the control code, that initiates the evaluation of e. If the function has to be clocked with a specific clock edge, this information belongs to the $guard$ as well.
  - $dst \in Out(M) \cup State$
  - $f ::= const \mid signal \mid var \mid var[f] \mid \circ(f_1, \ldots, f_n)$ with $\circ \in O$, $signal \in In(M)$, $var \in State$.

Furthermore the chip area of M in terms of x- and y-dimensions and the delay associated with each of the guarded assignments are parts of the module description. Bidirectional ports and busses are neglected in this description.

Figure 1 illustrates a simple example of a partial structure specification. The internal structure of the partial structure is not explicitly visible to the synthesis system, but the behavioural view describes the usage of the module sufficiently and completely.

```
PARTS
sadd:  MODULE shft_add
  (IN a,b:word;IN c:twobits; OUT f:word)
  BEHAVIOUR AtRtLevel IS
    ATTRIBUTE DimX IS ( 48*Lambda);
    ATTRIBUTE DimY IS (256*Lambda);
  BEGIN
    f <- CASE c OF
      0:  SHIFTLL(a) AFTER 4;
      1:  SHIFTLL(a + b) AFTER 12;
      2:  a + b AFTER 10;
      3:  a - b AFTER 10;
    END;
  END;
```

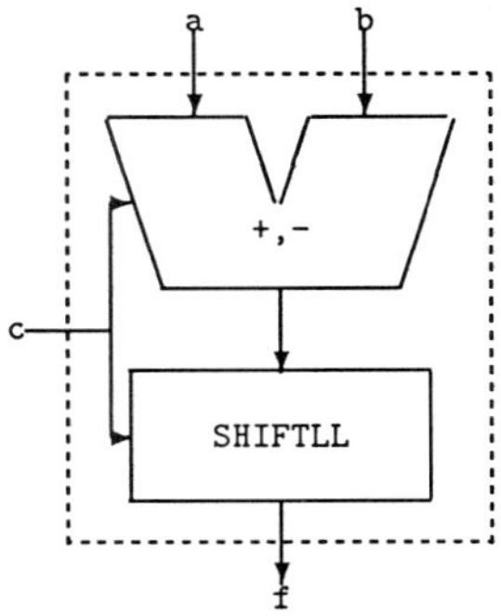

Figure 1: ALU with subsequent shifter

2.3 The Matching of System and Component Behaviour

In order to select suitable components from the module library, the synthesiser has to determine whether or not a module can implement a given expression. If there is a way of controlling the module to induce the desired functionality, the matching process generates an activation. The expression $SHIFTLL(2+3)$ may be implemented for example by using the activation $shft\_add(c \leftarrow 1, a \leftarrow 2, b \leftarrow 3)$.

The function $Match'$ implements an expression e using a function $f \in Fct(M)$ of module M. It returns a list of input ports with associated values (signal assignments), or $fail$, if M cannot do e.

$$Match'(e,M,f) = \begin{cases} X & if \quad e, f \in const, e = f \\ f \leftarrow e & if \quad f \in In(M) \\ a_1, \ldots, a_n & if \quad e = o_e(e_1, \ldots, e_{n_e}), f = o_f(f_1, \ldots, f_{n_f}) \\ & \qquad o_e = o_f, n_e = n_f \\ & \qquad \forall i \in \{1, \ldots, n\} : a_i = Match'(e_i, M, f_i) \neq fail \\ fail & otherwise \end{cases}$$

The input values represent a set of remaining subexpressions of e, which still have to be implemented.

The function $Match$ returns a complete set of activations that implement e on M.

$$Match(e,M) = \bigcup_{f \in Fct(M)} \{(M(ctr \leftarrow guard(f)), Match'(e,M,f)) \mid Match'(e,M,f) \neq fail\}$$

where $ctr \in In(M)$ is the control input of M.

The matching process actually implemented is extended by some algebraic equivalences. Indeed the activation $shft\_add(c \leftarrow 1, a \leftarrow 3, b \leftarrow 2)$ is another legal match with $SHIFTLL(2+3)$. The swapping of inputs, commutativity of operations, and the reduction of strength are mentioned in [Mar90].

3 Synthesis by Successive Approximation

The SuccAss algorithm starts with a maximum parallelised RT-program. The degree of parallelisation is restricted by the data dependencies among the statements of a parallel block. Further restrictions are the maximum cycle time for the chaining of operations, and the number of memory ports for the number of simultanious read and write accesses.

An initial structure is generated, which is large enough to execute each parallel block in a single microinstruction. A floorplan is generated from the initial structure, which usually exceeds the maximum allowed size. As long as the current structure is too large, the algorithm applies area decreasing transformations to it (see fig. 2).

Since the required behaviour of the design has to be kept, the hardware transformations preserve its functionality. The relationship between structure and behaviour is represented by a set of bindings.

Operation to module binding. Each operation may be performed by different modules. This kind of binding is refined by the concept of partial structures. Instead of a single operation, we consider a set of complex expressions a module may implement. To each expression in the program the set of module activations is attached. Moreover, a module may be activated in several ways to evaluate the given expression. That is, for each expression e we determine the set $Vers(e) = \bigcup_{M \in Lib} Match(e,M)$ of activations $Vers(e)$, which match e. The members of $Vers(e)$ are called versions of e. From compiler construction techniques, we adopted the value numbering approach, in order to get a fixed set of expressions from the behavioural specification. Common subexpressions are detected and numbered consistently. The numbers represent the fixed set of expressions belonging to the given design problem. Thus the set of versions is calculated only once.

Operation to part binding. This binding again is represented by means of activations. In order to avoid resource conflicts, the synthesiser assures that no part is used in different ways at the same time. An activation from $Vers(e)$ bound to a part of the current structure is called an alternative

```
Generate Initial Structure;
WHILE structure does not fit on the chip DO
  BEGIN
    apply transformations not generating new control steps
        in order to reduce area;
    IF more transformations applicable THEN
      apply transformation with least increase in control steps
          and largest decrease in chip area;
    IF size of control storage exhausted THEN
        EXIT design constraints could not be met;
  END;
Generate hardwired constants;
Generate Busses;
PRINT structure;
```

Figure 2: The SuccAss algorithm

of expression e. For the example above, the bound activation is $sadd(c \leftarrow 1, a \leftarrow 2, b \leftarrow 3)$. It refers to the part in a structure rather than to a library entry. A set of conflict free bound activations for the statements of a parallel block is called an alternative for that block, regardless whether or not additional interconnections are necessary. The synthesis algorithm maintains the set of alternatives for all parallel blocks in the program. It keeps track, that at least one alternative exists for each block.

Operation to control step binding. The notion of parallel blocks defines a set of statements executed in the same microinstruction. The data dependencies between two statements in a basic block define either:

- the strict execution sequence,
- or the possible parallel or sequential execution,
- or an arbitrary execution sequence.

The statements of a basic block have to be scheduled into a minimum number of parallel blocks. Since the SuccAss algorithm always works on a complete structure, this type of scheduling reduces to the microcode compaction problem under fixed resources. From an ASAP (as soon as possible) and an ALAP (as late as possible) schedule the scheduling space of each operation is determined. If a parallel block has to be split, the scheduling space of its operations is considered for composing a new parallel block.

We distinguish two modes the transformation are applicable in. In *restricted* mode parts of the structure are removed by rebinding operations bound currently to them. The structure is optimised without changing the operation to control step binding (scheduling) of the program. The cycle time may increase, but no additional microinstructions are generated.

Whenever no transformation decreases the chip area without increasing the number of parallel blocks, the SuccAss algorithm continues in *unrestricted* mode. Even the rescheduling of operations is allowed in *unrestricted* mode. A part may be removed, if all operations bound to it can be rebound to control steps, where idle hardware resources are available. If no rebinding to other microinstructions is possible, a new parallel block is introduced.

3.1 Local Transformations

A local transformation detects a part in the current structure which may be removed. This type of transformation is called local, because only a single part is considered, and no additional module is selected from the library. For all parallel blocks, which use the considered part, the set of alternatives is calculated.

The transformations create all possible alternatives that reduce chip area. Since the rebinding of operations may require additional interconnections and multiplexers, the algorithm trades between the saved area, and the spare area.

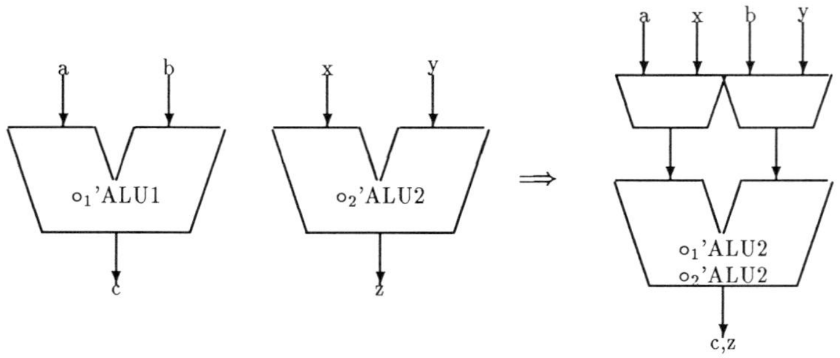

Figure 3: Removal of an ALU

Figure 3 illustrates a local transformation that removes $ALU1$. The expression $a \circ_1 b$ is previously performed by $ALU1$. The transformation detects the alternative $ALU2$ for the expression, and determines the new multiplexer and wiring costs. The length of the additional interconnections is estimated based on the most recent floorplan.

In restricted mode, the assignments $c \leftarrow a \circ_1 {}'ALU1\, b$ and $z \leftarrow x \circ_2 {}'ALU2\, y$ may not occur in the same parallel block. When applied unrestricted the transformation tries to reschedule one of the assignments to a microinstruction, that belongs to its scheduling space and where $ALU2$ is idle.

If no substructures can be removed further, a locally optimised structure is achieved. At this point the algorithm applies global transformations to the structure.

3.2 Global Transformations

The main task of the global transformation is the selection of a new module from the library, that enhances the current structure. The selection of the library entry is based on the following policy:

- The module is a multiple function unit.
- The module replaces some functionality of the structure; i.e., an incarnation of the module makes further local transformations applicable.

For each candidate the floorplanner is invoked in order to select the transformation with the largest decrease in chip area. The selected module is instantiated and the local transformations are applied in unrestricted mode. The resulting structure becomes current. It occupies less area, but probably needs more control steps.

4 Results

In this section we present the results from two examples available in the literature. The host system is a DN4000 APOLLO workstation with 4 MB memory.

PDP8: Table 1 shows the results of the synthesis of four PDP8 compatible processors. We examined four designs of the processor, each of them from a specification with an additional set of user defined bindings:

1. Variables to storage bindings.
2. The degree of parallelization is fixed by means of sequential and parallel executable blocks.
3. Definition of key components

4. Binding of operations to key components

The quality of the four designs is expressed in the chip area occupied, the number of static microinstructions, and the cycle time. In addition the number of wires is included.

manual binding	area	wires	#μ-instr.	cycl. time
1.	113	778	110	42
2.	265	846	88	45
3.	122	830	88	39
4.	172	832	90	33

Table 1: PDP8 processor design

The more information has been added to the description, the better is the resulting design. The turn around time decreases from 4000 cpu seconds for the first specification downto 1000 seconds with all sets of bindings added.

5th Order Elliptic Filter: This example is adopted from [Kun85]. It is commonly considered as a benchmark for scheduling algorithms, because there is a large scheduling space, and a rich set of possible sharable resources. The module library is restricted to single operation ALU's, and single port memories.

The smallest design is made from 5 registers, 1 adder, 1 multiplier, and 33 microinstructions are generated for control. We get the fastest design from a specification, where each variable is manually bound to a register, and a cycle time small enough to prevent the chaining of modules is chosen. The resulting structure consists of 8 registers, 3 adders, 1 multiplier, and is controlled by 14 microinstructions.

5 Future Work

Until now the synthesis algorithm creates multiplexers, tri-state drivers, and busses without referring to the module library. With the notion of generic modules we will allow for parametrised module descriptions.

Another extension of the module description will allow for multiple clocked operations. At the very end we will be able to incorporate complete processors into high level synthesis.

6 Conclusion

We have described the SuccAss algorithm. The most remarkable features are the use of partial structures and the integration of floorplanning into the design process.

This approach is based on the selection and application of hardware transformations. Starting with a maximum parallel structure, the exploration of the design space proceeds towards optimality. The transformation actually applied is chosen based on a floorplanner, which gives estimates of high confidence of the chip area.

The synthesis system supports a rich set of user defined bindings. With the restricted design space, the algorithm puts attention to the interesting solutions, not wasting time in unwanted alternatives. The designs are generated faster, and meets the users intention more closely.

References

[Bro89] Broß, O. *"RT-Synthese mittels fortgesetzter Annäherung"*, master's thesis, Inst. f. Informatik, University of Kiel, Aug. 1989.

[Bro89a] Broß, O.; Marwedel, P.; Schenk, W. *"Incremental Synthesis and Support for Manual Binding in the MIMOLA Hardware Design System"*, High Level Synthesis Workshop 1989, Kennebunkport (Maine) 1989.

[Jöh89] Jöhnk, R.; Marwedel, P. *"MIMOLA Reference Manual, Version 3.45"*, Report 8902, University of Kiel, Institute f. Informatik, 1989.

[Kun85] Kung, D.E.; Whitehouse, H.J.; Kailath, T. *"VLSI and Modern Signal Processing"*, Prentice Hall Information and System Series, 1985, pp. 258-264.

[Mar89] Marwedel, P.; Schenk, W. *"Improving the Performance of High-Level Synthesis"*, Proc. Euromicro 1989, Köln (W. Germany) 1989.

[Mar90] Marwedel, P. *"Matching System and Component Behaviour in MIMOLA Synthesis Tools"*, Proc. EDAC'90.

Logic and Architecture Synthesis I

Logic and Architecture Synthesis
P. Michel and G. Saucier (Editors)
Elsevier Science Publishers B.V. (North-Holland)

OASIS®: A Silicon Compiler for Semi–Custom Design

Gershon Kedem

Computer Science Department
Duke University
Durham, NC 27706
(919) 660–6555

Franc Brglez and Krzysztof Kozminski

MCNC
P. O. Box 12889
Research Triangle Park, NC 27709
(919) 248–1925

Abstract

OASIS is a cell–based silicon compiler system that supports rapid implementation of digital Application Specific ICs (ASICs) by taking a high–level language description and producing a layout of a testable IC. The language unifies functional descriptions of finite state machines (FSMs) and decoding logic with structural/algorithmic descriptions of data paths and regular structures such as adders or multipliers. The OASIS system integrates the language compiler with logic synthesis, simulation, verification, testability, and layout tools, providing the necessary control to explore the speed–versus–area tradeoffs.

Our experience with a language–based design capture system shows a marked increase in productivity when compared to schematic capture. This paper describes the system with its unique features and presents some of the results of using OASIS in designing semi–custom ASICs.

Introduction

The notion of a silicon compiler, introduced in the last decade [1], was originally used to describe a program that could lay out a parameterized set of macros. Later, this notion was extended to describe systems that could generate families of ICs implementing a fixed architecture with a parameterized set of modules [2] and systems that could translate functional descriptions into structural ones [3,4]. The most general notion of a silicon compiler is that of a translation program that converts a functional description of an algorithm into an efficient silicon layout implementing this algorithm. In general, this translation process appears to be ill–defined and too hard to implement. Only in some restricted domains was the compilation process carried out successfully from a functional description to the structural implementation [5]. Generally, a functional specification is not sufficient to describe all the constraints and design considerations that are needed for translating function into structure. Moreover, human designers can contribute knowledge, experience, and understanding of the specific problem at hand to help them design an efficient system implementation. In general, automating this process is very hard.

Our paper describes a pragmatic approach that has worked well in practice. We describe the overall system, its major components, and the design flow management first. Next, we introduce the LOGIC–III programming language. We describe the major components: Logic Synthesis and Optimization, Simulation and Verification, Test Pattern Generation and Redundancy Identification, Placement and Routing, and Standard Cell Generation and Maintenance. We conclude with a short description of four ASICs designed with the OASIS system; all were manufactured and tested successfully.

OASIS is a registered trademark of MCNC.

The OASIS concept

The Open Architecture Silicon Implementation Software, OASIS, was created to address several needs:

- Establish a platform for research and education in VLSI design;
- Provide a way of producing chips that are highly likely to work on first silicon, and with a very fast turnaround time;
- Focus collaboration between MCNC consortium, the participating universities, and industrial affiliates;
- Leverage productivity of a small semi-custom design team with a highly automated design system.

The limited resources provided an incentive for a design system with an open architecture: we have integrated a variety of public domain tools into OASIS in addition to developing our own. The modular design of OASIS provides for easy addition and substitution of any tool in the compilation sequence, thereby facilitating research and development of new algorithms as well as benchmarking of the new approaches versus the old ones. Figure 1 illustrates the major components of OASIS and its top-down approach to design. The four major components of the system are the compiler and logic synthesizer, the simulator, the automatic test pattern generator and the automatic layout generator.

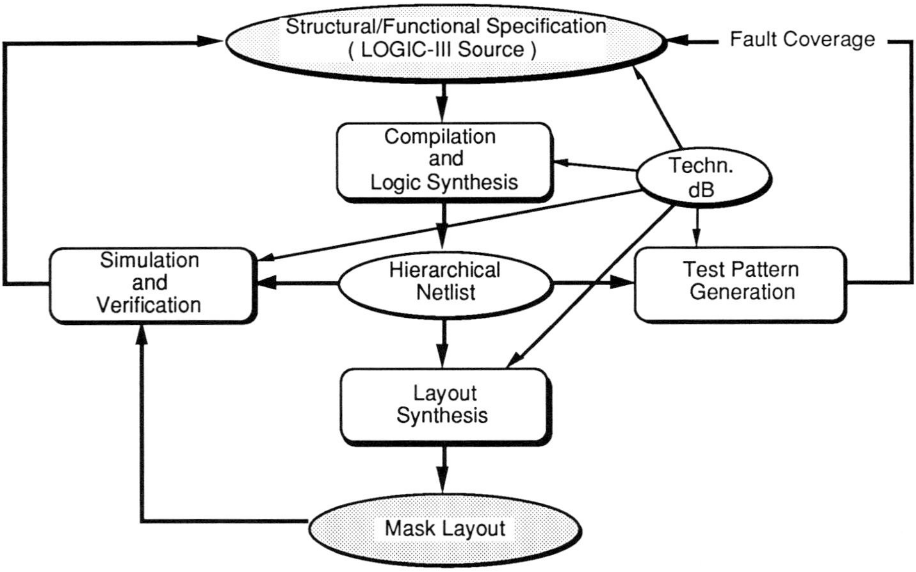

Figure 1: The OASIS concept

The design process starts by specifying/modelling the major functional blocks of the system in a high-level programming language. Once the design has been simulated and verified at the top level, the designer may simulate lower-level details. The designer's knowledge, insights and decisions are indispensable for specifying the structure of data path circuits. On the other hand, efficient synthesis of FSMs and combinational logic can be done automatically by software that is able to explore the flexibility of functional specifications. We rely on the process of compilation and logic synthesis to find a structural implementation of the specified function.

When the design is sufficiently verified, the gate-level netlist is examined for testability. Tools for redundancy identification and removal can be used to dispose of the untestable faults at this stage. If the design meets the specifications, the automatic placement, routing and layout are performed. There is a final performance verification step that uses the circuit extracted from mask data. The manufactured chip is tested with a combination of patterns used by the simulator and those obtained with the test pattern generator.

We have developed a language, LOGIC–III [11,23], that combines both the functional and structural descriptions. The former is reserved for finite state machines and control logic that can be readily translated into circuit structures [6–10]. All other components are described on the structural level. Either type of the description can use the usual programming language constructs.

The mixture of structural and functional descriptions proved to be a successful one. Most designers find the system easy to use. Moreover, the use of a programming language for hardware design proved to be far easier and more efficient than the more traditional approach using schematic capture. The language is flexible enough to accommodate module generators and custom components. In addition, it is possible to import portions of design that were done outside the system.

Design Management and Control

Each of the four major components of OASIS shown in Figure 1 links several tools to perform the required function. In order to help the designer manage a large collection of complex tools, and in order to relieve the designer from the need to remember all the possible files and options that each collection of tools require, we have developed a Design Control Language DECOL and a design management system [42] that automates most of the information flow. The ***DECOL*** supervisor program generalizes and expands facilities available from the UNIX *make* utility. The principle of matching a program usage template to the existing design data makes ***DECOL*** a powerful tool for describing concisely complex data dependencies and interactions between programs. ***DECOL*** provides version control and history mechanisms as well.

To a novice designer, OASIS appears as a single compiler system driven by simple commands. By typing "`oasis layout`", the designer translates a high-level specification into a layout, using the default design strategies throughout. By invoking "`oasis all`" the compiler will not only produce the layout, but will also generate scan-based test vectors, and the compiled code for the simulator. On the other hand, an experienced designer can control every step of the compilation by setting options, modifying defaults, choosing alternate programs and, if needed, using manual intervention. In order to maintain an open environment, we have chosen not to implement a unified database but rather to use the UNIX file system to capture the design hierarchy and DECOL to manage the processing and information flow.

A Structural/Functional Design Language

LOGIC–III is a PASCAL-like language that combines functional description of finite state machines and combinational logic with structural/algorithmic descriptions of computational circuits. The expressive power of a programming language is effective in describing the structure of large digital systems. The description is short and easy to understand and/or modify. An algebraic programming language such as PASCAL is also suitable for describing finite state machines. Moreover, the process of translating functional descriptions of state machines to efficient circuits is well understood. For data path-type structures, however, a functional description alone is not sufficient. Functional units like multipliers could be mapped onto many different hardware realizations, spanning a wide spectrum of trade-offs between, size, speed, and power. Without a hierarchical structure, some circuits, such as multipliers, cannot be effectively described as Boolean functions. The only practical way to express such circuits is in terms of their structure. By combining functional and structural descriptions in a single programming language, we are able to proceed in a consistent and effective manner with the compilation all the way to the mask level.

The language supports two module types: LOGIC_MODULE, used to describe a single FSM or combinational logic, and a NET_MODULE, used to describe the structure of a circuit as an interconnection of LOGIC_MODULEs and NET_MODULEs. The language retains the expressive power of PASCAL so that arbitrary connections can also be programmed and/or parameterized.

A LOGIC_MODULE begins with a list of formal parameters that can be either INPUT, CLOCKED_INPUT, OUTPUT, CLOCKED_OUTPUT, STATE, INTEGER, or BOOLEAN and a declaration section defining local variables and symbolic states. The main body of the module is a list of instructions describing the state machine. IF-THEN-ELSE and CASE statements are used for describing logical implications, and vector assignments can be used to treat sets of Boolean variables as integers. Also, loops and conditionals can be used to direct the flow of execution and to describe repetitive functions. Each LOGIC_MODULE is synthesized into either a netlist of standard cells [11,12] or a PLA.

Each NET_MODULE has a list of formal parameters, INPUT, OUTPUT, BIDIRECTIONAL, INTEGER, or BOOLEAN, followed by a declaration section describing the nodes and local variables. The body of a NET_MODULE consists of statements that connect other modules, LOGIC_MODULEs, NET_MODULEs, or even primitive elements such as transistors or capacitors. Statements such as IF-THEN-ELSE, FOR, WHILE, and UNTIL control the way the connections are made.

Addition of the programming language constructs to the netlist descriptions enables one to define parameterized macros, write algorithms to choose different hardware structures for different sets of parameters and to define the hardware at various levels of detail. To give the reader some flavor of the LOGIC-III language, we show below fragments of LOGIC-III code for a typical ASIC. `FSM1` and `FSM2` are examples of state machines, giving a functional description of the machine in terms of state transitions, and outputs as a function of the current state and inputs. The net module `Sreg` is a parameterized macro for N-bit shift register. The top level circuit module `MyASIC` (corresponding to a main program) is the top level of the description hierarchy. The `PIN` macro in this example provides an interface to the testability and placement and routing software.

```
LOGIC_MODULE FSM1(A,B,C: INPUT; D,E,F: OUTPUT)
VAR
states: [0..4] array of STATE ;
i, j,k, computing,idle, active,: INTEGER ;
Not_busy: LOGICAL ;
BEGIN
. . .
. . .
Not_busy := A or (B and C) ;
. . .
CASE:
. . .
states= computing:
      IF (Not_busy) THEN
      E :=B
      states := idle
      ELSE
      states := active ;
.     End. (* state = computing *)
. . .
END (* FSM1 *)

LOGIC_MODULE FSM2(...)
. . .
. . .
END. (* FSM2 *)
```

```
NET_MODULE Sreg(N: INTEGER; In, Ctrl : INPUT;
      Out : OUTPUT;)
VAR
i : INTEGER;
Temp : [0 .. N] array of NODE;
BEGIN
      DFF(In,Ctrl, Temp[0]);
      for i:= 1 to N-1 do
      DFF(Temp[i-1],Ctrl, Temp[i]) ;
      DFF(Temp[N-1],Ctrl, Out) ;
END.
. . .
. . .
Circuit MyASIC ;
VAR
. . .
BEGIN
      PIN( "Bus_1" , TOP, 13, 13, INPUT);
      FSM1( . . . );
. . .
      Sreg(17, Main_in, Shift_it,Delay_out);
. . .
END ;
```

The user controls the level to which the compiler expands each module. The choice depends on the target application. The simulator would typically expect a variety of representations beginning with the switch level. Any module that is not a transistor or a capacitor must have a corresponding functional model in a high-level language. It is not uncommon to have several functional models of the same macro, each at a different level of detail. For example, the layout system would typically expect the modules at the standard cell-level, and the test generation system expects them at the gate-level.

Logic Synthesis and Optimization

The process of compilation begins by interpreting the source code written in LOGIC-III. Depending on the options specified by the designer, a single pass by the compiler produces a netlist at the level intended for the application: switch-, gate- and functional-level for the simulator, gate-level for the test pattern generator, and cell-level for the layout generator. The functional description of LOGIC_MODULEs is translated into two-level Boolean equations which are minimized with ***espresso*** [6] and then factored and mapped with ***decaf*** [12] into a netlist of standard cells. Alternatively, the compiler can direct the module generator to produce a PLA. Further improvements, such as Partitioning with Resynthesis, [32,33], Technology Remapping with ***McMap*** [25], and Redundancy Removal with ***crisp*** [15,22] are done by post-processing the netlist.

Simulation and Verification

The role of simulation and verification is essential throughout the design cycle, from the early conceptual stages of the design to the last checkpoint before the masks are submitted for chip fabrication. The simulation/verification system must span a broad spectrum of representations, from the high-level during the conceptual stage to the switch-level during the final verification of the mask layout. In order to be effective and efficient, the simulator/verifier must simultaneously support various design representations.

LEONARDO [37,38] is a mixed-mode event-driven timing simulation and verification system. The core of the system is an event-driven switch-level timing simulator [13]. To this core one can link simulation modules that can model any desired component. We have satisfied the need for high-level modeling and simulation by developing **HLC**, an extended

C. This allows us to model gates and high-level functional units using a modern programming language. We have developed a set of models for our standard cell library that are an integral part of the OASIS system. In addition to simulation modules the simulator has a programmable interface. By writing **HLC** modules, user can add new instructions to the interactive simulation driver or drive the simulation in a batch mode using the **C** programming language to provide a simulation environment for their design.

To support the high-level functional modeling, ***LEONARDO*** provides tools for translating PLA descriptions and standard-cell netlists into compiled code and thereby speeding up simulation by orders of magnitude. In addition, the system has a hierarchical interface to the ***MAGIC*** layout system [14], removing the need to simulate the whole layout at the transistor level. A set of TTL models and tools for generating JDEC files and simulation models for field programmable devices make it easier to design board level systems and to verify that the ASIC devices will fit into their intended environment.

In addition to the simulation tools, two verification programs are used to aid the designer verify his design. The network comparison program ***geminiII*** [39] is used to compare the equivalence of two standard cells or transistor networks. A boolean function equivalence verification program, ***verify*** [40] compares the functional equivalence of truth tables, multilevel boolean networks, and transistor networks. The latter program is based on Binary Decision Diagrams. It is used to compare different levels of design abstractions, thus verifying the integrity of the design after it have been passed through many transformations.

Test Pattern Generation

Large-scale automatic generation of test patterns is currently feasible only for scan-based designs. Our test pattern generation system [15] consists of five basic components: scan-based audit to verify the compliance with scan-based rules, a testability assessment program that projects fault coverage before any patterns are generated, a fast fault simulator MIKE2 [34,35] that grades random and user-supplied patterns, an algorithmic test pattern generator [24,26] that produces tests for faults that resist detection with random patterns, and a test pattern compactor used to reduce the size of the test pattern set. An important byproduct of test pattern generation is the identification of redundant faults. These redundant faults are removed by ***crisp*** [15,22] during the resynthesis process.

The system continues to evolve within the cell-based OASIS system. It supports boundary scan [16], interfaces to the program generator for the the Mega_One tester [17], and serves as a platform for further research into hierarchical methods [41], automated test point insertion, improved random test pattern generation, cost-effective redundancy identification etc.

Placement and Routing

Our placement and routing software, ***VPNR***, is an open subsystem for implementing and testing placement and routing algorithms. VPNR is also the name of a language for describing all phases of the cell placement and routing process. The language provides a clear and concise interface that simplifies the communication among different tools in the system.

We have implemented a standard cell combined placement and routing algorithm based on quadrisection [18]. Benchmarks of the ***VPNR*** performance against other systems show that our implementation gives excellent results and very high performance [18,19]. In addition, VPNR includes a scan path wiring program, a choice of two channel routers, and a translator to the ***MAGIC*** layout system.

The VPNR system has a parser with a simple user interface. All tool developers are required to use the same parser. ***VPNR*** includes simple audit programs that check the consistency of the transformations made by each tool at each stage. As a result, we could readily import existing placement and routing tools, as, e.g., ***Timberwolf*** [20] and compare the performance of different algorithms and implementations.

Recently, we have developed algorithms for combined floorplanning and partitioning heuristics for mixed standard cells and large macro–block layouts [30]. A new software, ***MONTAGE***, for combined layout of standard cells and macro–blocks is under development [31]. ***MONTAGE*** enhances our ability to incorporate cell generators and custom design into OASIS.

Standard Cell Compilation

The rapid changes of technology bring forth the need to automate the standard cell library generation and maintenance. While our standard cell layouts are still manually optimized, all other data necessary for the operation of OASIS are computer–generated. Associated with each cell in the library are many different representations: mask data, cell outlines, transistor network, delay information, simulation models, etc. The only two representations that are entered manually are the LOGIC–III description of each cell and the layout, the rest of the representations are derived by running various tools for circuit extraction, simulation, and verification. This collection of tools, named the ***CellWizard***, enables a sophisticated user to add new cells to the library. Since this collection of tools is also controlled by DECOL, adding a new cell in to the library is highly automated and easy to do.

ASIC Design with OASIS

About ten ASICs were designed, manufactured and tested using the OASIS design system as described in the paper. Here, we highlight four of such chips. Two of the chips were designed for biomedical applications and two are parts for the Ray–Casting Machine [21]. The first two designs started with well–defined architectures and with architectural level simulation in place.

The QRS chip was designed as a single chip monitor for the detection and prevention of Sudden Death Syndrome [29]. The device analyzes ECG data and will display a warning if the patient's heart rate goes out of a designated region based on his/her physiological condition. The device is capable of operating at 20 MHz which allows sampling at up to 32k samples/sec and also incorporates Built-In Self-Test. The design time was on the order of 3.5 months to complete a 45k transistor device using MCNC's 1.2 micron CMOS technology.

The FAN chip [28] implements in silicon a real–time data compression algorithm for electrocardiogram data. The chip is made of two components. A 20x20 Booth multiplier is implemented as a full custom component and the rest of the circuitry is implemented with standard cells. The chip was designed to operate at 25 MHz using MCNC's 1.2 micron CMOS technology. About 45,000 of the transistors are in standard cells and about 16,000 transistors were laid out manually. Each portion of the design was done in about 4 man–months. Most of the standard cell design (about 3.5 man–months) was devoted to simulation. Only about two weeks were spent on layout verification and test generation. On the other hand, most of the multiplier design time was devoted to layout and verification.

The PC processor is a pipelined, bit–serial, floating point processor that solves a sequence of related quadratic equations [21]. The PC uses a mixture of double and single precision floating point operations. The arithmetic is done bit–serially at the rate of two double precision additions, one double precision square root, and two single precision additions every 38 clock cycles. The PC is designed to work at 40 MHz using 0.8 micron CMOS technology. Each PC chip has 10 processors. Each processor contains about 60,000 transistors for a total of more then 600,000 transistors. The PC was designed by a graduate student over a period of ten months.

The CC chip is a bit–serial pipelined processor that computes set operations (union, intersection, and difference) on sequences of interval segments. The CC processors are connected as a $\log N \times N$ array and can be programmed to form a very deep pipeline in the shape of a binary tree. Each CC processor is a single pipeline stage communicating with its sons and parent processors [21]. The CCs are designed to work both in fixed point and floating point arithmetic. Each CC processor is relatively small, containing some 1000 standard cells (or about 12000 transistors). The current generation CCs are designed to run at 40 MHz using .8 micron CMOS technology. We have integrated 40 such processors on a single die, of which only 32 will have to be fully functional. The CC processor was designed as a modification of an existing CC in about one man–month.

Summary

We have presented an overview of the OASIS system. The major components of the system are the language compiler combined with a logic synthesis and optimization, a mixed–mode multi–level timing simulation and verification system, a test pattern generation system, and an IC placement and routing layout system. This complex set of tools is integrated into a coherent design environment with the aid of a sophisticated design management and control software. The OASIS system successfully integrates existing public domain tools while adding major new components. We have described several Application Specific ICs that were designed with OASIS. It is our experience that the OASIS system dramatically increases designer productivity. We were able to design large and complex ICs in a few man–months only. The use of a high–level language combined with logic synthesis and automatic placement and routing relieved the designer from the the chores of dealing with low–level details, thus enabling him/her to concentrate on the more creative aspects of the design.

Acknowledgements

Most of the OASIS research is being supported by MCNC and NSF, with earlier grants from RTI, NT and BNR. Major contributors to OASIS include MCNC staff: David Bryan, Robert Lisanke (now with AutoDesign), Mike Lorenzetti, Margaret Nifong, Tom Krakow and John Rose, faculty consultants: John Ellis (Duke U.) and Michael Schulz (Techn. U. Munich, now with Siemens); and graduate students: Hala Abdalla (UNC–CH, now with SUN), Jitendra Apte (Duke U.), Jack Briner (Duke U.), John Calhoun (NCSU, now with IBM), Sujit Dey (Duke U.), Clay Gloster (NC A&T U., now NCSU), Douglas Maltais (NCSU), Dieter Pellkofer (Techn. U. Munich), Tan Phan (NCSU), Chris Ryan (NC A&T, now Virginia Tech.), Bernhard Seiss (Techn. U. Munich), Peter Suaris (Duke U., now in Sri Lanka), and Mimi Wong (UNCC).

References

[1] Johannsen, D.,"Bristle Blocks: A Silicon Compiler", *16th Design Automation Conf.*, 1979.

[2] Southard, J. R., "McPitts: An Approach to Silicon Compilation", *Computer*, Vol. 16, No. 12, pp. 74–82, December, 1983.

[3] Blackman, T. and J. Fox and C. Rosebrugh, "The SILC Silicon Compiler: Languages and features", *22nd Design Automation Conf.*, June 1985.

[4] Brewer, F. and D. Gajski, "An Expert System Design Paradigm", *23rd Design Automation Conf.*, pp. 62–68, June 1986.

[5] J. M. Rabaey *et al.*, "An Integrated Automatic Layout Generation System for DSP Circuits", *IEEE Trans. on Computer Aided Design*, CAD–4, July, 1985.

[6] Brayton, R. K. *et al.*, "*Logic Minimization Algorithms for VLSI Synthesis*", Kluwer Academic Publishers, Boston, MA, 1984.

[7] G. De Micheli, "*Synthesis of Control Systems, Design System for VLSI Circuits*", Martinus Nijhoff Publishers, Boston, 1987.

[8] R. Brayton and C. McMullen, "Synthesis and Optimization of Multi–Level Logic", *IEEE Intl. Conf. on Computer Design*, pp. 23–28, October, 1984.

[9] C. Shaw, *et al.*, "RESCUE: A Comprehensive Control Logic Layout System", *IEEE Intl. Conf. on Computer–Aided Design*, pp. 260 – 263, November, 1986.

[10] G. Kedem and K. Kozminski, "A Standard–Cell Based Silicon Compiler", IEEE 1986 *Custom Integrated Circ. Conf.*, pp. 120–124, May, 1986.

[11] H. Abdalla *et al.*, "LOGIC II: A High–Level Language for a Cell–Based Design System", Technical Report TR87–16, MCNC, Research Triangle Park, NC, August, 1987

[12] R. Lisanke *et al.*, "DECAF: Decomposition and Factoring for Multi–Level Logic Synthesis", Technical Report TR87–15, MCNC, Research Triangle Park, NC, August, 1987

[13] C. J. Terman, "Simulation Tools for Digital VLSI Design", Ph.D. Thesis, Massachusetts Institute of Technology, Cambridge, MA 1983

[14] Ousterhout, *et al.*, "A Collection of Papers on MAGIC", UC Berkeley, Berkeley, CA, 1983

[15] F. Brglez, *et al.*, "Automated Synthesis for Testability", IEEE Tr. on Industrial Electronics, vol. 36, No. 2, May 1989, pp. 263–277.

[16] C. Gloster, F. Brglez, "Boundary Scan with Built–In Self–Test", *IEEE Design & Test of Computers*, Feb. 1989, pp. 36–44.

[17] Ji–en M. Chang *et al.*, "Generic Test Program Translator (GTPT) for tester–per–pin ATE", *IEEE Intl. Test Conf.*, pp. 1044–1051, 1987.

[18] P. R. Suaris and G. Kedem, "A Quadrisection based Combined Place and Route Scheme for Standard Cells." IEEE Trans. CAD Vol 6 Num. 3 pp. 234–244. March 1989.

[19] A. E. Dunlop and B. Preas and K. Roberts, "Cell–Based Layout Benchmarks", *24th Design Automation Conf.*, pp. 318, 1987.

[20] Sechen, C. and Sangiovanni–Vincentelli, A., "The TimberWolf Placement and Routing Package", *IEEE Journal of Solid–State Circuits*, Vol SC–20, No.2, pp. 510–522, April, 1985.

[21] G. Kedem and J. L. Ellis, "The Ray Casting Machine Prototype", in *Parallel Processing for Computer Vision and Display*. P. M. Dow, R A. Earnshaw and T. R. Heywood, Eds, Addison Wesley Reading MA , pp378–401 1989.

[22] D. Bryan, *et al.*, "Redundancy Identification and Removal", *Intl. Workshop on Logic Synthesis*, Research Triangle Park, NC, May 1989.

[23] M. Wong, G. Kedem, F. Brglez, "A Guide to Logic–III", in preparation.

[24] J. Calhoun, *et al.*, "Automatic Test Pattern Generation for Scan Based Digital Logic". Technical Report TR87–15 MCNC August 1987.

[25] R. Lisanke, F. Brglez, G. Kedem, "McMap: A Fast Technology Mapping Procedure for Multi–Level Logic Synthesis". *IEEE Proc. Intl. Conf. on Computer Design* October 1988.

[26] F. Brglez, *et al.*, "A Modular Scan–Based Testability System," *IEEE Proc. Intl. Conf. on Computer Design* , October 1988.

[27] C. S. Gloster and F. Brglez, "Boundary Scan With Cellular–Based Built–In Self–Test". *Proc. Intl. Test Conf.*, September 1988.

[28] S.C. Roy, *et al.*, "The Design and Verification of a VLSI Chip for Electrocardiogram Data Compression", *Third IEEE Symp. on Computer–Based Medical Systems*, June 1990.

[29] S.C. Roy, *"A Testable CMOS QRS Detector and Arrhythmia Monitor"*, Master's Thesis, University of North Carolina at Chapel Hill, 1990.

[30] J. Apte and G. Kedem "Combined Floorplanning and Partitioning Heuristics for Mixed Standard Cells and Macro Block Layout", *The Sixth MIT Conf. on Advanced Research in VLSI*, April 1990.

[31] J. Apte and G Kedem, "MONTAGE: An Automatic Layout System for Mixed Standard Cells and Macro Blocks". *Intl. Workshop on Layout Synthesis*, Research Triangle Park, May 1990.

[32] S. Dey, F. Brglez, G. Kedem, "Circuit Partitioning and Resynthesis". *IEEE Proc. Custom Integrated Circuit Conf.*, May 1990

[33] S. Dey, F. Brglez, G. Kedem, "Corolla Based Partitioning and Resynthesis", *26th Design Automation Conf.*, June 1990.

[34] M. H. Schulz and F. Brglez, "Accelerated Transition Fault Simulation." *24th Design Automation Conf.*, pp. 237–243, June 1987.

[35] K. J. Antreich amd M. H. Schulz, "Accelerated Fault Simulation and Fault Grading in Combinational Circuits". *IEEE Trans. on CAD*, CAD–6(5), pp. 704–711, September 1987

[36] R Lisanke, *et al.*, "Testability–Driven Random Pattern Generation", *IEEE Trans. CAD*, CAD–6(5) pp. 704–711, September 1987.

[37] J. V. Briner Jr. "LDVSIM: A Mixed Level System Simulator", Technical Report 89–29, Computer Science Department, Duke University, Durham, NC, 1989.

[38] J. V. Briner Jr., J. L. Ellis, G. Kedem, "The LEONARDO Simulation System", to be presented at the *21st Annual Pittsburgh Conf. on Modelling and Simulation*, May 1990

[39] C Ebeling, "GeminiII: A Second Generation Layout Validation Program", *Proc. IEEE Intl. Conf. on CAD* November 1988.

[40] D. S. Reeves and M. J. Irwin, "Fast Method for Switch Level Verification of MOS Circuits", *IEEE Trans. on CAD*, Vol6, No. 5 pp. 766–779, September, 1987.

[41] J. Calhoun and F. Brglez, "A Framework for Hierarchical Test Generation", *IEEE Proc. Intl. Test Conf.* pp. 480–490, August, 1989.

[42] K. Kozminski, "Design Control in MCNC's Open Architecture Silicon Implementation System OASIS", Technical Report TR89–54, MCNC, Research Triangle Park, NC, December, 1989.

Logic and Architecture Synthesis
P. Michel and G. Saucier (Editors)
Elsevier Science Publishers B.V. (North-Holland)

High-level Synthesis and Optimization Strategies in Hercules and Hebe

David Ku Giovanni De Micheli

Center for Integrated Systems
Stanford University
Stanford, CA 94305

Abstract

In this paper, we present an approach to automated synthesis of digital circuits from behavioral specifications. The system, called *Hercules* and *Hebe*, offers many advantages to the designer. First, the system supports constraint-driven synthesis where timing and resource constraints are applied to guide the synthesis decisions. Second, systematic design space exploration is possible, where the designer explores the tradeoff between area and performance to meet the design objectives. Third, logic synthesis techniques are uniformly incorporated within the synthesis framework to provide estimates to guide high-level decisions. Along with a synthesis oriented hardware description language called *HardwareC*, *Hercules/Hebe* provides an environment for the design of general synchronous digital circuits, with specific attention to the requirements of ASIC designs. The system has been applied to complex ASIC designs, including the the Digital Audio I/O, MAMA, and Bi-Dimensional DCT chips.

1 Introduction

Computer-aided synthesis of digital circuits from behavioral specification can greatly improve the complexity and quality of hardware design. By providing a synthesis framework of tools, a circuit can be specified by its functional description, timing requirements and interface constraints. The description can then be optimized by the high-level and logic synthesis systems to produce a logic level implementation that satisfies the user constraints. The benefits of a high-level synthesis-based design methodology include standardization, self-documentation and ease of modification of the hardware specifications, shortened design time, increased productivity of the designers, and better quality of the synthesized design due to the use of optimization techniques at the functional level.

While logic synthesis techniques have been established as standard steps in the design methodology for digital circuits [1], high-level synthesis techniques have been lagging behind for several reasons. First, most proposed and used hardware description languages (HDLs) have been conceived for hardware simulation and documentation. For a circuit description to be effective for synthesis, it is necessary to have a well-defined hardware synthesis semantic for the language. A consensus on a behavioral HDL platform for synthesis has not been reached yet. Second, given the diversity of the approaches to digital circuit designs, it is difficult to encode all implementation decisions in terms of algorithms or rules that can be universally applied. Therefore, practical high-level synthesis techniques need to support *automatic* and *user-driven* synthesis modes to leverage off the designer's knowledge and experience. Third, as designs increase in size and complexity, system integration issues often dominate a design. Hardware interfacing and design constraints on timing and area need to be addressed at both the design specification and design synthesis levels.

Previous work on high-level synthesis system addressed primarily microprocessor and digital signal processing designs [2], [3], [4], [5], [6], [7], [8], [9], [10], and [14]. We believe that high-level synthesis can be especially effective in the synthesis of Application Specific Integrated Circuits (ASICs). For this reason, we have developed a hardware description language and a high-level synthesis system for the synthesis and simulation of general-purpose digital circuits, with specific attention to the requirements of ASIC designs. The system is divided into two parts, called *Hercules* and *Hebe*, and has the following features:

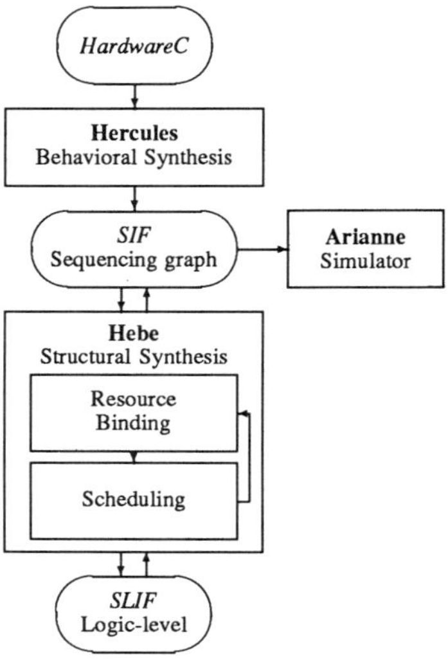

Figure 1: Block diagram of the *Hercules* and *Hebe* system.

- *Uses synthesis-oriented HDL. HardwareC* is a hardware description language designed for synthesis of digital circuits [12]. It supports hardware descriptions with both procedural and declarative semantics, as well as constraints on the hardware implementation. *HardwareC* also provides constructs to support the description of libraries of generic hardware resources.

- *Supports constraint-driven synthesis with partial binding.* Timing and resource requirements that are specified in the high-level description are used to guide the results of the synthesis optimizations. The system supports the synthesis of partially bound hardware descriptions, where certain operations are pre-bound to specific hardware modules, and performs synthesis on the remaining operations. It provides a flexible framework for systematic exploration of the design space of tradeoffs between area and performance, and guarantees *optimal* scheduling of each point in the design space.

- *Incorporates logic synthesis techniques.* To meet the area requirements, resource sharing is a necessary part of the synthesis system. Since resources correspond to models that are described and invoked in the high level description, the characterization of resources to evaluate sharing feasibility is carried out using logic synthesis techniques to provide estimates on timing and area. This methodology is particularly suited for ASIC designs that tend to rely on application-specific logic functions. The use of logic synthesis for estimates improves the quality of the synthesized designs, and avoids erroneous high-level decisions due to insufficient data or inappropriate assumptions.

- *Supports automatic and user-driven synthesis.* The synthesis flow can be fully automated, transforming an input *HardwareC* description directly to a logic-level implementation. The system also supports user-driven synthesis, where a designer can intervene and drive high-level decisions based on an evaluation of the possible design tradeoffs.

We have implemented two programs for the high-level synthesis of synchronous digital circuits, called **Hercules** [11] and **Hebe**. The two programs transform a behavioral description of hardware in *HardwareC*, through a series of translations and optimizations, to a synchronous logic implementation that satisfies the timing and resource constraints that are imposed on the design. *Hercules* performs the front-end parsing and behavioral optimizations. It generates an implementation-independent description of the hardware behavior in a graph-based representation, called the *Sequencing Intermediate Form* (SIF). The SIF can be simulated by *Arianne* to provide feedback to the designer on the functional correctness of the input description. *Hebe* maps the SIF into a logic-level implementation, described in the *Structural/Logic Intermediate Form* (SLIF). A block diagram of the system is shown in Figure 1.

2 Hardware Modeling

HardwareC is a high-level hardware description language with a C-like syntax. The language has its own hardware semantics, and it differs from the C programming language in many respects. *HardwareC* supports both declarative semantic (e.g. interconnection of modules) and procedural semantic (e.g. set of operations ordered in time) in the modeling of hardware. There are four fundamental design abstractions, corresponding to *block*, *process*, *procedure*, and *function models*. At the topmost level, a design is described in terms of a block, which contains an interconnection of logic and instances of other blocks and processes. A process consists of a hierarchy of procedures and functions, and represents a functionality that executes repeatedly, restarting itself upon completion. Since a process executes concurrently and independently with respect to the other processes in the system, it allows the modeling of *coarse-grain* parallelism at the functional level. A procedure or function is an encapsulation of operations, and may contain calls to other procedures and functions.

In contrast to micro-architectural synthesis systems that use a predefined set of library elements as building blocks, *Hercules* and *Hebe* treat each *model* in the input description as a *resource* that can be allocated and shared among the calls to the models (either procedures or functions). For example, two calls to a model `A` can be implemented either by a single resource corresponding to the hardware implementation of `A`, where both calls share the use of the resource; or by two resources, where each call is implemented by a different resource. Operators such as + or − can either be converted into calls to the appropriate library models, or by default be implemented in terms of logic expressions.

HardwareC supports the usual iterative and branching constructs, including both fixed-iteration and data-dependent looping constructs. Data-dependent loops can be used to detect signal transitions, which are important in describing external interfaces. For example, the construct `while (data==0);` will wait until the rising transition of the signal `data`. In addition, there are several features of *HardwareC* that support hardware specification and synthesis:

- *Interprocess communication* – To support communication and synchronization among the concurrent processes, *HardwareC* supports both *parameter passing* and *message passing*. The former assumes the existence of a shared medium (e.g. shared bus or memory) that interconnects the hardware modules implementing processes. The handshaking protocols are described in the *HardwareC* description. The latter uses a synchronous *send/receive* mechanism that can be used for synchronization or data transfer. The corresponding hardware for communication, as well as its protocol, are automatically synthesized.

- *Explicit instantiation of models* – Hierarchical designs are supported through the use of model calls. A call to a model can be either *generic* or *instantiated*: a generic call invokes a model without specifying the particular instance that is used to implement the call, whereas an instantiated call identifies also a specific instance of the model which will implement the call. Through explicit instantiation of model calls, *HardwareC* supports resource constraints and partial bindings of operations to resources. The designer can constrain the synthesis system to explore a subset of the possible structures corresponding to a behavioral model to satisfy a particular architectural requirement.

- *Template models* – Templates are models that take one or more integer arguments, and support *polymorphism* in the language by modeling several behaviors with a single description. As an example, a single template can be used to describe a family of adders of different size. Templates are therefore very useful in describing libraries of hardware operators at a high level.

- *Degree of parallelism* – For procedural semantic models, *HardwareC* offers the designer the ability to adjust the degree of parallelism in a given design through the use of *sequential*, *data-parallel*, or *parallel* groupings of operations. In the first case, operations are executed sequentially. In the second one, all operations are executed in parallel, unless data dependency requires serialization. In the last case, all operations execute in parallel unconditionally.

```
process gcd ( xin, yin, restart, result )
    in port xin[8], yin[8], restart;
    out port result[8];
[
    boolean x[8], y[8];
    tag a, b;

    /* set output to zero during computation */
    write result = 0;

    /* wait for restart to go low */
    while ( restart )
        ;

    /* sample inputs */
    <
        constraint mintime from a to b = 2 cycles;
        constraint maxtime from a to b = 2 cycles;

        a: x = read(xin);
        b: y = read(yin);
    >

    /* Euclid's algorithm */
    if ( (x != 0) & (y != 0) ) {
        repeat {
            while (x >= y)
                x = x - y;
            /* swap values */
            < y = x; x = y; >
        } until (y == 0);
    } else
        x = 0;

    /* write result to output */
    write result = x;
]
```

Figure 2: Example of a *HardwareC* description to find the greatest common divisor of two values.

- *Constraint specification* – Timing constraints are supported through tagging of operations, where lower and upper bounds are imposed on the time separation between the tags. Timing constraint are useful in interface specification by constraining the time separation between I/O operations. Resource constraints limit the number of and binding of operations to resources in the final implementation.

An example of a *HardwareC* description that computes the greatest common divisor of two numbers is given in Figure 2. The model `gcd` waits until the `restart` signal is low, samples the inputs, then performs Euclid's algorithm iteratively. The read operations are tagged, and timing constraints are applied on the tags to ensure that the reading of y occurs exactly 2 cycles after the reading of x. Note that any statement in the description can be tagged.

3 Hercules – Behavioral Synthesis

The objective of *behavioral synthesis* is to identify the maximal parallelism that exists in the input description. This gives an indication of the *fastest* design that the system can produce, assuming that in the design implementation each operation is implemented by a dedicated hardware component. While this assumption may not be realistic in some cases due to area and interconnection costs, it is important to compute the related performance as a limiting bound for a given behavior.

The input *HardwareC* description is parsed and translated first into an abstract syntax tree representation, which provides the underlying model for semantic analysis and behavioral transformations. The transformations are categorized into *user-driven* and *automatic* transformations. User-driven transformations are optional, and allow the designer the capability of modifying the model calls and hierarchy of the input description. They include the following:

- *Selective in-line expansion of model calls*, where a call to a model is replaced by the functionality of the called model. Once expanded, the optimization algorithms can be applied across the call hierarchy.

- *Selective operator to library mapping*, where operators, such as "+" or ">=", in the input description are mapped into calls to specific library template models. Although an operator can be synthesized in a variety of different implementation styles, the designer is often constrained to elements of a particular library. With such mapping, the designer has the flexibility to select the specific implementation for the operators. If no mapping is given, then by default the operators are implemented as combinational logic expressions.

Automatic transformations optimize the behavior by performing transformations similar to those found in optimizing compilers. The automatic transformations are carried out without human intervention, and include the following:

- `For`-*loop unrolling*, where fixed-iteration loops are unrolled to increase the scope of the optimizations.
- *Constant and variable propagation*, where the reference to a variable is replaced by its last assigned value.
- *Reference stack resolution*, where multiple and conditional assignments to variables are resolved by creating *multiplexed values* that can be referenced and assigned [11].
- *Common sub-expression elimination*, where redundant operations that produce the same results are removed.
- *Dead-code elimination*, where operations whose effects are not visible outside the model are removed.
- *Collapse conditional*, where conditionals with branches containing only combinational logic are collapsed to increase the scope in which logic synthesis can be applied.
- *Data-flow analysis*, where data and control dependencies among the operations are identified.

Upon completion of the automatic transformations, the behavior is optimized with respect to the data-dependencies that exist among the operations. At this point, *combinational coalescing* is performed to group together combinational logic operations into *expression blocks*. The expression blocks define the largest scope (without crossing register boundaries) in which logic synthesis can be applied, and identify the critical combinational logic delays through the design. They are directly passed to logic synthesis for minimization, the results of which are fed-back as estimates on area and timing that are used to refine the design. Combinational coalescing is important particularly for ASIC designs because of their extensive use of logic expressions in the hardware specification.

Sequencing Graph Model. The optimized behavior resulting from behavioral synthesis is translated into a graph abstraction called the *sequencing intermediate form* (SIF). The sequencing graph is a concise way of capturing the partial order among a set of operations, and it is modeled as a polar, directed acyclic graph. The vertices represent the operations to be performed, and the edges represent the dependencies that are either explicit in the hardware specification, or represent dependencies due to *data-flow* restrictions or hardware *resource-sharing* considerations. A vertex is enabled when all its predecessors have completed execution. Since a vertex may have multiple predecessors and successors, the model supports *multiple threads of concurrent execution flow*.

The vertices are categorized as either *simple* or *complex* vertices. Simple vertices include primitive computations in the language, such as arithmetic or logic expressions and message passing commands. Complex vertices allow groups of operations to be performed, and include model calls, conditionals, and loops. The complex vertices induce a hierarchical relationship among the graphs. A call vertex invokes the sequencing graph corresponding to the called model. A conditional vertex selects among a number of branches, each of which is modeled by a sequencing graph. A loop vertex iterates its body until the exit condition is satisfied; the body of the loop is also a sequencing graph. The sequencing graph is *acyclic* because only structured control-flow constructs are assumed (no `goto`), and loops are broken through the use of hierarchy.

The sequencing graph model can be simulated by **Arianne**. Therefore, *HardwareC* descriptions can be simulated at the functional level by transforming them first into the SIF representation. Since behavioral synthesis in Hercules is very efficient and fast, validation of the *HardwareC* models via simulation can be effectively supported. The SIF is the underlying representation for structural synthesis.

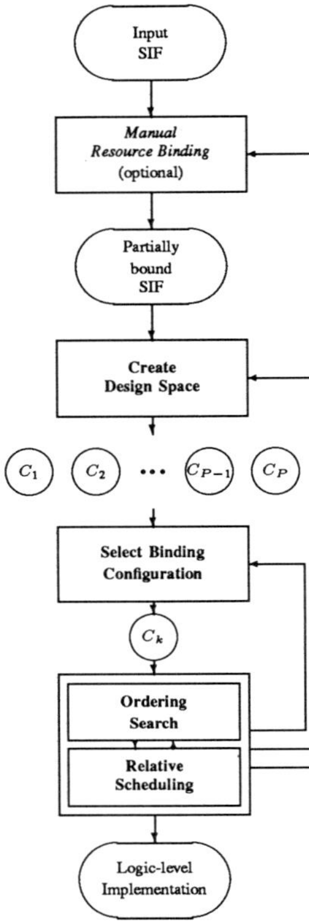

Figure 3: Block diagram of the *Hebe* structural synthesis system.

4 Hebe – Structural Synthesis

The objective of the *Hebe* structural synthesis system is to explore the design tradeoffs to obtain a suitable structure that satisfies the user constraints on area and timing. The constraints can either be specified in the input *HardwareC* description, or entered interactively by the designer. *Hebe* provides a flexible underlying representation of the design space that serves as the basis both for algorithmic exploration as well as for user-driven synthesis. In particular, the designer can either let *Hebe* explore the design tradeoffs automatically, or manually guide the direction of synthesis by imposing constraints on the design.

Structural synthesis involves performing two tasks – *resource binding*, where operations are assigned to hardware resources, and *scheduling*, where operations are assigned to control states. An effective strategy is to perform resource binding before schedule to provide scheduling with detailed interconnection delays, as in *Caddy* [14] and *BUD* [15]. This basic approach is extended in *Hebe* to provide closer interaction and guidance to the designer, and is shown in Figure 3. The flow of structural synthesis in *Hebe* is to first bind operations to specific resources, then perform scheduling to find a schedule that satisfy the timing constraints. The process repeats for different possible binding alternatives.

A **resource pool** is a set of hardware resources (e.g. implementations of models) with an upper-bound on the number of instances of each type of hardware resources that the user allows in the final implementation. A **binding configuration** is a matching of the operations (i.e. the vertices of the sequencing graph) with specific resources in the resource pool. The **design space** is the entire set of binding configurations. Examples of

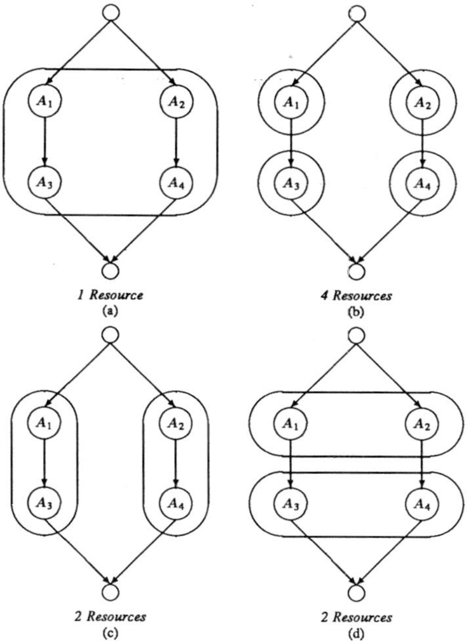

Figure 4: Examples of binding configurations, where operations within a group are bound to the same resource instance.

binding configurations for a sequencing graph containing four calls to model A are shown in Figure 4. All operations that are grouped together share the same resource instance in the final implementation, e.g. the binding configuration of Figure 4(a) utilizes one resource instance, the binding configuration of Figure 4(b) utilizes four resource instances, etc.

An important aspect of the design space formulation is that it is a *complete* characterization of the entire set of possible design tradeoffs for a given allocation of resources, and offers two important advantages:

- *Uniformly incorporates partial binding information.* In some circuits the designer may wish to bind certain operations to resources in order to achieve high-level design goals. This information can be used to limit the design space such that the synthesis system focus on the remaining unbound operations. At the extreme, if all operations are bound, then the design space trivially reduces to a single point.
- *Optimal scheduling under timing constraints.* Since *Hebe* decouples resource binding from scheduling, the latter problem can be solved exactly and efficiently, even under timing constraints, for each binding configuration.

The size of the design space may be large, because it grows exponentially with the number of shareable resources. However, it is often the case in ASIC designs that the number of shareable resources is sufficiently small to make systematic exploration of all binding configurations practical. Furthermore, bounding techniques based on evaluation of various *cost criteria* can be used to prune the design space, and speed the search for a suitable implementation.

Select Binding Configuration. Given a design space, *Hebe* supports both *exact* and *heuristic* search of the binding configurations. The system supports a set of *cost criteria* that are used to evaluate the design space.

The cost criteria represent the effect of a particular binding configuration on the area, interconnection, and delay of the final implementation, and include:

- *Area Cost*: A binding configuration implies a certain degree of resource utilization and sharing. The area cost corresponds to the area costs of the resources in the resource pool.
- *Interconnection Cost*: The interconnection structure is the steering logic that guides the appropriate values to their proper destinations in the final implementation. Since a binding configuration is a complete assignment of operations to resources, the interconnection structure is completely specified. The interconnection cost is proportional to the size and delay of the interconnection structure.
- *Width Cost*: The *width* of a binding configuration is the number of threads of parallelism that *need to be serialized* in order to avoid resource contentions. In particular, all operations bound to the same resource instance should not execute simultaneously, e.g. either there exists sequencing dependencies among the operations, or the operations occur in mutually exclusive branches of a conditional. The width cost is a heuristic measure of the effect of the binding configuration on the performance of the design.

The decision of whether one alternative is favorable with respect to another depends on the relative importance of these criteria, which is determined by the value of a *weight* associated with each criterion. *Hebe* provides a flexible framework in which the designer can experiment with different design goals by adjusting the values of the weights. The design goals indicate the emphasis of the final implementation with respect to area and/or performance, and include:

- Find the *minimum area* configuration that satisfies the timing constraints, or
- Find the *maximum performance* configuration that satisfies the area constraints

The designer can focus the synthesis efforts on the binding configurations with acceptible costs. For example, if the goal is to minimize the area, then the area and interconnection costs can be used to identify the binding configurations with minimal area. Likewise, if the goal is to maximize performance under area constraints, then the area and interconnection costs can bound the search to those configurations that meet the area constraints, while the width cost can provide further pruning of the design space.

Ordering Search. Once a resource binding is selected, the operations bound to the same resource component are *ordered* to resolve any resource conflicts that may occur. For example, three calls to model A executing in parallel but bound to the same resource must be serialized to ensure that they cannot execute simultaneously, as shown in Figure 5. The goal of the *ordering search* is to find a set of dependencies among the operations that resolves the resource conflicts. A branch-and-bound approach is used to explore the ordering alternatives. Since resource sharing can occur across the sequencing graph hierarchy, and dependencies may exist among the operations, not all possible orderings are valid. Therefore, significant pruning of the search space can be achieved. Once an ordering is found, the sequencing graph is serialized accordingly and scheduling is applied. If no schedule exists under timing constraints, then another ordering is tried. If no schedule exists for any valid ordering, then the binding configuration is discarded.

Relative Schedule. Given an ordering, the sequencing graph is free from resource conflicts. However, scheduling is necessary to define the detailed temporal relationships among the operations to satisfy the imposed timing constraints. A complication arises in that some of the operations may have *unbounded execution delays*, corresponding to synchronization primitives and data-dependent loops. The unbounded delay operations invalidate the traditional scheduling formulation where operations are statically assigned to specific time slots. We use a novel technique called *relative scheduling* that uniformly supports operations with fixed and unbounded delays [16].

The scheduling problem under timing constraints is modeled by means of a *constraint graph*. The vertices of the polar directed graph represent the operations, and the weights on the edges represent the timing constraints between pairs of operations. We define a subset of the vertices, called *anchors*, that serve as reference points for specifying the start times of the operations. The anchors consist of the source vertex and the set of unbounded delay vertices. *Offsets* are then defined with respect to each anchor of the graph. In particular, the *anchor set*

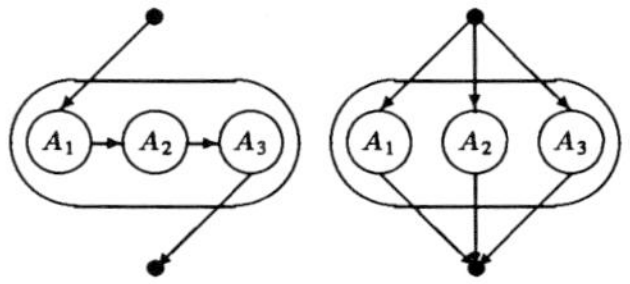

Figure 5: Example of (a) no resource conflict, and (b) resource conflict, among operations bound to the same resource.

of a vertex is the set of anchors that are predecessors to the vertex, and represents the *unknown* factors that affect the activation time of the vertex. The start time of a vertex is then generalized in terms of fixed time offsets from the completion of each anchor in its anchor set. Note that if there are no unbounded delay vertices in the graph, then the start times of all operations will be specified in terms of offsets from the source vertex, which reduces to the traditional scheduling formulation.

An important consideration during scheduling is whether the timing constraints can be satisfied for any value of the unbounded delay operations. We use the concept of *well-posed* verses *ill-posed* timing constraints in the presence of unbounded delays [16]. Specifically, a timing constraint is well-posed if its satisfiability does not depend on any unbounded delays. Given an ordering of shareable operations to resolve resource conflicts, additional serialization may be required to make the constraints well-posed. If no consistent serialization can be found, or if the constraints are not satisfiable, then the ordering is rejected as unfeasible. Otherwise, a *minimum* relative schedule can be computed by means of an *iterative incremental scheduling* algorithm. The time complexity of making the constraints well-posed and the scheduling algorithm are both polynomial. This allows relative scheduling to be effectively integrated within the ordering search.

Once a satisfactory structure with respect to both resource and timing constraints is obtained, a logic-level description is generated for the data-path, the interconnection, and the control circuitry to activate the data-path according to a given schedule. We use logic synthesis to perform combined synthesis of both control and data-path, the result of which can be mapped to a library, or implemented as macro-cells.

5 Implementation and Design Experiences

Hercules and *Hebe* have been implemented in C, with approximately 100,000 lines of code. They have been tested on the benchmark circuits for high-level synthesis. In addition, the system has been used to design three ASIC circuits at Stanford University, namely a Bi-dimensional Discrete Cosine Transform (BDCT) chip [17], a Digital Audio Input Output (DAIO) chip [18], and a decoder chip for the Multi-Anode Microchannel Array (MAMA) detector for the space telescope [19].

The BDCT chip is used for video compression applications. An 8×8 BDCT architecture was synthesized by *Hercules* and implemented in a compiled macro-cell design style [13] as a 9×9 mm^2 image in $2\mu_m$ CMOS technology. The DAIO chip provides an interface, following the Audio Engineering Standard (AES) protocol, between a standard 16/32 microprocessor bus with audio devices, such as compact disk or digital audio tape player. The DAIO specification in HardwareC was compiled and mapped into a logic netlist suitable for implementation in LSI Logic 9K-series sea-of-gates technology. The logic specification had about 6000 equivalent gates. The MAMA chip is designed to discriminate the information generated by a multi-anode detector in a space telescope. Also described in HardwareC, it was synthesized and fabricated with LSI Logic 9K-series sea-of-gates technology.

6 Acknowledgments

Thomas Truong implemented the SIF simulator *Arianne*. The authors would like to thank R. Gupta and T. Truong for helpful comments. This research was sponsored by NSF, under grant No. MIP 8719546, by AT&T and DEC jointly with NSF, under a PYI Award program, and by a fellowship provided by Phillips/Signetics.

References

[1] A. de Geus, *Logic synthesis Speeds ASIC Designs*, IEEE Spectrum, Vol 26, No. 8, August 1989, pp. 27-31.

[2] M. Crastes de Paulet, C. Duff, R. Leveugle, F. Poirot, G. Saucier, P. Sicard, *ASYL: A Logic and Architecture Design Automation System*, Proceedings of EuroAsic 89, Jan 1989.

[3] J. Rabaey, H. De Man, et. al., *Cathedral II: A Synthesis System for Multiprocessor DSP Systems*, in *Silicon Compilation*, Ed. D. Gajski, Addison Wesley 1988, p. 311-360.

[4] J.Huisken, H.Janssen, P.Lippens, O.McArdle, R.Segers, P.Zegers, A. Delaruelle and J. van Meerbergen, *Efficient Design of Systems on Silicon with PYRAMID*, in *Logic and Architecture Synthesis for Silicon Compilers*, North Holland, Amsterdam, 1989.

[5] R. Walker, D. Thomas, *The Systems Architect Workbench*, Kleuer Academic Press, 1989.

[6] A. Parker, J. Pizarro, M. Mlinar, *MAHA: A Program for Data Path Synthesis*, Proceedings 23th Design Automation Conference, June 1986, p. 461-466.

[7] F. Brewer, D. Gajski, *Knowledge Based Control in Micro Architecture Design*, Proceeding 24th DAC, p. 203-209, June 1987.

[8] P. G. Paulin, J. P. Knight, E. F. Girczyc, *HAL: A Multi-Paradigm Approach to Automatic Data-path Synthesis*, Proceedings 23th Design Automation Conference, June 1986, p. 263-270.

[9] R. Brayton, R. Camposano, G. De Micheli, R. Otten, J. van Eijndhoven, *The Yorktown Silicon Compiler System* in *Silicon Compilation*, Ed. D. Gajski, Addison Wesley 1988, p. 204-310.

[10] G.Zimmermann, *The MIMOLA Design System: Detailed Description of the Software System*, Proc 16th Des Autom. Conf, 1979, pp 56-63.

[11] G. De Micheli, D. Ku, *HERCULES - A System for High-Level Synthesis* Proceedings 25th Design Automation Conference, June 1988, p. 483-488.

[12] D. Ku, G. De Micheli, *HardwareC - A Language for Hardware Design* Stanford Technical Report, CSL-TR-88-362, August 1988, and CSL-TR-90, April 1990 (Version 2.0).

[13] F. Mailhot, G. De Micheli, *Automatic Layout and Optimization of Static CMOS Cells* Proceedings of ICCD, Rye, New York, pp. 180-185, 1988.

[14] R. Camposano, W. Rosenstiel, *Synthesizing Circuits from Behavioral Descriptions*, IEEE Trans. on CAD, p. 171-180, Feb 1989.

[15] M. J. McFarland, *Using Bottom-Up Design Techniques in the Synthesis of Digital Hardware from Abstract Behavioral Descriptions*, Proceedings 23rd Design Automation conference, pp. 474-480, 1986.

[16] D. Ku, G. De Micheli, *Relative Scheduling Under Timing Constraints*, Stanford Technical Report, CSL-TR-89-401, November 1989, and Proceedings of 27th Design Automation Conference, Orlando, Florida, June, 1990.

[17] V. Rampa, G. De Micheli, *The Bi-dimensional DCT Chip*, Proceedings of ISCAS, 1988.

[18] M. Ligthart, A. Bechtolsheim, G. De Micheli, A. El Gamal, *Design of a Digital Audio Input Output Chip*, Custom IC Conference, May 1989.

[19] D. B. Kasle, *High resolution decoding techniques and single-chip decoders for multi-anode microchannel arrays*, Proceedings of SPIE, Vol. 1158, 1989, pp. 311-318.

Logic and Architecture Synthesis
P. Michel and G. Saucier (Editors)
Elsevier Science Publishers B.V. (North-Holland)

CASTOR: STATE ASSIGNMENT IN A FINITE STATE MACHINE SYNTHESIS SYSTEM

Gerd RIETSCHE and Michael NEHER

Computer Science Research Centre at the University of Karlsruhe (FZI)
Automation of Circuit Design (Prof. Dr. D. Schmid, Dr. W. Rosenstiel)
Haid-und-Neu-Straße 10 - 14, D-7500 Karlsruhe

This paper describes some of the state assignment heuristics, which are applied in the finite state machine synthesis system CASTOR. It concentrates on algorithms for two-level logic implementations, while work on algorithms for multi-level logic implementations is in progress. Input to the system is a description of the FSM in form of a state table. CASTOR generates an appropriate controller consisting of pre- and postprocessing structures around a kernel FSM, which may be a PLA, a ROM or random logic. In the case of a PLA or random logic controller coding constraints are extracted for the kernel FSM. Then these coding constraints have to be satisfied by adequate state assignment algorithms in order to minimize the area of the physical implementation.

1. INTRODUCTION

A general view of the finite state machine synthesis system CASTOR is given in figure 1. CASTOR is starting from a symbolic state table as a description of the FSM. After some usual consistency checks a state reduction can be performed. Then an appropriate controller is generated. Pre- and postprocessing structures like multiplexers and decoders are grouped around the kernel FSM, which may be a PLA, a ROM or random logic. Some of CASTOR's structure options are especially adapted to the needs of the control part synthesis in the Carlsruhe Digital Design System (CADDY) [4], [6]. Thereafter adequate coding constraints are extracted for PLA and random logic controllers. In the case of a PLA FSM a symbolic minimization [3] is performed, in the case of a random logic controller similar techniques have to be applied. Then these coding constraints have to be satisfied by adequate state assignment algorithms. After a final logic minimization we get a kernel FSM with pre- and postprocessing structures. In the case of a multi-level logic implementation, the structure of the kernel FSM has to be generated from the two-level description.

2. DEFINITIONS

Let S be the set of states to be encoded and let $s = |S|$ be the number of states. Let n be the number of coding constraints and let b be the number of Boolean variables used to represent S, i. e., the encoding length.

Coding constraint: A coding constraint is a set of states, which have a transition to the same next state, while the values of the primary inputs as well as the values of the primary outputs are compatible. It can be represented by a vector $c \in \{0, 1, -\}^s$. $c_i = 1$ indicates, that state i is an element of constraint c. Accordingly $c_i = 0$ indicates, that state i is not an element of constraint c and a "-" entry in position i indicates, that state i can be an element of constraint c or not.

Constraint matrix: The constraint matrix $C \in \{0, 1, -\}^{n \times s}$ is a matrix, whose rows are the coding constraints.

Encoding matrix: The encoding matrix $E \in \{0, 1\}^{s \times b}$ is a matrix, whose rows are the encodings of the states.

Face: The face of a constraint is the minimal Boolean subspace, which contains the encodings of all the states, which are an element of this constraint.

Face matrix: The face matrix $F \in \{0, 1, -\}^{n \times b}$ is a matrix, whose rows are the faces of the constraints.

Satisfaction: A coding constraint is satisfied, if its face doesn´t contain encodings of states, which are not an element of this constraint. If a coding constraint is satisfied, the corresponding state transitions can be implemented by one common product term.

Compatibility: Two coding constraints c_1 and c_2 are compatible, if either ($c_{1j} = c_{2j}\ \forall\ j$ s. t. $c_{1j} \neq$ '-' and $c_{2j} \neq$ '-') or ($c_{1j} \neq c_{2j}\ \forall\ j$ s. t. $c_{1j} \neq$ '-' and $c_{2j} \neq$ '-').

Encoding problem: The encoding problem is to find an encoding matrix E, which satisfies the coding constraints represented by a given constraint matrix C.

3. STATE OF THE ART

Algorithms for state assignment for two-level logic implementations have been published in [3], [9], and [10]. For the solution of an encoding problem, there are two basic approaches. Row based encoding algorithms compute the encoding matrix E row by row, i. e., one state is encoded after the other. Column based encoding algorithms compute the encoding matrix E column by column. They compute one code bit for each state in each step. Both approaches have been presented in [3], while ASYL [9] and NOVA [10] are row based algorithms.

In the row based algorithms published in [3] the number of computation steps is equal to s (number of states), while in column based algorithms only b (encoding length) computation steps have to be performed. Especially for larger FSMs, row based algorithms are significantly slower than column based algorithms. Column based algorithms, however, have an other problem. For better understanding of this problem we first want to solve the encoding problem represented by the following constraint matrix C as an example:

$$C = \begin{pmatrix} 1 & 1 & 1 & 0 & 0 & 0 \\ 0 & 0 & 1 & 1 & 0 & 0 \\ 0 & 0 & 0 & 1 & 1 & 0 \\ 1 & 0 & 0 & 0 & 0 & 1 \end{pmatrix}$$

The encoding matrix $E = C^T$ is a valid solution of this encoding problem. In this case each of the constraints in the constraint matrix C is satisfied by the corresponding column in the encoding matrix E. However, the encoding length b = 4 is not minimal.

Therefore during the algorithm single coding constraints or combinations of compatible coding constraints are chosen as encoding columns. Compatible coding constraints can be satisfied by one column in the encoding matrix E. If compatible coding constraints can be detected, the column cardinality in the encoding matrix E, i. e., the encoding length b can be decreased.

In the first step there is no pair of compatible coding constraints. In this case one of the coding constraints is chosen as first encoding column, for example the first one. Then the face matrix is computed. So we get the following constraint matrix C, face matrix F, and transposed partial encoding matrix E^T:

$$C = \begin{pmatrix} 1 & 1 & 1 & 0 & 0 & 0 \\ 0 & 0 & 1 & 1 & 0 & 0 \\ 0 & 0 & 0 & 1 & 1 & 0 \\ 1 & 0 & 0 & 0 & 0 & 1 \end{pmatrix}, \quad F = \begin{pmatrix} 1 \\ - \\ 0 \\ - \end{pmatrix}$$

$$E^T = \begin{pmatrix} 1 & 1 & 1 & 0 & 0 & 0 \end{pmatrix}$$

Now the coding constraints can be reduced. Since all the states which are elements of the first constraint (states 1, 2, and 3) are encoded with "1" the face of this constraint is "1". The states which must not intersect this face (states 4, 5, and 6) are all encoded with "0". Therefore this constraint is completely satisfied by the partial encoding matrix E. The faces of the second and forth constraint are "-", because some of the states, which are elements of these constraints are encoded with "0" and some are encoded with "1". In this case no reduction is possible. The face of the third constraint is "0". Some of the states which must not intersect this face (states 1, 2, and 3) are encoded with "1". This constraint is only partially satisfied, because the encoding of state 6 is still intersecting its face. The constraints, which remain to be satisfied can be represented by the following reduced constraint matrix C:

$$C = \begin{pmatrix} 1 & 1 & 1 & - & - & - \\ 0 & 0 & 1 & 1 & 0 & 0 \\ - & - & - & 1 & 1 & 0 \\ 1 & 0 & 0 & 0 & 0 & 1 \end{pmatrix}$$

In the second step the last two coding constraints are compatible. They can both be satisfied by one single encoding column $e_2^T = (\,1\,0\,0\,0\,0\,1\,)$.

In the third step the encoding column $e_3^T = (\,0\,0\,1\,1\,0\,0\,)$ can satisfy the second coding constraint, so that the encoding problem is completely solved by the following minimal transposed encoding matrix E^T, whose columns are the encodings of the states:

$$E^T = \begin{pmatrix} 1 & 1 & 1 & 0 & 0 & 0 \\ 1 & 0 & 0 & 0 & 0 & 1 \\ 0 & 0 & 1 & 1 & 0 & 0 \end{pmatrix}$$

A problem of column based algorithms is the choice of the first encoding column. Especially for larger FSMs the coding constraints normally contain much more "0" entries than "1" entries. In the first step one of these coding constraints has to be chosen as an encoding column and many states are encoded with "0", while only few states are encoded with "1" or vice versa. Assume we have to solve an encoding problem with eight states and the maximal coding constraint cardinality is two. Then in the first step of the algorithm six states are encoded with "0" and only two states are encoded with "1". Only to distinguish the encodings of the six states, which are encoded with "0", three additional encoding columns are necessary, so that the minimal possible encoding length is exceeded by one.

4. CASTOR'S STATE ASSIGNMENT

In CASTOR several state assignment algorithms are implemented. In this paper we want to introduce column based state assignment heuristics, which are founded on the column based algorithm mentioned above. They comprise the speed advantages of the column based algorithms, but try to avoid or limitate the problem, we mentioned above.

In the sequel we describe the major steps of our algorithm:

Step 1: Partition the set C of coding constraints into a maximal number m of disjoint subsets $C_1, C_2, ..., C_m$, so that coding constraints, which are elements of different subsets, don´t intersect. Compute adequate encoding columns to partition the Boolean space into disjoint subspaces, so that each of the subsets $C_1, C_2, ..., C_m$ is related to one of these subspaces.

The following two steps have do be performed for each of the subsets C_k, k = 1, ..., m.

Step 2: Partition the set C_k of coding constraints into a minimal number n of disjoint subsets $C_{k1}, C_{k2}, ..., C_{kn}$, so that coding constraints, which are elements of the same subset, don´t intersect. Partition each of the subsets C_{ki}, which contains more than one element itself into two subsets $C_{ki}{}^0$ and $C_{ki}{}^1$ of approximately equal cardinality. Compute a partial encoding column for each of the subsets C_{ki} with more than one element. Encode states, which are element of a coding constraint $c \in C_{ki}{}^0$, with a "0" and states, which are element of a coding constraint $c \in C_{ki}{}^1$, with a "1". Assign unspecified code bits in order to improve the efficiency of the algorithm.

Step 3: Compute additional encoding columns according to the above mentioned column based algorithm in order to satisfy remaining unsatisfied constraints until all constraints are satisfied or the minimal possible or a preset encoding length is achieved, then distinguish identical encodings and STOP.

While step 3 is invoked in any case, steps 1 and 2 can be suppressed by option. Default is the use of all three steps.

As an example we want to solve the encoding problem represented by the following constraint matrix C:

$$C = \begin{pmatrix} 1 & 1 & 1 & 0 & 0 & 0 & 0 & 0 & 0 & 0 & 0 & 0 & 0 & 0 \\ 0 & 0 & 1 & 1 & 0 & 0 & 0 & 0 & 0 & 0 & 0 & 0 & 0 & 0 \\ 0 & 0 & 0 & 1 & 1 & 0 & 0 & 0 & 0 & 0 & 0 & 0 & 0 & 0 \\ 1 & 0 & 0 & 0 & 0 & 1 & 0 & 0 & 0 & 0 & 0 & 0 & 0 & 0 \\ 0 & 0 & 0 & 0 & 0 & 0 & 1 & 1 & 0 & 0 & 0 & 0 & 0 & 0 \\ 0 & 0 & 0 & 0 & 0 & 0 & 0 & 0 & 1 & 1 & 0 & 0 & 0 & 0 \\ 0 & 0 & 0 & 0 & 0 & 0 & 0 & 0 & 0 & 0 & 1 & 1 & 0 & 0 \\ 0 & 0 & 0 & 0 & 0 & 0 & 0 & 0 & 0 & 0 & 0 & 0 & 1 & 1 \\ 0 & 0 & 0 & 0 & 0 & 0 & 1 & 0 & 1 & 0 & 1 & 0 & 1 & 0 \end{pmatrix} = \begin{pmatrix} c_1 \\ c_2 \\ c_3 \\ c_4 \\ c_5 \\ c_6 \\ c_7 \\ c_8 \\ c_9 \end{pmatrix}$$

In step 1 the coding constraints are partitioned into a maximal number of subsets, so that coding constraints, which are elements of different subsets don´t intersect. In our example we get the two subsets $C_1 = \{c_1, c_2, c_3, c_4\}$ and $C_2 = \{c_5, c_6, c_7, c_8, c_9\}$. Then the Boolean space is partitioned into two subspaces. States, which are an element of a coding constraint $c_i \in C_1$, are encoded with a "0" and states, which are an element of a coding constraint

$c_i \in C_2$, are encoded with a "1". After the reduction of the coding constraints, we get the following reduced constraint matrix C and transposed encoding column e:

$$C = \begin{pmatrix} 1 & 1 & 1 & 0 & 0 & 0 & - & - & - & - & - & - & - & - \\ 0 & 0 & 1 & 1 & 0 & 0 & - & - & - & - & - & - & - & - \\ 0 & 0 & 0 & 1 & 1 & 0 & - & - & - & - & - & - & - & - \\ 1 & 0 & 0 & 0 & 0 & 1 & - & - & - & - & - & - & - & - \\ - & - & - & - & - & - & 1 & 1 & 0 & 0 & 0 & 0 & 0 & 0 \\ - & - & - & - & - & - & 0 & 0 & 1 & 1 & 0 & 0 & 0 & 0 \\ - & - & - & - & - & - & 0 & 0 & 0 & 0 & 1 & 1 & 0 & 0 \\ - & - & - & - & - & - & 0 & 0 & 0 & 0 & 0 & 0 & 1 & 1 \\ - & - & - & - & - & - & 1 & 0 & 1 & 0 & 1 & 0 & 1 & 0 \end{pmatrix}$$

$$e^T = (0 \; 0 \; 0 \; 0 \; 0 \; 0 \; 1 \; 1 \; 1 \; 1 \; 1 \; 1 \; 1 \; 1)$$

With the exception of two diagonal blocks, the reduced constraint matrix C constists of "-" entries only. These two blocks can now be seen as two encoding subproblems, which can be solved independently of each other. This means that we have to solve the encoding problems represented by the following two constraint matrices C_1 and C_2:

$$C_1 = \begin{pmatrix} 1 & 1 & 1 & 0 & 0 & 0 \\ 0 & 0 & 1 & 1 & 0 & 0 \\ 0 & 0 & 0 & 1 & 1 & 0 \\ 1 & 0 & 0 & 0 & 0 & 1 \end{pmatrix}, \quad C_2 = \begin{pmatrix} 1 & 1 & 0 & 0 & 0 & 0 & 0 & 0 \\ 0 & 0 & 1 & 1 & 0 & 0 & 0 & 0 \\ 0 & 0 & 0 & 0 & 1 & 1 & 0 & 0 \\ 0 & 0 & 0 & 0 & 0 & 0 & 1 & 1 \\ 1 & 0 & 1 & 0 & 1 & 0 & 1 & 0 \end{pmatrix}$$

The encoding problem represented by constraint matrix C_1 was already solved above by the following transposed encoding matrix E_1^T:

$$E_1^T = \begin{pmatrix} 1 & 1 & 1 & 0 & 0 & 0 \\ 1 & 0 & 0 & 0 & 0 & 1 \\ 0 & 0 & 1 & 1 & 0 & 0 \end{pmatrix}$$

Now we have to solve the encoding problem, which is represented by the constraint matrix C_2, and for which the above algorithm cannot find a minimal solution.

$$C_2 = \begin{pmatrix} 1 & 1 & 0 & 0 & 0 & 0 & 0 & 0 \\ 0 & 0 & 1 & 1 & 0 & 0 & 0 & 0 \\ 0 & 0 & 0 & 0 & 1 & 1 & 0 & 0 \\ 0 & 0 & 0 & 0 & 0 & 0 & 1 & 1 \\ 1 & 0 & 1 & 0 & 1 & 0 & 1 & 0 \end{pmatrix} = \begin{pmatrix} c_1 \\ c_2 \\ c_3 \\ c_4 \\ c_5 \end{pmatrix}$$

In step 2 the coding constraints are partitioned into a minimal number of disjoint subsets, so that coding constraints, which are elements of the same subset don´t intersect. In our example we get the two subsets $C_1 = \{c_1, c_2, c_3, c_4\}$ and $C_2 = \{c_5\}$. Then subset C_1, which contains more than one element is itself partitioned into two disjoint subsets $C_1^0 = \{c_1, c_2\}$ and $C_1^1 = \{c_3, c_4\}$. Thereafter states, which are an element of a constraint $c_i \in C_1^0$, are encoded with a "0" and states, which are an element of a constraint $c_i \in C_1^1$, are encoded with a "1". After the reduction of the coding constraints, we get the following reduced constraint matrix C_2, face matrix F_2 and transposed partial encoding matrix E_2^T:

$$C_2 = \begin{pmatrix} 1 & 1 & 0 & 0 & - & - & - & - \\ 0 & 0 & 1 & 1 & - & - & - & - \\ - & - & - & - & 1 & 1 & 0 & 0 \\ - & - & - & - & 0 & 0 & 1 & 1 \\ 1 & 0 & 1 & 0 & 1 & 0 & 1 & 0 \end{pmatrix}, \quad F_2 = \begin{pmatrix} 0 \\ 0 \\ 1 \\ 1 \\ - \end{pmatrix}$$

$$E_2{}^T = (0 \quad 0 \quad 0 \quad 0 \quad 1 \quad 1 \quad 1 \quad 1)$$

Now the first four coding constraints are compatible. In step 3 they can be satisfied by a second encoding column $e_2{}^T$ = (1 1 0 0 1 1 0 0), while the last coding constraint can be satisfied by the encoding column $e_3{}^T$ = (1 0 1 0 1 0 1 0). So we get the following transposed encoding matrix E^T as a minimal solution of the encoding problem represented by the constraint matrix C_2:

$$E_2{}^T = \begin{pmatrix} 0 & 0 & 0 & 0 & 1 & 1 & 1 & 1 \\ 1 & 1 & 0 & 0 & 1 & 1 & 0 & 0 \\ 1 & 0 & 1 & 0 & 1 & 0 & 1 & 0 \end{pmatrix}$$

In step 1 the initial encoding problem has been partitioned into two encoding subproblems, which have then been solved in steps 2 and 3. Now we have to merge the encoding column e, which we computed in step 1, and the encoding matrices E_1 and E_2, which solve the encoding subproblems. in order to receive the solution of our initial encoding problem, which is the following transposed encoding matrix E^T:

$$E^T = \left(\begin{array}{c|c} \multicolumn{2}{c}{e^t} \\ \hline E_1{}^T & E_2{}^T \end{array} \right)$$

$$E^T = \left(\begin{array}{cccccc|cccccccc} 0 & 0 & 0 & 0 & 0 & 0 & 1 & 1 & 1 & 1 & 1 & 1 & 1 & 1 \\ \hline 1 & 1 & 1 & 0 & 0 & 0 & 0 & 0 & 0 & 0 & 1 & 1 & 1 & 1 \\ 1 & 0 & 0 & 0 & 0 & 1 & 1 & 1 & 0 & 0 & 1 & 1 & 0 & 0 \\ 0 & 0 & 1 & 1 & 0 & 0 & 1 & 0 & 1 & 0 & 1 & 0 & 1 & 0 \end{array} \right)$$

5. BENCHMARK RESULTS

Table 1 and table 2 show results for some of the MCNC FSM benchmarks [8]. In both tables the first four columns give the basic properties of the examples. Table 1 shows a comparison between CASTOR and NOVA [10] for two-level logic implementations, while table 2 shows a comparison between CASTOR and JEDI [7], a state encoder for multi-level logic implementations.

In table 1 the fifth column presents the number of product terms achieved by CASTOR's state assignment without initial processing and postprocessing optimizations to allow a direct comparison with other state assignment systems and a logic minimization. The sixth column shows the corresponding results of NOVA.

In table 2 the fifth column presents the number of literals achieved by CASTOR's state assignment without respect to any multi-level coding constraints and a multi-level synthesis by MIS 2.0 [2]. The last column shows the results of JEDI's state assignment and a multi-level synthesis by MIS 2.1.

PI : number of primary inputs
PO : number of primary outputs
S : number of states
PT : number of product terms in the original symbolic state table
CL : minimal possible codelength

Name	PI	PO	S	PT	CL	CASTOR	NOVA
bbara	4	2	10	60	4	22	24
bbsse	7	7	16	56	4	29	30
bbtas	2	2	6	24	3	8	8
beecount	3	4	7	28	3	11	11
cse	7	7	16	91	4	46	45
donfile	2	1	24	96	5	27	26
ex1	9	19	20	138	5	44	44
ex2	2	2	19	72	5	27	27
ex3	2	2	10	36	4	17	18
keyb	7	2	19	170	5	47	48
lion	2	1	4	11	2	6	6
lion9	2	1	9	25	4	8	8
mark1	5	16	15	22	4	18	18
mc	3	5	4	10	2	8	9
planet	7	19	48	115	6	86	87
s1	8	6	20	107	5	73	63
s1a	8	6	20	107	5	68	65
styr	9	10	30	166	5	94	94
tbk	6	3	32	1569	5	48	57
train11	2	1	11	25	4	9	9

CASTOR : number of product terms after CASTOR's state assignment and logic minimization
NOVA : number of product terms after NOVA's state assignment and logic minimization

table 1: two-level benchmark results

Name	PI	PO	S	PT	CL	CASTOR	JEDI
bbara	4	2	10	60	4	60	57
bbsse	7	7	16	56	4	133	111
bbtas	2	2	6	24	3	21	21
beecount	3	4	7	28	3	51	39
cse	7	7	16	91	4	212	200
donfile	2	1	24	96	5	67	76
ex1	9	19	20	138	5	220	250
ex2	2	2	19	72	5	107	122
ex3	2	2	10	36	4	74	66
keyb	7	2	19	170	5	188	140
lion	2	1	4	11	2	16	13
lion9	2	1	9	25	4	28	13
mark1	5	16	15	22	4	109	66
mc	3	5	4	10	2	22	23
planet	7	19	48	115	6	519	547
s1	8	6	20	107	5	304	152
s1a	8	6	20	107	5	223	94
styr	9	10	30	166	5	445	508
tbk	6	3	32	1569	5	179	278
train11	2	1	11	25	4	31	27

CASTOR : number of literals after CASTOR's state assignment and multi-level optimization
JEDI : number of literals after JEDI's state assignment and multi-level optimization

table 2: multi-level benchmark results

6. CONCLUSIONS

We introduced some of the state assignment algorithms in the finite state machine synthesis system CASTOR. We chose a subset of the MCNC FSM benchmark set, in order to make CASTOR's state assignment algorithms comparable to other systems. Although only a part of CASTOR's capabilities could be applied, very encouraging results have been achieved. Even for multi-level implementations CASTOR's results are in the most cases comparable to the results of a real multi-level state encoder, although we do not yet consider special multi-level coding constraints.

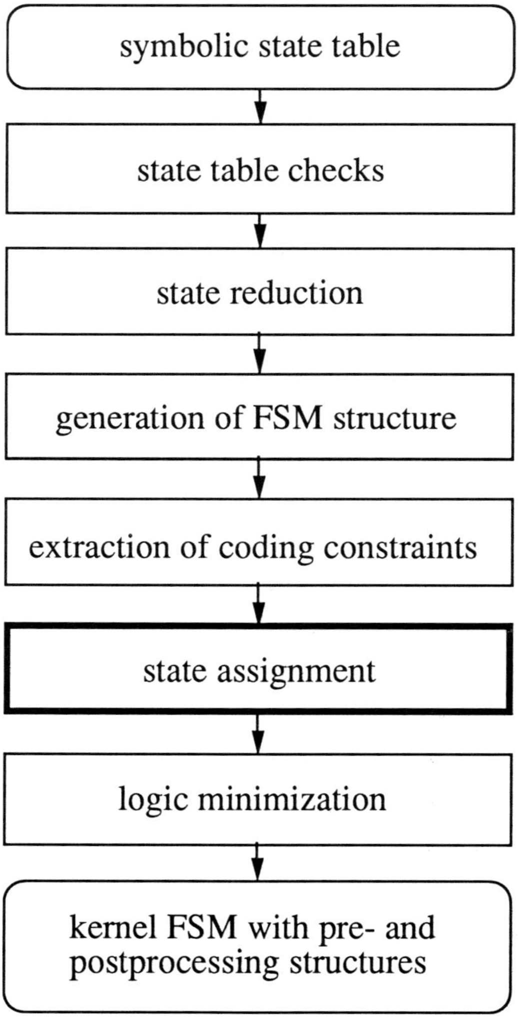

figure 1: the FSM Synthesis System CASTOR

REFERENCES

[1] R. K. Brayton, G. D. Hachtel, et al.: "Logic Minimization Algorithms for VLSI Synthesis", Kluwer Academic Publishers, Boston, 1984.
[2] R. K. Brayton et al.: "MIS: A Multiple Level Logic Optimization System", IEEE Trans. on CAD, vol. CAD-6, 1987.
[3] G. De Micheli: "Symbolic Design of Combinational and Sequential Logic Implemented by Two-Level Macros", IEEE Trans. on CAD, vol. CAD-5, 1986.
[4] P. Duzy, H. Krämer, M. Neher, M. Pilsl, W. Rosenstiel and T. Wecker, "CALLAS - Conversion of Algorithms to Library Adaptable Structures", VLSI 89.
[5] W. Grass, "Steuerwerke", Springer-Verlag, 1978.
[6] H. Krämer and W. Rosenstiel, "System Synthesis using Behavioural Descriptions", EDAC 1990.
[7] B. Lin and A. R. Newton: "Synthesis of Multiple Level Logic from Symbolic High-Level Description Languages", VLSI 89.
[8] R. Lisanke: "Introduction of Synthesis Benchmarks", Int. Workshop on Logic Synthesis, North Carolina, 1989.
[9] G. Saucier et al.: "State Assignment Using a New Embedding Method Based On an Intersecting Cube Theory", Proc. of the DAC, 1989.
[10] T. Villa and A. Sangiovanni-Vincentelli: "NOVA: State Assignment of Finite State Machines for Optimal Two-Level Logic Implementations", Proc. of the DAC, 1989.

Logic and Architecture Synthesis II

Logic and Architecture Synthesis
P. Michel and G. Saucier (Editors)
Elsevier Science Publishers B.V. (North-Holland)
IFIP, 1991

MULTI-LEVEL SYNTHESIS ON PROGRAMMABLE DEVICES IN THE ASYL SYSTEM

Gabrielle SAUCIER, Pascal SICARD

Institut National Polytechnique de Grenoble - Laboratoire CSI
46 avenue Felix Viallet
38031 GRENOBLE Cedex _ FRANCE
Tel: 76574687 _ FAX: 76503421

Laurent BOUCHET

Laboratoire d'Annecy-le-vieux de Physique des Particules
Chemin de Bellevue BP 110
74941 ANNECY-LE-VIEUX Cedex _ FRANCE
Tel: 50233245 poste 232 _ FAX: 50279495

ABSTRACT: Starting from Boolean equations or, at a higher level, from a control flowchart, the automatic synthesis tool presented here, looks for an optimized mapping on a network of programmable modules. For the PALs, the output will be a network of PALs and a netlist ready for a Jedec fusemap; for the Xilinx PGAs the system delivers a network of Xilinx blocks and a netlist. For the last target a place and route phase is necessary.

1. INTRODUCTION

A lot of developments have concerned in the last decade automatic synthesis of Boolean functions or finite state machines on PLAs or standard cells. Optimized synthesis methods relying on significant result in global minimization for PLAs [2], in factorization and technology mapping [1, 17, 18], in state assignment [7, 11, 16] have led to effective tools in commercial silicon compilers.

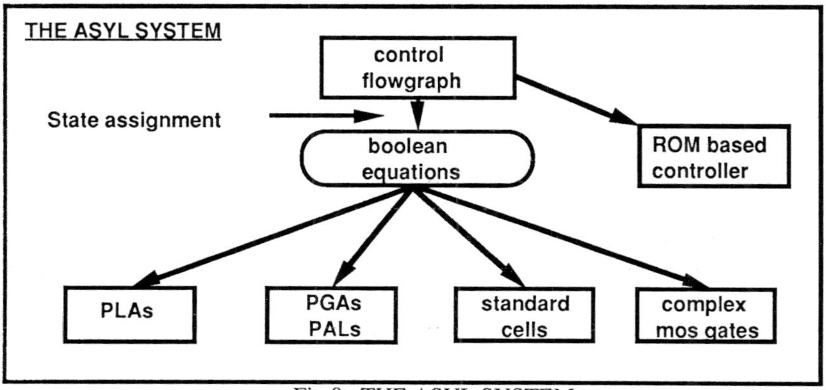

Fig.0 : THE ASYL SYSTEM

These fantastic progresses in logic synthesis move now slowly to board design and one of the first target was PAL synthesis[13, 14, 9, 15]. Nevertheless, the results recently exhibited in this field [6] show that this subject is not yet mature and that significant progresses will be achieved in the near future.

The approach presented here, is based on algebraic division techniques [3] decomposing Boolean functions in portions which are mappable in the considered programmable devices. A phase called "reinjection phase" optimizes this decomposition aiming at avoiding any unnecessary decomposition. A final phase performs splitting of functions or product terms which have not been decomposed at a sufficient extent in the previous phases.

The positive and original aspect of this approach is that the final target is considered in an early stage during the factorization and not afterwards, in the technology mapping phase. This approach has already been applied successfully to PALs [10] and is presently extended to Xilinx PGAs.

This tool is a part of the ASYL system (Fig.1) wich is a complete system dedicated to the synthesis of digital circuits [12].

The paper is organized as follows. In a first section the decomposition of Boolean equations by algebraic division is presented. In a second section, the splitting of functions which cannot or can no more be decomposed by algebraic division is explained. In a last section the final mapping is shown.

2. DECOMPOSITION OF BOOLEAN FUNCTIONS FOR SYNTHESIS ON PROGRAMMABLE DEVICES

2.1. MAPPABLE AND NON MAPPABLE FUNCTIONS

The Boolean functions are described by a set of equations minimized locally. That means that the functions are minimized individually without looking for common product terms. It has been shown, at least for the PAL target, that global minimization has a negative effect. This may be easily understood as in the PAL structure there is no sharing of product terms between functions.

A function is mappable if it can be implemented at least in one element of the considered library. For the PAL target, the number of product terms and the number of variables of a product term should be smaller than those of the largest PAL [8]. For the Xilinx, the number of variables has to be smaller than five (family Xc3000) [4, 5].

A preliminary phase investigates if a function or its complementary form is more interesting. If none of the two forms is mappable, the decomposition phase is invoked. The decomposition is necessary if the number of variables is too large (this is true for both PAL and Xilinx target) or if the number of product terms of a function is too large for the PAL target. If a decomposition is required, a kernel search is performed first. If at least one kernel exists, a decomposition through algebraic division is realized.

2.2. DECOMPOSITION OF BOOLEAN FUNCTIONS BY ALGEBRAIC DIVISION

The decomposition process is based on algebraic division and kernel extraction. A kernel algebraic division $F = K.Q + R$ is interesting as:

- the number of variables in the kernel K and in the quotient Q is inferior to the number of variables in the function F.
- the number of product terms in the kernel K, in the quotient Q and in the remainder R is inferior to the number of product terms in the function F.
- the number of product terms in the remainder R is not always inferior to the number of product terms in the function F but it decreases with the number of product terms.

This decomposition phase starts by computing all the kernels of ~F. ~F represents either a function or its complementary form according to the choice of the preceeding step.

The best kernel is chosen and an algebraic division is performed. A kernel will be interesting if the associated quotient, remainder and the kernel itself are easy to be implemented, i.e. if their implementation requires a minimal number of further decompositions. If so, the number of recursive decompositions will be limited and therefore the critical path for this function. With regards to the propagation delay, the length of the longest path in the decomposition tree has bo be reduced, producing a well-balanced decomposition tree. The selected kernel has to minimize simultaneaously the number of variables it contains and the number of variables in r. So the chosen kernel minimizes the MAX (number of variables in k, number of variables in r). The decomposition process is iterated until q, k and r can be mapped or there is no more division capability.

Considering the following example and suppose that:
F=v1.v2^v3+v4.^v5.v6+v1.v2^v4.v6+v6.v7.^v8.v9.^v10.
The kernels are: ^v3 + ^v4.v6;
v4.^v5 + v1.v2.^v4 + v7.^v8.v9.^v10;
and the function itself.
Kernel ^ v3 + ^ v4.v6 is selected as it minimizes at the same time the number of variables it contains and the number of variables in the remainder. Therefore F is rewritten as F = k.q + r where:

k = ^ v3+^ v4.v6; q = v1.v2;
r=v4.^ v5.v6+v6.v7.^ v8.v9.^ v10;

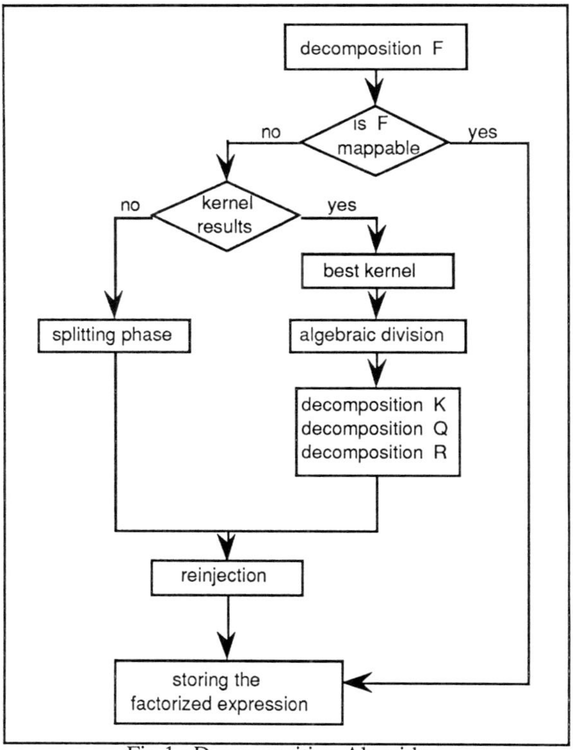

Fig.1 : Decomposition Algorithm

2.3. DECOMPOSITION OF BOOLEAN FUNCTION BY SPLITTING METHODS

This technique is applied when no more division possibilities exist for the generated subfunctions and these are still not mappable. All the variables appear then at most twice, once in the direct form and once in the complementary one. Otherwise, a kernel would exist. Two cases are considered:
(i) There are product terms containing too many variables. A splitting of these product terms into two balanced product terms is iterated until the product terms become mappable.
(ii) The current functions are not mappable because they contain too many variables (PAL and Xilinx target) or too many product terms (PAL). These functions are then split into subfunctions, grouping the pairs of product terms depending on the same variable. Remember that such a variable appears once in its direct form, once in its complemented form. This partitioning is done by creating balanced subfunctions until they are mappable on the target.

2.4. REINJECTION PHASE

According to the previous algorithm, we obtain a decomposition tree but some of these decompositions may be unnecessary.

Once the decomposition is done, we try to eliminate some subfunctions by reinjecting them in the upper level of the decomposition. As the efficiency of this reinjection is order-dependent, a heuristic to determine a near-optimal list of successive reinjections is needed. In the decomposition tree, we will first try to reinject subfunctions in the longest path of the tree and then in the other paths. As the propagation delay depends on the depth of the tree, this heuristic minimizes the propagation delay in the final result. This reinjection is guided by the final target.

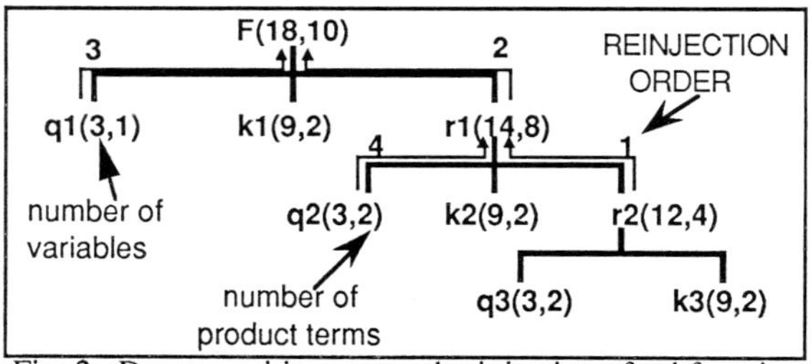

Fig. 2 : Decomposition tree and reinjection of subfonctions

Consider, for instance, the decomposition tree for a given function F for a PAL target (Fig. 2). The numbers between brackets are respectively the number of variables and the numbers of product terms of the current function. Suppose also that the PALs used have a maximum number of 10 input variables and of 6 product terms.

The initial decomposition leads to the decomposition given in Figure 3a (4 levels and 8 subfunctions). Reinjections 1,2,3, and 4 lead to the decomposition given in Figure 3b (2 levels and 4 subfunctions). The information pointed by the arrows in Figure 3a, 3b indicate the number of input variables used by each "connection".

At the end of the recursive decomposition step, all the subfunctions are mappable at each level of decomposition and they are stored for definitive mapping.

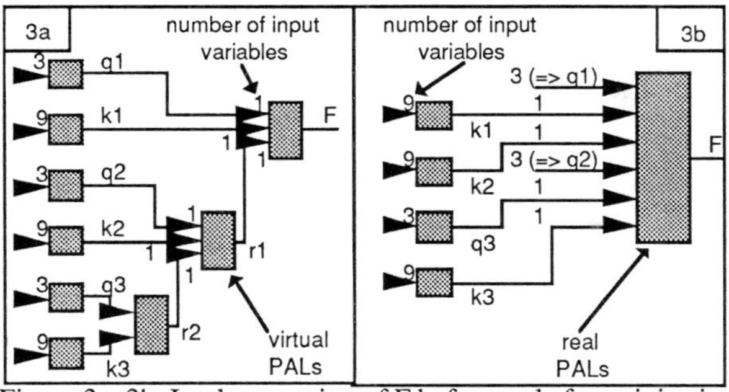

Figure 3a, 3b: Implementation of F before and after reinjection

3. DEFINITIVE MAPPING

3.1. MAPPING ON PAL

At this stage of the algorithm, all the subfunctions can be assigned to a device of the library. As either the subfunctions or their complements can be in most of the cases mapped on the devices, the assignment step will consist, for each function, of selecting the form of the function to be implemented. The goal of this step is to minimize at first the total number of devices and next the cost. The first difficulty is that each function can be assigned to a large set of devices, and the second one is that each device may usually be used to implement several functions. Therefore, several functions with various characteristics have to be gathered on the same device. The algorithm tries all the possible assignments for each function, trying to use already existing devices.

Once an initial solution has been computed, other solutions are searched by backtracking on the successive assignments. Backtracking in the search space is limited by a branch and bound heuristic. The tree of the possible assignments has to be reduced in order to reach a near-optimal solution in reasonable time. The first point is that we will first assign the functions that can be assigned to a minimum number of devices. The second point is that the functions will be assigned to the devices with a maximum number of outputs, thus trying to get a maximum number of functions on the same device. This is done while bearing in mind that if the more constrained functions are assigned to devices with a maximum number of outputs, then the remaining functions will be probably assignable to the same devices.

Before assigning the functions to devices, the algorithm first classifies both the functions and the devices. Let #Fi(resp.#F^i) the number of devices to which the function Fi (resp.^ Fi) can be assigned. The functions are classified according to the following criteria given in decreasing order of significance:
- (#Fi + #^ Fi) increasing,
- minimum of the number of product terms of the function and of its complement decreasing,
- number of variables decreasing.

For each function, the direct and complemented forms (Fi, ^ Fi) will be ordered according to the two following criteria, given in decreasing order of significance:
- #~Fi increasing,
- number of product terms in ~Fi decreasing.

As usual, the number of variables is the same in Fi and ^ Fi, we will first try the form which can be assigned to the maximum number of devices. In this way, we have an increased probability to use a device already selected.

The devices are classified according to the following criteria given in decreasing order of significance:
-number of outputs decreasing,
- number of variables decreasing,
- number of product terms per output decreasing,
- programmable polarity.

The same classification will be applied when selecting a device from the list of already selected devices during the assignement step.

This heuristic has proved to be efficient on practical examples: the first solution is quite good and the search space is reduced, leading to a limited number of backtrackings.

3.2. MAPPING ON XILINX

For the XILINX family each CLB can have at most five input variables and can generate either a single function of five variables or two independent logic functions of up to four variables (the total number of variables for this two functions is at most five).

Starting from this properties the functions are classified into 4 classes:

(a) functions of 5 variables
(b) functions of 4 variables
(c) functions of 3 variables
(d) functions of 2 variables.

Each function of the first class is mapped directly because it can't be paired with another function.

The other ones are paired into the same CLB using a special pairing procedure. The pairing algorithm is the following: for each function f belonging to (b) we search a function g belonging also to (b) so that f and g may be paired, this means that the set of variables used by f and g is of cardinality 5. If we don't find any solution, we try to apair again the same f but with g belonging to (c), and if we don't succeed we try with g belonging to (d). If no solution appears, we map f alone. When all the set (b) has been considered, we start again with f belonging to (c) and g belonging to (c), next to (d) if necessary. We stop this last step when the number of functions in (c) is equal to the number of functions belonging to (d) as all pairs (f,g), f belonging to (c) and g belonging to (d), may be mapped on a CLB. For each function of set (c) the number of variables is 3 and for each function of set (d) the number of variables is 2, when we paired them the set of variables used is of cardinality 5.

4. PRACTICAL RESULTS

4.1. PAL RESULTS

4.1.1. Benchmark results (table 1a and 1b)

The presented algorithms have been implemented using a prototype version of the system written in Prolog (compiled version of DELPHIA) on SUN workstations under UNIX, a VAX under VMS and a UNIGRAPH workstation under UNIX.

The library used is made up of 23 PALs from MMI: pal10h8, pal12h6, pal14h4, pal10l8, pal12l6, pal16l8, pal16r4, pal14l4, pal16r6, pal16p8, pal16r6, pal16rp4, pal12l10, pal14l8, pal16l6, pal18l4, pal20l2, pal20l10, pal20l8, pal20s10, pal20rs4, pal20rs8, pal20rs10.

It could be easily extended to other family like ALTER, INTEL, LATTICE, as the description language is flexible enough. The algorithms have been validated on a set of both combinational and sequential examples which are standard benchmarks. Computation times are indicated on a UNIGRAPH 3000 workstation. A reasonable backtracking time has been allocated to reach a good solution.

NAME	O	I	NB-FCT	T-REP	T-DEC	NB-PAL	T-PLAC
9sym	1	9	6	2	5	**2**	0.7
bw	28	5	0	1	2.7	**4**	1.8
c1	1	5	2	2	1.3	**1**	0.3
con1	2	8	0	1	0.2	**1**	0.3
duke2	29	22	8	2	27	**14**	13.9
f2	4	4	0	1	0.3	**1**	0.3
missex1	7	8	0	1	0.5	**1**	0.6
vg2	8	25	19	2	35	**11**	7.6

table 1a: combinational benchmarks

NAME	O OL	I	NB-FCT	T-REP	T-DEC	NB-PAL	T-PLAC
planet	19 6	13	20	2	24	**10**	8.3
sm1	6 5	13	10	2	11.5	**10**	7.7
bbsse	7 4	11	0	1	1.6	**2**	3
keyb	2 5	12	11	2	12.9	**4**	1.9
tav	4 2	6	0	1	0.7	**1**	0.5
dk14	5 3	6	0	1	0.9	**1**	0.6
mark1	14 4	8	0	1	2.8	**3**	1.5

table 1b: sequential benchmarks

O, OL : Number of non-registered (resp. registered) functions
I : Number of input variables
NB-FCT : Number of functions created after decomposition
T-REP : Maximum number of levels for the decomposed functions (critical path)
T-DEC : CPU time (s) for the decomposition
NB-PAL : Number of PALs after placement
T-PLAC : CPU time (s) for the placement
COST : Total cost
TIME : Running time

4.1.2. Quality estimation (Table 2)

The mapping heuristic gives a good result in a very short time. This result is generally equal or practically equal to the quasi-optimal solution found with extensive backtracking and is always faster. This is due to the decomposition phase. The mapping phase follows what has been prepared by this previous phase. As extensive backtracking in Prolog is available, it was possible to compare a fast result, obtained in a short time with a result obtained after a very long backtracking procedure which is, therefore, quasi-optimal.

NAME	FIRST SOLUTION			QUASI OPTIMAL SOLUTION		
	NB-PAL	COST	TIME	NB-PAL	COST	TIME
9SYM	2	48	0.7	2	48	0.7
BW	4	84	1.8	3	68	37.8
C1	1	24	0.3	1	24	0.3
CON1	1	20	0.3	1	20	0.3
DUKE2	14	336	13.9	14	328	35
F2	1	20	0.5	1	20	0.5
MISSEX1	2	40	0.9	1	20	3.6
VG2	11	256	7.6	11	252	27

table 2 : mapping quality

4.1.3. Interest of complementary computation (Table3)

The following table 3 shows that the solution is equal or better for 2 functions when choosing between direct and complementary function. This confirms the interest of complementary computation in our preliminary phase.

NAME	solutions WITH complementary function computation	solutions WITHOUT complementary function computation
9SYM	1	1
BW	3	4
C1	1	1
CON1	1	1
DUKE2	14	15
F2	1	1
MISSEX1	1	1
VG2	11	11

table 3 : number of PALs

4.1.4. Importance of an initial global or local minimization (Table 4)

It is proven that a global minimization does not bring a better result than a local minimization, and in some cases (eg: "BW" and "DUKE2") global approach has a negative effect on the number of PALs. This can be explained by the increasing number of product terms of each function. If the global minimization reduces the number of global set of product terms, this number may increase for each function. This confirms the comment of the introduction.

NAME	Local minimisation results	global minimisation results
9SYM	2	2
BW	4	5
C1	1	1
CON1	1	1
DUKE2	15	17
F2	1	1
MISSEX1	1	1
VG2	11	11

table 4 : number of PALs

4.1.5. Interest of large libraries (table 5)

We wand to explore the gain provided by a large library. Some examples have been processed with reduced and full libraries.The results prove the interest of a large library. Moreover, if the user knows very well the PAL library, he can choose the most adequate for his application. The result obtained can be very good. In "table 5" PAL20110 is adapted for "misex 1" and "c1".

NAME	solution		using
	all library	PAL20l2	PAL20l10
BW	4	15	7
C1	1	7	1
F2	1	2	1
MISSEX1	2	4	2

table 5 : number of PALs

4.2. XILINX RESULTS (table 6)

The presented algorithms have been implemented in C language. The number of CLBs for the three benchmarks are given in table 6. The other benchmarks results will be given at the conference.

Since there is no published result for xilinx architecture we can not compare our results with anyone else.

NAME	# out	# in	# sf	#CLB	Crit.P	CPU
misex 1	7	8	11	14	3	1.3
misex 2	18	25	25	34	3	4.4
sao2	4	10	44	38	6	61
5xp1	10	7	19	23	4	5.2
duke2	22	29	324	292	11	115
con1	2	7	1	3	2	0.2
9sym	1	9	81	64	7	102

table 6 : combinational benchmarks

5. CONCLUSION

Both key ideas of this synthesis tool have proven their efficiency: the kernel algebraic division allows easy and quick decrease of the number of variables of a function which has to be partitioned; the achieved reinjection avoids useless decompositions. More generally, the system has led to excellent results both in terms of number of PALs and of critical path.

As far as PGAs, it is difficult to appraise our work because of the lack of comparison with other tools.

REFERENCES

[1] R.K. Brayton et al, "MIS: a multiple-level logic optimization system", IEEE trans. on CAD, pp. 1062-1081, November 1987.

[2] R.K. Brayton et al, "Logic Minimization Algorithms for VLSI synthesis", Kluwer Academic Publishers, 1984.

[3] R.K. Brayton and C.T. McMullen, "The decomposition and factorisation of booleann expressions", ISCAS Proceedings, April 1982.

[4] XILINX Programmable Gate Arrays 'User's Guide", XILINX.INC., 1988.

[5] XILINX Programmable Gate Arrays "Technical Data, XC3000 - Logic Cell Array Family", XILINX.INC., 1988.

[6] Rajeev P. Goré and Kotagiri Ramamohanarao, "Automatic Synthesis of Boolean Equation Using Programmable Array Logic", 26nd DAC, 1989.
[7] S. Devadas et al, "MUSTANG: State assignment of finite state machines for optimal multi-level logic implementations" ICCAD 87, pp. 16-19.
[8] Monolithic Memories Inc., Advanced Micro Devices, "PAL device handbook 1988".
[9] Ronald R. MUNOZ and CHARLES E. STROUD, "Automatic Partitioning of Programmable Logic Devices", VLSI Systems Design, October 1987.
[10] G. Saucier, L. Bouchet, P. Sicard, "Multi-level synthesis on PALs", EDAC 90.
[11] G. Saucier, C. Duff, F. Poirot, "State assignment for controllers for optimal area implementation", EDAC 90.
[12] G. Saucier, M. Crastes, P. Sicard, "ASYL: a rule-based system for controller synthesis", IEEE trans. on CAD, pp. 1088-1097, November 1987.
[13] P.G. SCHRIER, "New Logic Technologies Inpact Desktop Programming Systems", Personal Engeneering and Instrumentation News, June 1987.
[14] Charles E. STROUD and Ronald R. MUNOZ, "CONES: a system for automated synthesis of VLSI and programmable logic from behavioral models", ICCAD, 1986.
[15] P. SICARD, "Nouvelles methodes de synthese logique", PhD Dissertation, CSI Lab. , INPG, FRANCE, September 1988.
[16] G. Saucier, C. Duff, F. Poirot, "State assignment using a new embedding method based on an intersecting cube theory", DAC 89.
[17] P. Abouzeid, K. Sakouti, G. Saucier, F. Poirot, "Multi-level synthesis minimizing the routing factor", DAC 89.
[18] K. Sakouti, G. Saucier, "A fast and effective technology mapper on an autodual library of standard cells", IFIP 90, Paris, FRANCE.

Logic and Architecture Synthesis
P. Michel and G. Saucier (Editors)
Elsevier Science Publishers B.V. (North-Holland)

A Methodology for Performance Driven Data-Path Compilation.*

Stefaan Note[1] Francky Catthoor[1] Gert Goossens[1]
Hugo De Man[1,2]

[1]IMEC Laboratory, Kapeldreef 75, 3030 Leuven, Belgium
[2]Professor at Katholieke Universiteit, K. Mercierlaan 94, 3030 Leuven,Belgium

Abstract

In this paper, a methodology is described for performance driven compilation of data-paths suited for high throughput real time DSP applications. The emphasis lies on the designers point of view i.e. on the effect of the different performance optimizations, and the sequence in which to apply them. The resulting methodology has been implemented in a tool called CHOPIN.

1 Introduction

In front-end audio, medium-level image, and video DSP systems, the number of available clock cycles to complete the computation of a pixel or a frame is typically ranging between 1 and 20. Hardware resources can therefore not be shared or reused for many different operations of the algorithm. In this case, the use of application specific units, or customized data-paths, is unavoidable. Predefined, optimized units like multipliers and arithmetic units is not any longer feasible for this target domain. The dedicated, customized data-paths are hardwired or slightly programmable and closely resemble the algorithmic signal flow graph. Typically, the number of blocks in these data-paths vary between 3 and 40. The architectural style advocated for this kind of applications is referred to as lowly multiplexed co-operating data-paths [1]. In our applications , the clock frequencies at which the different data-paths operate range from 10 to 60 Mhz.
Furthermore, also for data-paths in the medium throughput range, optimizing the clock frequency is important. For a given sample frequency, higher clock frequencies increase the available number of cycles ($= f_{cl}/f_s$) to execute the algorithm at hand.

The driving force, in generating or assembling these data-paths, is the stringent performance requirement imposed by the application. Compared to the currently available

*Research supported by the "ESPRIT 97" and 2260 (SPRITE) program of the E.C.

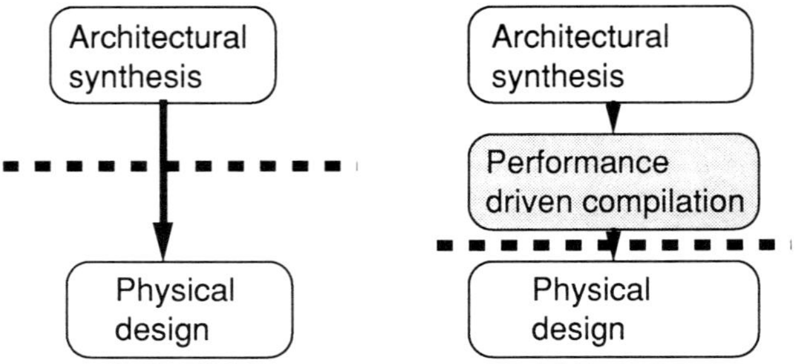

Figure 1: Assembly versus Performance Driven Compilation approach.

data-path assemblers [2][3][4], which only *assemble* in a straightforward way the different blocks which constitute the data-path, the need for *performance driven data-path generation* is required. Transformations have to be applied, and certain selections have to be made in such a way that the final data-path fulfills the high throughput requirements. Furthermore, the incorporation of this performance aspect during the data-path generation process is essential in order to reduce the number of design iterations, and the overall design time. Figure 1 shows the previous assembly, physical design, approach and the current performance driven approach.
In the design of a data-path different degrees of freedom can be exploited, to achieve the goal of satisfying the high performance constraint. In the following, only optimizations which guarantee the bit-true behaviour are considered.

The degrees of freedom which are exploited correspond to optimizations at different hierarchical levels :

- *Register transfer level.* This transformation involves the exploitation of the freedom how to redistribute or where to insert additional registers in the data-path.

- *Logical level.* This optimization deals with the selection of different hardware alternatives. For example, how to select either a ripple-adder or a carry-select adder in a data-path.

- *Circuit level.* This task refers to the selection of cells so that their driving capability is matched to the load at the output of the cells.

- *Topological level.* Wiring delays can be reduced if a proper relative placement of the different blocks of the data-path can be found.

One contribution of our work consists in the *inventarization* of the different performance optimizations, their effect on the final performance, and the dependencies between them.

A second major contribution is the *strategy for the combination* of these optimizations. This strategy resulted in a script determining the sequence of the different optimizations, and the development of a compiler called CHOPIN [5]. Starting from a data-path description, and a constraint on the minimal clock frequency, the objective of the compiler is to generate an optimized netlist. This netlist serves as input to a layout generation program which performs compaction, and detailed routing. The term compiler, instead of assembler, is adopted to emphasis the performance driven optimizations which are involved.
In literature, a number of data-path assemblers have been described [2][3][4]. However to our knowledge, no performance driven optimizations are included in these systems.
In section 2, the terminology which will be required throughout the rest of the paper will be introduced. Section 3 describes the different optimizations and their effect on the performance. In section 4, the overall strategy of invoking the different throughput optimizations is described. Finally, in section 5, a few test-vehicles are used to demonstrate the effect of these optimizations on the performance and area of the data-path.

2 Terminology

We assume that after high-level architectural synthesis, every operation in the signal flow graph description of the algorithm has been assigned to an *abstract building block* (ABB). An abstract building block is a building block for which different hardware realizations exist. Such a building block either corresponds to a combinatorical block, e.g. an adder, a memory block, e.g. a register, or an interconnection block, e.g. a multiplexer. Abstraction is made from the detailed hardware implementation.
The *data-path description* consists of a number of ABBs, and the interconnection between them. The *specification* of a data-path consists of its description and a constraint on the maximal clock period.
During the data-path compilation process, every ABB is mapped on a *hardware building block* (HBB). A hardware building block is defined as a reusable entity of hardware which implements a certain ABB. A software procedure describes in a parameterized way how the HBBs are composed out of different leafcells. Examples of HBBs corresponding to the abstract building block "adder" are ripple adders, carry-select adders, carry- bypass adders.
Leafcells are the basic cells from which the HBBs are hierarchically composed, and which are laid out as one entity. Furthermore, these cells are characterized with respect to their timing behaviour.
Finally, a *path* in a data-path is defined as a sequence of HBBs $(h_1, h_2, ...h_n)$ such that every HBB h_i in that sequence is connected to its predecessor h_{i-1} and successor h_{i+1}. Furthermore, the first HBB of the sequence must either be a register or an external input, and the last HBB of the sequence must either be a register or an external output.

3 Performance driven data-path generation

In this section, different transformation and selection tasks which affect the perfor-

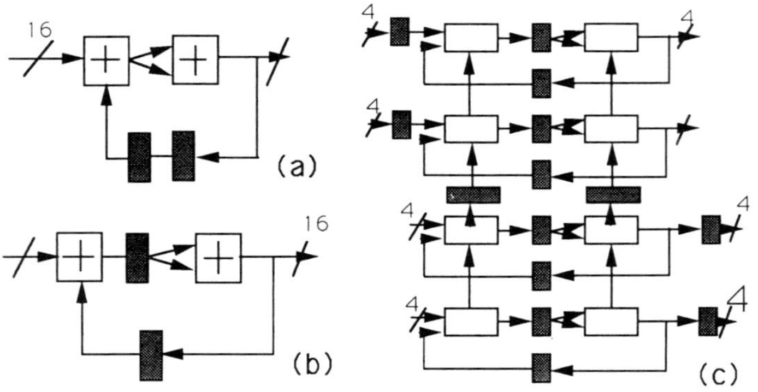

Figure 2: Pipelining at the HBB and leafcell level.

mance of a data-path are discussed. A brief survey has already been given in the introduction.

3.1 Pipelining at the Register Transfer Level

This optimization step involves an automatic pipelining methodology of complex data-paths based on the concept of retiming [6]. Whereas retiming only considers the distribution of the available registers, pipelining allows the insertion of additional pipeline registers. Starting from the specification of the data-path, and timing models for the leafcells of the HBBs, a behaviorally equivalent and pipelined description of the data-path is generated. This description is obtained by both moving and inserting pipeline registers in the original data-path. In most cases, this will result in an increase of the latency of the data-path. Based on the experience of several industrial chip designs, pipelining at 2 hierarchical levels was necessary :

- The HBB level. At this level, pipeline registers are inserted or redistributed between the HBBs. Figure 2b shows the redistribution of the available registers in the data-path.

- The leafcell level. If the throughput after pipelining at the HBB level is not sufficient, additional pipeline registers can be inserted between the different bit-slices of the HBBs. In figure 2c, registers are inserted after 4 full-adder stages. Performance could even further enhanced if registers are inserted between every full-adder.

The effect of this performance optimization on the throughput depends on the level of pipelining considered. For pipelining at the HBB level, the maximum throughput corresponds to a situation where the critical path contains only 1 combinatorical HBB. The relative gain in speed depends both on the word-length of the HBBs in the data-path,

and the number of HBBs in cascade. As the word-length increases, the effect on the performance decreases. If the non-pipelined data-path contains a large number of HBBs in cascade, the performance improvement is more important.
If we consider pipelining at the bit level, the throughput is only limited by the delay of one leafcell. However for recursive structures, the critical path can depend on the delay of more than 1 leafcell.
Compared to other pipelining algorithms like in [7], our approach can handle recursive structures, and is based on accurate timing models which is crucial for the applications envisioned.

3.2 Hardware Selection at the Logic Level

In contrast to non-recursive signal flow graphs, it is not always possible to introduce more registers into the loops of recursive structures : only a redistribution of the available registers is possible. However if redistribution is not sufficient, increasing the throughput can further be achieved by selecting faster hardware for some ABBs of the data-path. Moreover, in non critical parts of the data-path, less costly HBBs are selected in order to reduce the total area. The objective of this task is to select an optimal HBB for every ABB in the data-path description so that the performance constraint is met, and all paths in the data-path are balanced. The technique is based on accurate hierarchical timing models for the HBBs, and is formulated as an integer linear programming (ILP) problem. The improvement or impact of this optimization depends on the word-length. If large word-lengths occur in the data-path, selecting a faster structure has a significant improvement. Carry-bypass or carry-select adders only outperform simple ripple-adders if the word-length is large enough.
This task and the previous pipelining task are strongly connected. Indeed, depending on the hardware cost of a register compared to additional "speed-up circuitry", either a pipeline register will be inserted or a faster HBB will be selected. This trade-off is efficiently solved by combining the pipelining and hardware selection algorithm.

3.3 Cell Selection at the Circuit Level

This step involves the substitution of leafcells by more optimized, faster cells. A frequently occurring example is selecting a buffer which matches its output capacitance. Either these alternative cells can be selected from a discrete library of leafcells, or cells can be generated during the data-path compilation process using transistor sizing and automatic cell generation algorithms [8]. The impact of cell selection on the overall throughput depends on which transistors in the cell are considered. For example sizing the transistors related with the som generation of full-adder in a ripple adder has only a minor effect on the performance since the critical path is situated in the carry chain of the adder. Experiments indicate performance improvements of only a few %. Only if very large loads occur, sizing these transistors will have an impact. On the other hand, optimizing the transistors of the carry chain, will result in an important speed improvement because the carry chain is in the critical path of the data-path.

Currently, only the selection of alternative cells from a library is considered.

3.4 Linear Placement at the Topological Level

This task involves the linear placement of the different HBBs. The importance of the placement algorithm in a data-path compilation environment is twofold :

- An optimal placement of the HBBs could result in a higher performance due to the reduction of the capacitive wiring loads. Two kind of cost functions have been defined. A first function describes the total netlength which must be minimized. A second function describes that the maximal delay of a path must be minimized.
- Since the number of feed-throughs over the HBBs is limited, it is important to find a final placement which can be routed successfully within the limited number of feeds. This constraint is handled as a penalty function during the placement optimization.

Although wirelength is related with capacitance and thus to speed, it is important to mention is that experiments showed that for most of the data-paths encountered, no significant performance improvement was noticed ($< 5\%$), using the placement algorithm. A first reason is that through the use of the penalty function, a placement with an already restricted interconnection cost was resulting. A second reason is that our 3 $\mu Cmos$ library leafcells have a reasonable high driving capability. A last reason is that due to the word-lengths of the data-paths which are ranging between 8 and 24 bits, the effect of slowing down the circuit as a result of routing, is inversely proportional to the word-length. Because of the large word-lengths, the most time critical nets are the nets between the leafcells inside the HBBs. However these leafcells were preplaced and as a consequence optimized with respect to timing. For example, the net which corresponds to the carry signal is the most critical net of a ripple-adder.

4 Synopsis of the compiler Chopin

Starting from the specification of a data-path, the objective of the compiler is to generate an optimized netlist which serves as input to a layout generation program. The different performance optimization transformation and selection tasks have been described in the previous section. The goal of this section is to derive an order in which to apply all these transformations so that the number of iterations over the design is minimized. A script which contains the different steps is shown in figure 3. Each of these steps are implemented as a CAD tool. The first 3 tasks are the performance optimization tasks which have been described in the previous section. Remark that the first optimization combines both pipelining and hardware selection in order to effectively exploit the trade-off between register and "speed-up circuitry" cost. The linear placement task follows the pipelining task because due to pipelining, registers can be inserted which also need to be placed. The result of the optimizations is an optimized data-path description which serves as input for the data-path assembly step [9]. This task involves global routing and the

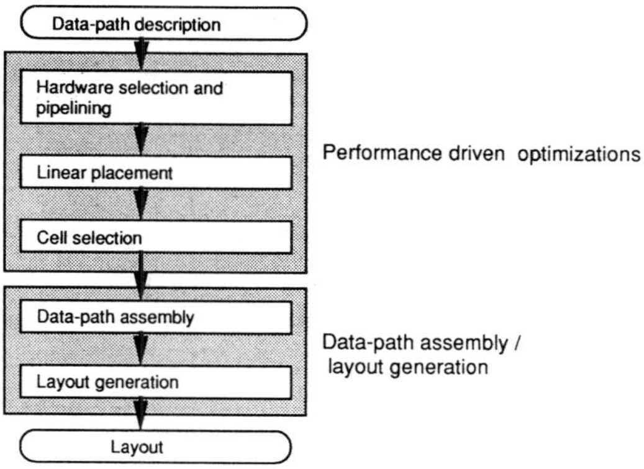

Figure 3: Synopsis of CHOPIN

conversion to a specific layout generation program. Currently, CHOPIN [10] can generate the input for 2 layout programs : the module generator environment [11] and a general cell place & routing tool. The first output corresponds to a full-custom layout style while the second refers to a semi-custom layout style.

From the designers point of view, different scenarios are possible using the compiler. For all video applications, and for most front-end audio applications, the sample frequency of the input streams is fixed. In addition, these sample frequencies are very high so that no hardware sharing is possible. This often means that the clock frequency equals the sample frequency. The objective during compilation is, given the clock frequency, optimize the data-path with respect to the performance constraints so that the area is minimized.

Another scenario is more tuned to image applications. For these applications, in a first step, the maximal clock frequency is derived from the most time critical data-path used. In a second step, all the other data-paths are optimized with respect to this clock frequency. In this way, the delays of the non critical data-paths are relaxed, which results in a reduced area occupation of all the data-paths.

5 Experimental Results

A first example is the recursive filter of figure 2a. Figure 4 gives an overview of the timing behaviour of a ripple and a bypass adder. The results of either pipelining and hardware selection are shown in table 1. A first conclusion is that the impact of word pipelining decreases as the word-length of the data-path increases. Performance improvements due to word-level pipelining, varying between 10% and 40%, can be obtained. A second conclusion is that more optimal adder structures like carry-bypass, carry-select adders are only effective for large word-lengths. The gains in performance obtained with

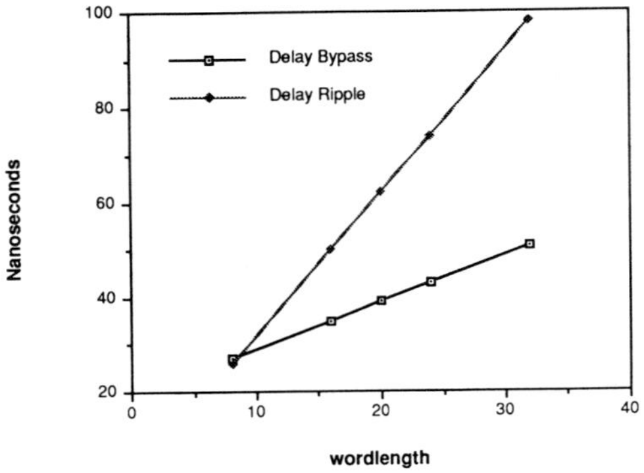

Figure 4: Timing information for a ripple and bypass adder.

Wordlength	*Not Pipel. ripple (nsec)*	*Pipel. ripple (nsec)*	*Improv.*	*Not Pipel. ripple (nsec)*	*Not Pipel. bypass (nsec)*	*Improv.*
8 bit	45	35	22%	45	45	0 %
16 bit	65	55	15%	65	51	21%
24 bit	85	75	11%	85	57	32%
32 bit	105	95	9%	105	43	59%

Table 1: Hardware selection and pipelining

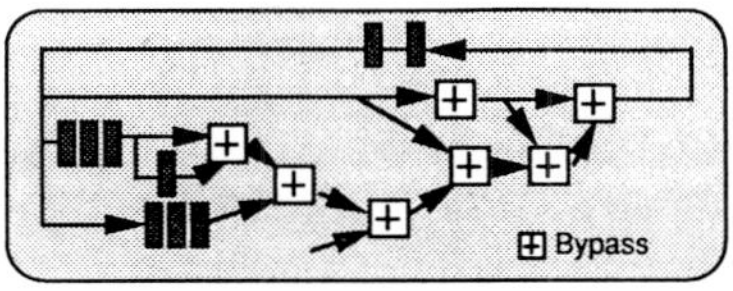

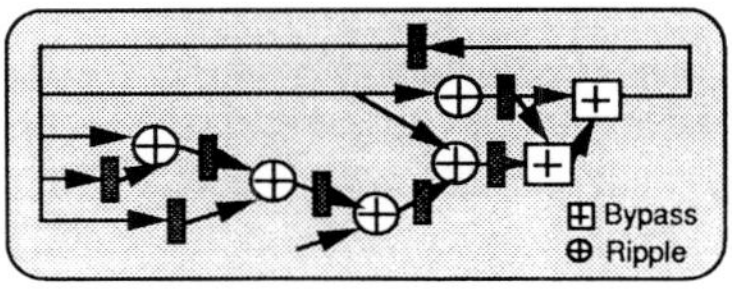

Figure 5: Adapted DPCM example.

Adpcm	*Hardware selection*			*Combination*			*Gain*	
Wordlength	*Area* (mm^2)	*Clock* *(nsec)*	*CPU* *(sec)*	*Area* (mm^2)	*Clock* *(nsec)*	*CPU* *(sec)*	*Area*	*Perf.*
8	1.95	76	3	1.95	36	11	-	52%
16	4.80	84	19	4.16	50	151	13%	40%

Table 2: Combination of hardware selection and pipelining

optimal hardware assignment are ranging between 10% and 60%. If pipelining at the bit level is considered, simple ripple adders are probably the most appropriate structures. Indeed, due to bit level pipelining, the word-length between registers is normally small (see figure 2).

Figure 5 shows a more complex example. Figure 5a shows the resulting data-path after hardware selection : all adders are assigned to bypass adders. Figure 5b shows the result after combined hardware selection and pipelining. Only 2 adders are now assigned to bypass adders, the rest are ripple-adders. On one hand, table 2 shows the increasing CPU times related with the combination of hardware selection and pipelining compared to only hardware selection. Secondly it shows the gain in area and performance if both pipelining and hardware selection are combined.

A last example is a finite impulse response filter (FIR) with symmetrical coefficients (figure 6). Table 3 summarizes the results. Because the larger number of adders in cascade, 4 compared to 2 in figure 2, the gain due to word level pipelining is higher.

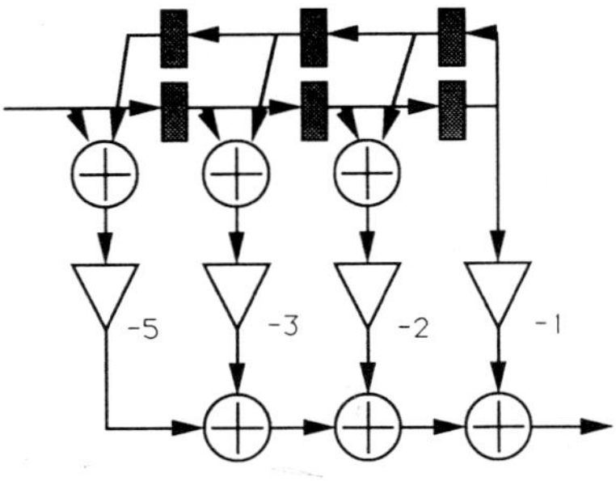

Figure 6: Symmetrical filter structure

	Clock period (nsec)	*Gain*
Original implementation	43.7	
After Word pipelining	26.0	40%
After Bit pipelining	13.0	70%

Table 3: Hierarchical pipelining

Furthermore, if pipeline registers are allowed between the bitslices of the HBBs, performance can still be increased with respect to word level pipelining. Remark the important performance improvement due to bit level pipelining.

6 Conclusions

Future high level synthesis systems will take more and more advantage of the automatic generation of complex data-paths. In this paper, an integrated approach for performance driven data-path compilation has been presented. The ultimate goal of the compiler is to reduce design time, and end up with solutions which are almost as efficient (performance and area) as manual designs. In this paper, the emphasis is on the performance driven transformation and selection tasks. Furthermore, their effect on real life examples is described.

Acknowledgement Thanks are due to Jef Van Meerbergen from Philips Natlab who has actively contributed to this work with many suggestions and valuable information.

References

[1] F.Catthoor, H.De Man, "Application-specific architectural methodologies for high-throughput digital signal and image processing", *accepted for publication in IEEE Trans. on Acoustics, Speech and Signal Processing*, 1990.

[2] C.S. Shung, "LagerIII, a framework for algorithm-specific IC design", IEEE solid-state circuits and technology committee fall workshop on application specificIC's for DSP, Napa Valley CA, Sep 2-4, 1987.

[3] Jim Rowson, Bill Walker and Suresh Dholakia,"A Datapath Compiler for Standard Cells and Gate Arrays",IEEE Custom Integrated Circuits Conference, 1987.

[4] R. Gordon, S. McNeary, T. Ng, Y. Rotblum, M. Tate, "An N-bus Data-path compiler for IC design", IEEE '89, CICC, San Diego, May 15-18, 1989.

[5] S. Note, J. Van Meerbergen, F.Catthoor, H.De Man,"Automated synthesis of a high speed Cordic algorithm with the Cathedral-III compilation system",IEEE International Symposium on Circuits and Systems, ISCAS '88 Helsinki, Finland, June 6-9, 1988.

[6] C. Leiserson, J. Saxe, "Optimizing Synchronous Circuitry by Retiming", Third Caltech Conference on VLSI, Computer Science Press, 1983.

[7] N. Park, A. Parker, "Sehwa : A Software Package for Synthesis of Pipelines from Behavioural Specifications", IEEE Transactions on Computer Aided Design, Vol. 7, No. 3, March 1988.

[8] H. Moscovitz, W. Bullman, J. Carelli, L. Davieau, J. Ardekami, K. Mondal, G. O'Donnell, "Automatic Full-Custom Design of Application Specific DSP VLSI." IEEE International Symposium on Circuits and Systems, ISCAS '88 Helsinki, Finland, June 6-9, 1988.

[9] H. Cai, P. Six, H. De Man, "A Data Path Layout Assembler for High Performance DSP Circuits", accepted for publication at DAC conference 1990.

[10] S. Note, J. Van Meerbergen, F. Catthoor, H. De Man, "Hardwired Data Path Synthesis for High Speed DSP Systems with the Cathedral-III Compilation Environment", International Workshop on Logic and Architecture Synthesis for Silicon Compilers, Grenoble, May 25-27, 1988.

[11] P. Six, I. Vandeweerd, H. De Man, "An Environment for creating Module Generators", ESSIRC, pp. 65-70, Sep 1986.

Logic and Architecture Synthesis
P. Michel and G. Saucier (Editors)
Elsevier Science Publishers B.V. (North-Holland)

Automatic synthesis of μprogrammed controllers

Laurent GERBAUX, Gabrièle SAUCIER IEEE Senior member

Institut National Polytechnique de Grenoble / CSI
46, Av Félix VIALLET, 38031 Grenoble Cedex FRANCE
Tel: (33) 76 57 46 87

This paper presents an optimal synthesis method of programmed controllers.Starting from a control flowchart, a synthesis tool generates automatically the content of the ROM and a standard cell implementation of the μsequencer. An optimization effort focuses on the ROM block. It is based on an optimized assignment of states to ROM addresses using embedding techniques in the hypercube.

Introduction

Automatic synthesis of controllers or finite state machines (FSMs) starting from a control flowgraph is presently available in most of the commercial silicon compilers. Their quality depends on the effectiveness of the state assignment which is the most critical step and allows to save 30% or 40% of the area with respect to random state assignment [4]. The basic architecture is made up of D flip flops and standard cells implementing next state and outputs equations.
Practically, only small controllers are implemented as flat finite state machines. As they become complex, hierarchical, pipelined structures are used as well as different basic styles. One of the basic styles uses as memory element counter/load devices, and relies on the decomposition of the control flowgraph in paths. It uses the incrementation primitive of the counter as next state sequencing primitive [1][5]. This target structure is adequate as long as paths are identified in the control flowgraph.
In this paper, we consider large controllers in which not only obvious paths exist but some of them are repeated several times in the control flowgraph. This has usually in the past led to μprogrammed implementation. We look here for automatic synthesis of μprogrammed controllers starting from a classical control flowgraph and leading to an automatic filling of the ROM and an automatic synthesis of the equation of the microsequencer on standard cells ; thus providing the same facilities for large μprogrammed controllers than for flat finite state machines.
In section I the target architecture is announced ; in section II the decomposition of the control flowgraph required by this approach is presented. In section III, emphasis is put on the optimisation techniques aiming at minimizing the global area. Finally in section IV, we show layouts of μprogrammed controllers generated by this approach.

I The target architecture

1.1 Basic style

Classically, two basic styles of μprogramming have been studied :
the horizontal μprogramming where μinstruction have a large number of bits as they

contain both the control signals sent by the controller and the information about next μinstruction address.

the vertical μprogramming where 2 types of μinstructions are distinguished by a tag bit :

(i) the μinstructions activating the control signals ; in this case the next μinstruction is at the next address of the ROM.

(ii) the branching micro-instructions specifying explicitly the next address in the ROM.

As a first target, in our tool, the horizontal μprogramming is considered. The micro-instructions are divided into 4 fields : the most left field sends the control signals, the next field defines the sequencing primitives used for next μinstruction address computation ; the third field specifies the test conditions if some conditions are involved in the next μinstruction address computation ; the last field gives an explicit address in the ROM for branching μinstructions.

In order to compute the address of the next state, 5 classical sequencing primitives are used :

- Incrementation
- Branching
- Mapping on opcodes ; this means that the next address depends on the opcodes.
- Branching to sub-μprogram
- Return from sub-μprogram

These 5 sequencing primitives are reduced to 4 as the incrementation is considered as a branching situation with an always false condition.

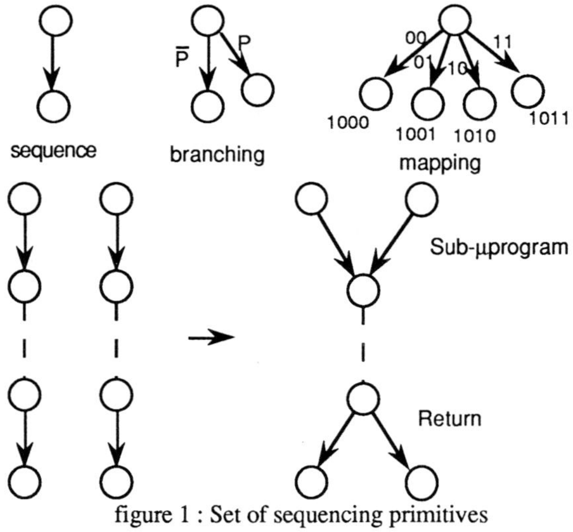

figure 1 : Set of sequencing primitives

1.2 The target architecture

The target architecture is shown in figure 2.

The ROM is surrounded by an address register called SCM (for Selection Control Memory). A multiplexer selects one among the 4 sequencing primitives. This selection is performed according to the 2 bits of the sequencing primitive field and the test conditions. The next address comes either from the stack value where return addresses of subμprograms are stored or from explicit next address field or from the control memory selection register through an incrementer or from the mapping logic.

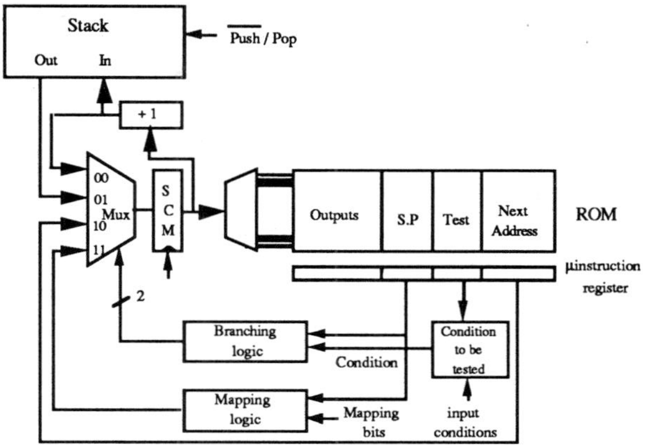

figure 2 : Target architecture

II Control flowgraph decomposition

The control flowgraph decomposition is based on a situation detection.

2.1 Branching and mapping situation

If a node has more than one successor such a situation is detected. If the number of successors is 2 then there is a choice between the implementation of a branching or a mapping on a 1 bit sequencing primitive. If the number of successors is greater than 2, then a mapping primitive is implemented. For a branching situation, one of the two successors will lead to an incrementation primitive. This choice is used to solve some conflicts. Consider for instance the pattern of figure 3. It is mandatory to consider state S3 as a state reached from S1 or S2 by a branching primitive as it cannot be reached on both side by incrementation.

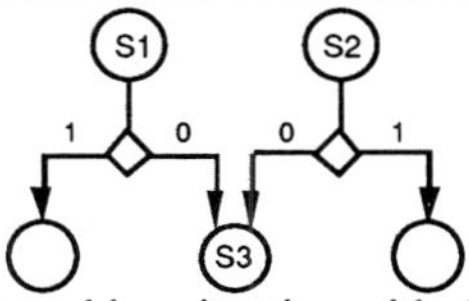

figure 3 : Branching situation with shared node

A mapping situation is made up of a node having 2^k successors. If this is not the case node replication is presently performed to simplify the approach.

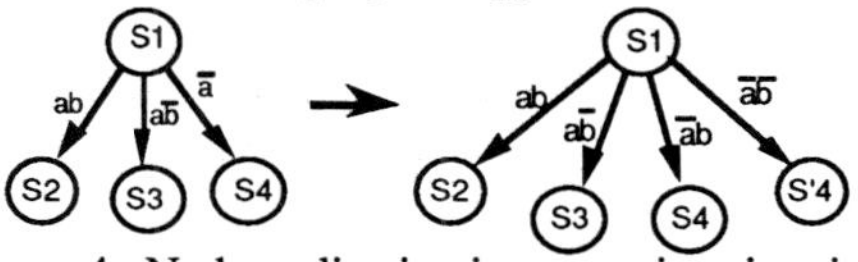

figure 4 : Node replication in a mapping situation

Similarly if two mapping situations have a node in common, this node is replicated (fig. 5).

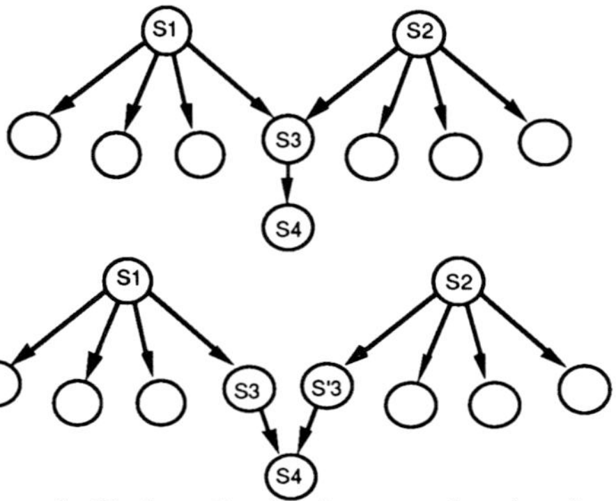

figure 5 : Node replication in a mapping situation

Finally, if a node of a mapping situation is already included in another situation (sub-micro-program return for instance) and thus has a constrained address, this node will be replicated (fig. 6).

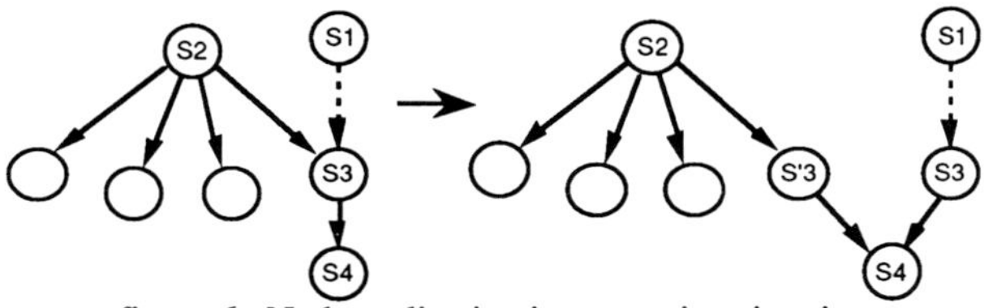

figure 6 : Node replication in a mapping situation

2.2 Subμprograms and paths identification

Each time a segment is repeated twice or more in the control flowgraph, it becomes a subμprogram. Nested subμprograms are similarly defined. Before the first state of a subμprogram there must be a state calling the sub-micro-program. Similarly after the subμprogram there must be a return state which is the successor of the call state. If this is not the case a phantom state with no output signal is added in the control flowgraph.
More generally for a path situation, the initial state is one of the following :

- the first state of a subμprogram.
- the successor states of a mapping situation.
- the successor of a branching situation which is not on the single path crossing the branching.
- a state having two or more predecessors. This state is reached, at least, from one of its predecessors by a branching primitive. Thus it is definitively considered as the beginning of a path.

2.3 Illustration on a small example

Consider the control flowgraph in figure 7.

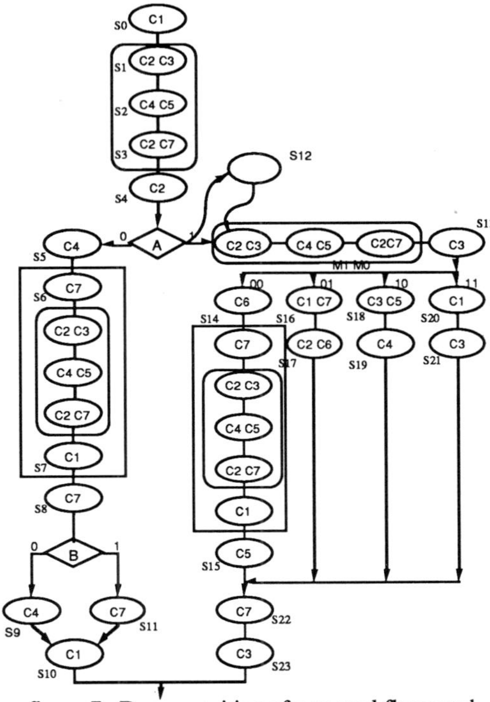

figure 7 : Decomposition of a control flowgraph

In this example, there are two identified sub-µprograms ; the first one with two states (S6,S7) the second one with three states (S1,S2,S3).

A phantom state S12 needs to be inserted before state S1

There is one mapping situation at state S13, involving the state : (S14,S16,S18,S20).

The following segments are identified as paths : ((S0,S4,S5,S8,S9),(S10),(S11),(S12,S13)).

III The synthesis method

3.1 General method

The synthesis method consists in assigning states to ROM addresses, filling the ROM, generating automatically the equations of the µsequencer and mapping them on standard cells. The most critical point is the minimization of the size of the ROM. This size will be of high importance for the global area optimization as it will be the hard block of the controller. A lot of work has been done in the past aiming at minimizing the output bit field (minimal compatibility classes of control signals) [2], [3]. These methods are still applied here. For the right part of the µinstructions, containing the branching addresses, a good way to compact this part is to look at these addresses as points on the hypercube and try to put them on a face of minimal dimension. This dimension will be the number of bits or width of this field.

This original approach will be more explained in section 3.2. As far as the standard cell parts of the micro-sequencer is considered, the logical expressions of the others blocks (branching logic, mapping logic) are obtained very easily and minimized by a classical standard cell minimizer.

3.2 ROM minimization

As said previously this step consists in placing the mapping situation and the starting nodes of the paths (corresponding to branching situation) in a minimal face of the hypercube. Concerning the mapping situation (figure 8), the size of the cube on which the head of the mapping segments have to be placed, is fixed. It is equal to $\log_2$[nb of mapping segments].

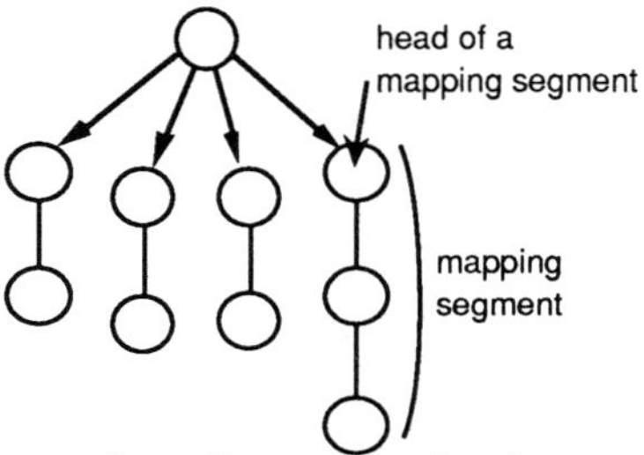

figure 8 : mapping situation

Moreover, a node corresponding to a head of a path or of a mapping segment, is followed by the other nodes of the path. Thus, the places of these nodes must be saved when placing the head. Figure 9a shows an hypercube of dimension 5 with boolean codes and figure 9b shows the decimal codes outpointing the incrementation and path existence.

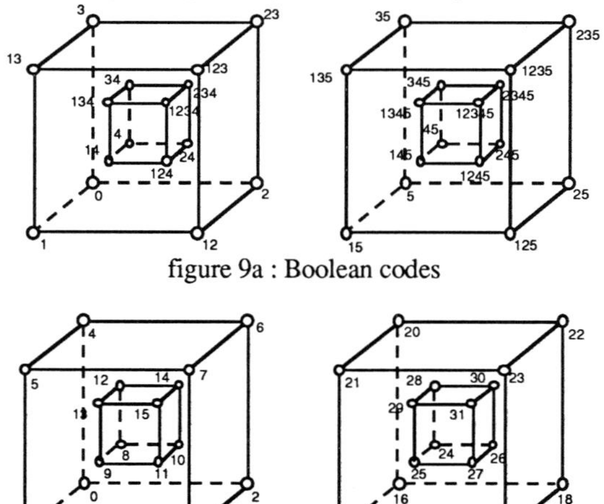

figure 9a : Boolean codes

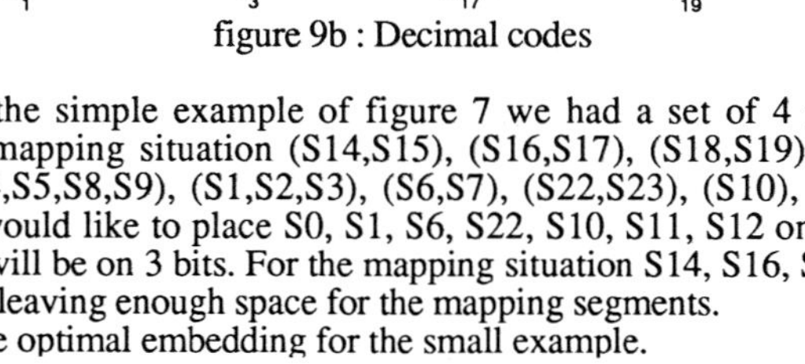

figure 9b : Decimal codes

If we go back to the simple example of figure 7 we had a set of 4 mapping segments associated to the mapping situation (S14,S15), (S16,S17), (S18,S19), (S20,S21), and 7 other paths (S0,S4,S5,S8,S9), (S1,S2,S3), (S6,S7), (S22,S23), (S10), (S11), (S12,S13). For the paths we would like to place S0, S1, S6, S22, S10, S11, S12 on a 3-cube, then the next address field will be on 3 bits. For the mapping situation S14, S16, S18, S20 have to be placed on a 2-cube leaving enough space for the mapping segments.
Figure 10 shows the optimal embedding for the small example.

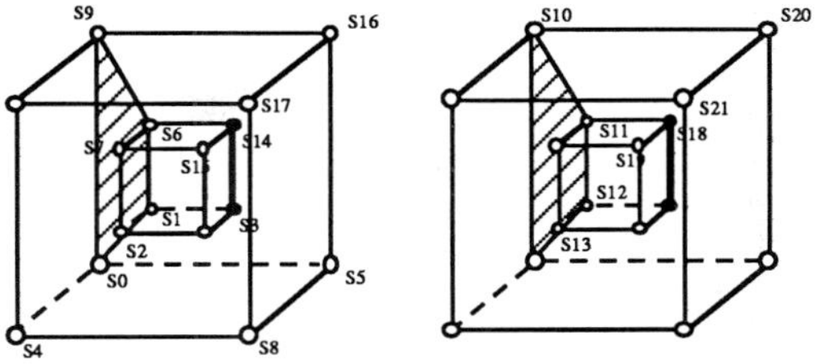

MAPPING CUBE
4 PATHS
(S14, S15), (S16, S17), (S18, S19), (S20, S21)

EMBEDDED IN THE CUBE [23, 45]

PATH CUBE
7 PATHS
(S0, S4, S5, S8, S9), (S1, S2, S3), (S6, S7), (S22, S23), (S10), (S11), (S12, S13)

EMBEDED IN THE CUBE [0, 345]

figure 10 : Embedding in the hypercube

Figure 11 shows a ROM filling corresponding to this example.

SMC dec	SMC bin	Commands 1	2	3	4	5	6	7	Br	Cond	Adr	State
0	00000	*							01		01000	S0
1	00001		*						00	A	11000	S4
2	00010						*		01		01100	S5
3	00011							*	00	B	11100	S8
4	00100				*				00	T	10100	S9
5	00101											
6	00110	*						*	00	F		S16
7	00111		*				*		00	T	10000	S17
8	01000		*	*					00	F		S1
9	01001				*	*			00	F		S2
10	01010		*					*				S3
11	01011											
12	01100							*	01		01000	S6
13	01101	*							10			S7
14	01110						*		01		01100	S14
15	01111					*			00	T	10000	S15
16	10000							*	00	F		S22
17	10001			*					00	T	00000	S23
18	10010											
19	10011											
20	10100	*							00	T	00000	S10
21	10101											
22	10110	*							00	F		S20
23	10111			*					00	T	10000	S21
24	11000								01		01000	S12
25	11001			*					11			S13
26	11010											
27	11011											
28	11100							*	00	T	10100	S11
29	11101											
30	11110			*		*			00	F		S18
31	11111				*				00	T	10000	S19

figure 11 : Mapping on the ROM

IV Example of a layout

As an example a small μprocessor has been designed with this technique.
The data path and the control flowgraph are given on figure 12a and figure 12b. There are 4 families of instructions :

- transfer instructions LD, ST
- instructions with 1 operand INC, DEC
- instructions with 2 operands ADD, SUB
- branching instructions BRA, BRZ

There are 4 classical addressing modes which are indirect, direct, Register and immediate addresses.

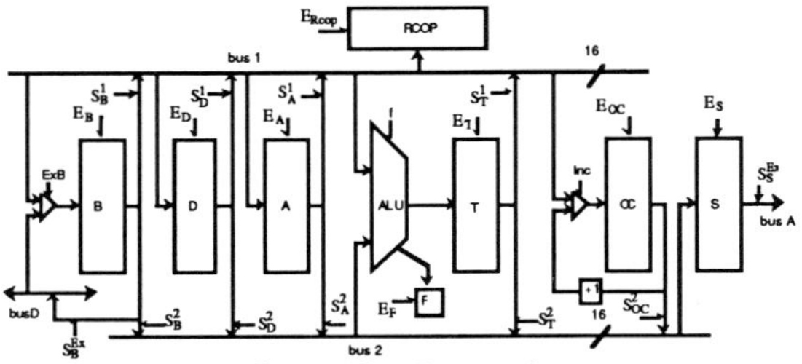

figure 12a : Data path

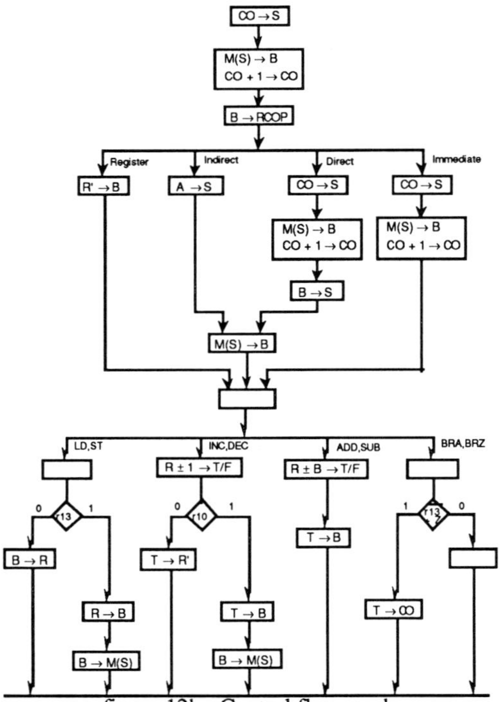

figure 12b : Control flowgraph

The control flowgraph decomposition is shown in figure 13.

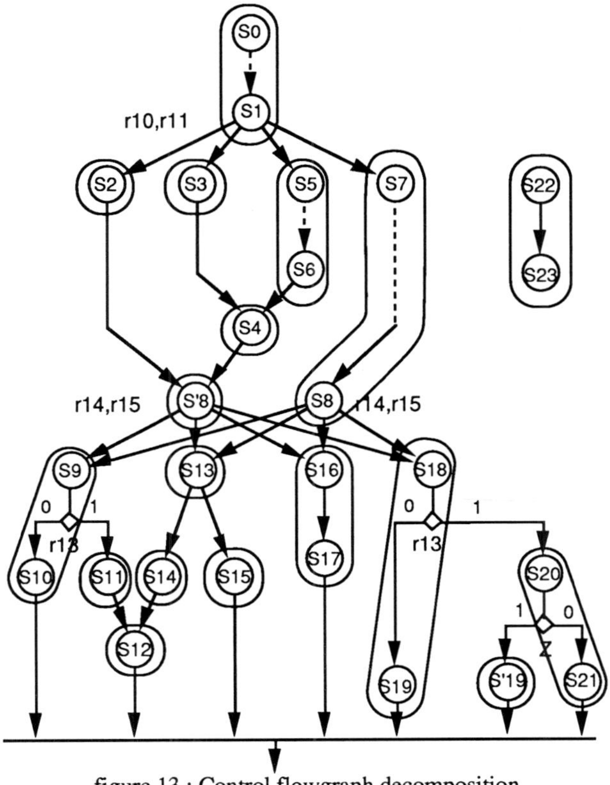

figure 13 : Control flowgraph decomposition

The mapping of the addresses is shown in figure 14.

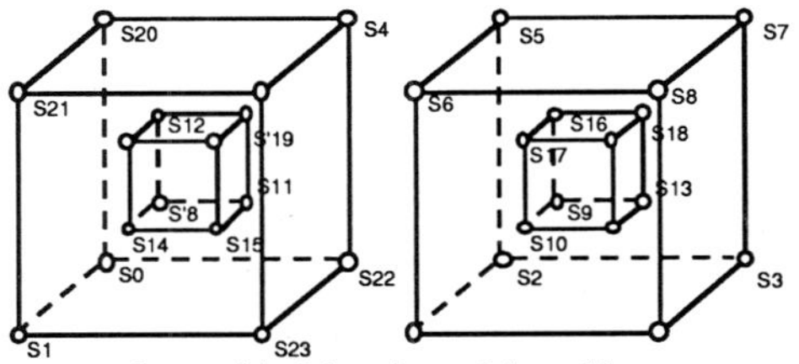

figure 14 : Mapping of the address

A layout have been generated in figure 15 on a 1.5u CMOS technology with the tool of VLSI technology. All the µsequencer logic, the selection control memory and the stack (which is composed by only one level in this exemple) have been implemented on standard cells. The data path compiler and the automatic synthesis tool on standard cell have been used.

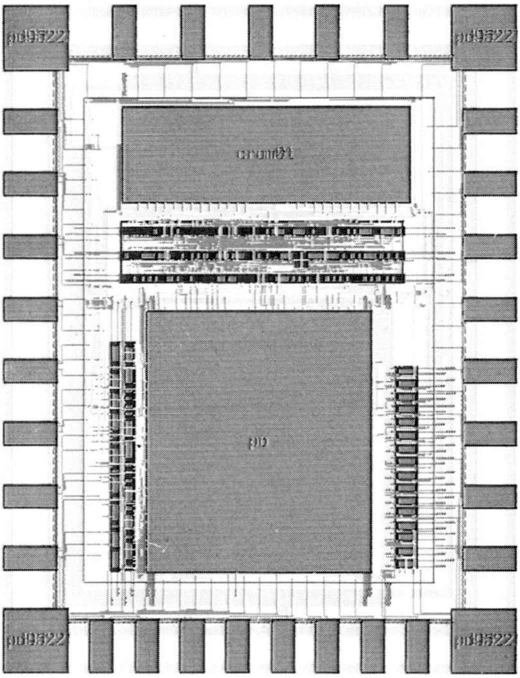

figure 15 : layout

References

[1] R. Amann, U.G. Baitinger "New State Assignment for Finite State Machines Using Counters and Multiple PLA/ROM Structures" Proceedings of the IEEE International Conf. on CAD, November 1987, pp. 20-23
[2] S.R. Das, D.K. Banerji and A. Chattopadhyay "On Control Memory Minimization in Microprogrammed Digital Computers" IEEE Trans Comp, vol C-22, September 1973, pp. 845-848
[3] T. Jayasri and D. Basu "An Approach to Organize Microinstructions which minimizes the Width of Control Store Words" IEEE Trans Comp, vol C-25, May 1976, pp. 514-521
[4] G. Saucier, F. Poirot and C. Duff "State Assignment Using a New Embedding Method Based on an Intersecting Cube Theory". 36th DAC, Las Vegas, June 1989, pp 321-326
[5] G. Saucier, F. Poirot and C. Duff "A New Embedding Method For State Assignment". MCNC Workshop on logic synthesis, North Carolina, May 1989.

Controller Synthesis

Logic and Architecture Synthesis
P. Michel and G. Saucier (Editors)
Elsevier Science Publishers B.V. (North-Holland)

The Structure of Constant Rank State Machines

Nico F. Benschop

Philips Research Laboratories
Eindhoven, The Netherlands

Abstract

Constant Rank (CR) state machines play an important role in the general structure theory of Finite State Machines. A machine is of constant rank if each input and input-sequence maps the state set onto the same number of next states. CR-machines are analysed, using their sequential closure (semigroup), which is shown to be a **simple semigroup,** that is: a direct product LxRxG of a left- and a right-copy semigroup, and a group. So in general a CR-machine is a parallel composition of: a **branch-**, a **reset-** and a **permutation** machine, which are three of the five basic types of state machines [1].

1 Introduction: Sequential Closure and Rank

A brief review of [1] is necessary to set up the required concepts. A state machine M(Q,A) with stateset Q and input alphabet A is a function M: Q x A $\rightarrow$ Q, which maps present state and input to next state. It is specified by a state transition table with $|A|$ columns and $|Q|$ rows. Each input 'a' in A is interpreted as a function a: Q $\rightarrow$ Q, mapping stateset Q into itself, called a **state transform**, or in short: a transform.

Sequential composition 'ab' of two transforms a and b is defined by q(ab)=(qa)b, for all q in Q. In other words, in state q first apply input 'a' to get state qa, then apply b, which yields state (qa)b = q(ab) = qab. Notice the left-to-right notation of this **function composition**, with stateset Q as domain and codomain. Two input sequences over A are defined **equivalent** if they yield the same Q-transform: a=b if qa=qb for all q in Q.

The **sequential closure of M**, called **semigroup S**, is the (finite) set of Q-transforms generated by all sequences over A, denoted S = A^+/Q. Here A^+ denotes the infinite semigroup of non- empty strings, length $\geq$ 1, over alphabet A, under string concatenation.

Closure S of machine M is a finite semigroup (of order $\leq n^n$, if M has n states) since transform composition is **associative**: a(bc) = (ab)c for all a,b,c in S, which is clear from above definition of transform composition. Input-strings with the same Q-transform are defined *equivalent* with respect to machine M, so the transform representation of each element of S is unique. State transform x: Q $\rightarrow$ Q is a function defined on a state set Q, which is both domain and co-domain. To state transform 'x' correspond:

- **range Qx** which is the set of function values (next states), and
- **partition Px** equivalences states that map onto the same next state.
- **rank r(x)** = the order $|Qx|$ of its range = the number of partition blocks.

Lemma 1 : Non-increasing Rank property

(a) Left composition (x.) does not increase range: $Qxy \subseteq Qy$ ($\subset$: subset of)

(b) Right composition (.y) does not refine partition: $Pxy \geq Px$ ($>$: coarser than)

(c) Rank does not increase under transform composition: $r(xy) \leq r(x)$ and $r(xy) \leq r(y)$

(d) All elements x with $rank(x) \leq k$ form a subsemigroup which is an ideal Z_k of S.

Proof: (a) $Qxy \subseteq Qy$ follows from set inclusion and associativity. $Qx \subseteq Q$, for all x, and right composition with y yields: $(Qx)y = Q(xy) \subseteq Qy$.
(b) $Pxy \geq Px$ follows from associativity and right composition of states i,j that are equivalent under x: ix=jx implies ixy=jxy for all y. Hence $i \equiv_x j$ implies $i \equiv_{xy} j$.
(c) This **monotone rank property** follows directly from (a) and (b),
because range ordering (a) implies rank ordering $|Q(xy)| \leq |Qy|$, so $r(xy) \leq r(y)$,
and partition ordering (b) implies rank ordering $|P(xy)| \geq |Px|$, so $r(xy) \leq r(x)$.
(d) It follows immediately that if x and y have rank $\leq$ k, then so does composition xy. This closure property means that all elements of rank not exceeding k form a subsemigroup Z of S. In fact, composition of any element z in Z with any element s in S yields zs with $r(zs) \leq r(z) \leq k$, sothat zs is in Z. The same holds for sz. Hence Z is both left- and right ideal, that is an ideal of S: $ZS \subseteq Z$ and $SZ \subseteq Z$ (see def-2 next section). ◇

Basically, this paper tries to render results from semigroup structure and their state representation better accessible for state machine decomposition purposes. In fact, the earliest known result in semigroup theory (Suschkewitch, 1928 [2, p207]) is on the structure of the minimal ideal of a semigroup, essentially our theorem 2.

2 Basic Machines and Simple Semigroups

Machine decomposition is seen as implementing a machine as a network of *smaller* machines. Semigroups, as the sequential closures of state machines, are essential for the **equivalencing** and **ordering** of machines. Two machines are defined to be equivalent if they have isomorphic semigroups. Two machines are ordered $M1 \leq M$ if their closures are ordered $S1 \leq S$, meaning that S1 is (isomorphic to) a subsemigroup of S. Naturally, a minimal or **basic machine** has a closure with *no proper subsemigroup* [1].

The **five basic state machines**, with semigroups of order two, are derived in [1], with their interpretation as the elementary digital functions: *logic*, *arithmetic* and *memory*. A semigroup S is also a state machine M(S,S) with itself as inputset and stateset. For unique representation by state transforms (distinct columns), one extra state suffices if some columns are equal in the SxS composition table, see tables Z2 and L2. Components C2 and Z2 have a single generator '1', the others have two invariant generators a^2=a. Their semigroup composition resp. state-transition tables are:

C2	1	0
1	0	1
0	1	0

Z2	1	0
1	0	0
0	0	0
2	1	0

H2	1	0
1	1	0
0	0	0

L2	1	0
1	1	1
0	0	0
2	1	0

R2	1	0
1	1	0
0	1	0

Closure-Tables

Fig.1a

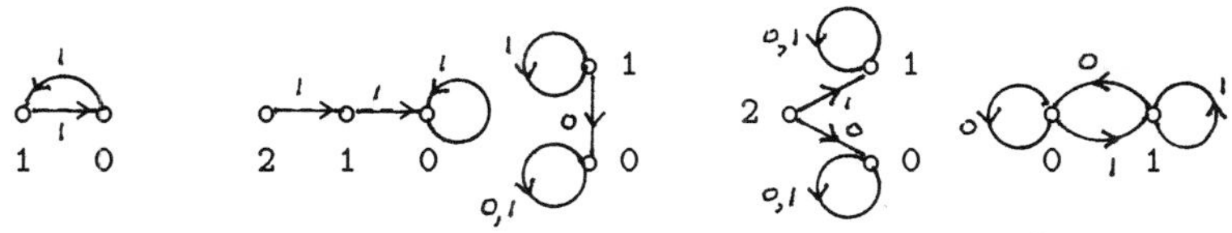


```
                                                                       State-
                                                                       Diagrams

2-counter     2-counter  | AND, OR      | 2-branch     set/reset       Component
 periodic     monotone   |  isomorph    |   mux          D-FF          Functions
add(mod 2)    converge   | mpy(mod 2)   | if-else      assign :=
    C             Z      |     H        |    L            R
<---- iterative a* ----> | <---- invariant : aa=a -------->             Algebraic
                         | <- LOGIC ->  | <-select-> <-store->         Properties
<--- ARITHMETIC : commutative -----> | <- MEMORY non-cmt ->
```

Fig.1b

Def 1: a semigroup is of **constant rank** (CR) if it can be represented by transforms of equal rank. A state machine is of constant rank if its closure is a CR-semigroup.

Three basic components are of constant rank, namely L2, R2 and C2. They are the smallest cases of the following three types of **constant rank semigroups**:
L : Left-copy semigroup with ab=a, ba=b for all a,b in S (n-branch, n+1 states)
R : Right-copy semigroup with ab=b, ba=a for all a,b in S (n-reset, n states)
G : Group (Permutation machine: permutes n states, $|G| \leq n!$)
All three are special cases of the following general type of semigroup [2, p5]:

Def 2: an **ideal** of a semigroup S is a subset Z with $SZ \subseteq Z$ and $ZS \subseteq Z$.
A semigroup is called **simple** if it has *no proper ideal.*

An ideal is like a multiplicative 'zero' (a.0=0 for all a) or 'trap'. Notice that Z2 (monotone counter with a final state) and H2 (hierarchy of two ordered invariants, see next section) are not simple semigroups, nor are they of constant rank. In general they model the *monotone sequential* aspects and *combinational logic* aspects of state machines respectively.

Corollary 1: a **simple** semigroup is of **constant rank**.

This follows directly from lemma 1d, since otherwise the elements of minimum rank would form a proper ideal. In fact, it will be shown that any simple semigroup is a direct product LxRxG of the three basic types of simple semigroups L, R, G. Hence, a general CR-machine is the parallel composition of a *branch* machine, a *reset* machine and a *permutation* machine. In a way, this is a conservation law of sequential logic.

3 Iterations: monotone, periodic, invariant

Iteration in a semigroup S is the repetition a^i of a single element. By virtue of associativity, the result is a unique element in S, independent of bracketing. The closure of a single element 'a' in S is the finite set of its **iterations** $a^+ = \{a^i, i=1..n\}$ which in general has a tail-cycle structure ('/Q' is omitted if no confusion can arise):

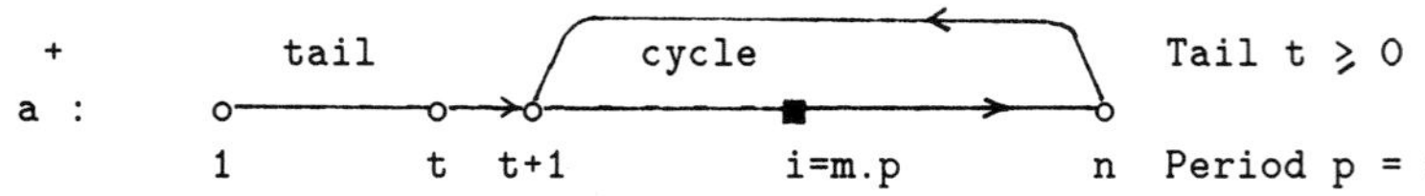

Fig.2

Since a^+ is finite, there is a smallest n for which $a^{n+1} = a^{t+1}$ with **tail(a)=t**, $0 \leq t < n$ and **period(a)=p** = n-t. There is precisely one **invariant** $a^i = (a^i)^2$ where i=m.p is the first and only multiple of p in the cycle, and $a^k = a^{k+p}$ for k>t.

An element of semigroup S is called **periodic** [**monotone**] if its closure has no tail, t=0 [no cycle, p=1]. Clearly, *invariants* aa=a are the only elements which have both properties. Elements which have a tail and a cycle are called *aperiodic*.

Def 3: a pair e,z of **commuting invariants**: ez=ze, e^2=e, z^2=z, is said to be **ordered** **e≥z** when ez=ze=z : e is identity for z. This is an ordering relation, since it is easily seen to be reflexive, anti-symmetric and transitive [2, p23].

3.1 Ordered Invariants: H

It will be shown that any *simple semigroup* S, being of constant rank, contains only *periodic* elements. Moreover, its invariants are not ordered but are all *equivalent* in some sense. So basic components of type Z2 (monotone iteration)and H2 (hierarchy of ordered invariants, or combinational logic) do not occur. In fact it turns out that S is a disjoint union of isomorphic groups (G), with identities forming a direct product of a left-copy (L) and a right-copy (R) semigroup.

Lemma 2 : Ordered Invariants
The ordering of commuting invariants z≤ e is their range ordering: Qz ⊆ Qe,
hence: – distinct commuting invariants have distinct ranges, and
– ordered invariants z<e have ordered ranks r(z) < r(e).

Proof : Let invariants z and e be ordered z≤e, then e is identity for z: ez=ze=z, so their ranges are ordered because Qz=Q(ze)=(Qz)e ⊆ Qe. Notice that ze=z suffices: e is right identity for z. Conversely, for commuting invariants: Qz ⊆ Qe implies z≤e. This follows from the *state transform structure of an invariant* e: qee=qe means that each state q maps to a state qe which is fixed under e. In other words, no state chains of length > 1 occur in the state transition diagram of e. Range **Qe** is the set of **fixed states of e**. Now, if Qz⊆Qe then z maps each state q into a fixed state of e: (qz)e=qz for all q, so ze=z. Since by assumption e and z commute, we have ez=ze=z, which means z≤e. Clearly, if Qe=Qz for commuting invariants e and z, then e≤z and z≤e, and hence e=z: commuting invariants with the same range are equal. ◇

Corollary 2: a simple semigroup S has *no ordered invariants*, and
no pair of invariants commutes (anti-commutative).

Proof: Ordered invariants have different ranks according to the previous lemma. Let k be the lowest rank of an ordered pair of invariants. Then, with lemma 1d, S has a proper ideal consisting of all elements with rank≤k which contradicts S being simple.

If invariants e,f commute: ef=fe, then their composition d=ef is also invariant: d^2=d since ef.ef= ef.fe= e.ff.e= e.fe= e.ef= ee.f= ef. Moreover d is ordered under e, since ed= eef= efe= de=d so d≤e, and similarly d≤f. In fact, d is the **greatest lower bound or**

meet of e and f, which is easily verified [2, p24]. So a commuting pair of invariants is either ordered, or their composition is ordered under both, contradicting simple S. Hence no pair of invariants commutes. ◇

So a semigroup of *commuting invariants* is partially ordered set where each pair has a meet (set intersection), called a lower semilattice, with a global zero. For n states, there are at most 2^n commuting invariants (Boolean lattice).

3.2 Equivalent Invariants: L, R

Consider now the invariants of a simple semigroup. They do not commute (cor 2). Invariants that do not commute may be equivalent in the following sense:

Def 4: Equivalent Invariants

- Invariants a,b that form a left- [right-] copy semigroup L2 [R2], are called **left- [right] equivalent**, denoted aLb [aRb].
- Invariants a,b are **equivalent**, denoted a~b, if they are left- or right equivalent: either *directly* so they form L2 or R2, or *indirectly:* alternating L- and R- equivalent via other invariants.

Lemma 3: Consider invariants a,b in any semigroup S, represented over stateset Q:
(a) Equivalent invariants have equal rank: a~b ⇒ $|Qa| = |Qb|$
(but equal rank is not sufficient for equivalence: see b)
(b) a and b are equivalent a~b iff aba=a and bab=b;
then {a,b,ab,ba} are 2 or 4 invariants of equal rank forming L2, R2 or L2xR2.

Proof:

(a) There are three cases of equivalence for invariants a,b: left-, right- and indirect equivalence. In the first two cases of 'direct' equivalence, rank-lemma 1 yields:

aLb implies r(a)=r(ab)≤r(b) and r(b)=r(ba)≤r(a), sothat r(a)=r(b);
aRb implies r(a)=r(ba)≤r(b) and r(b)=r(ab)≤r(a), sothat r(a)=r(b).

Hence left- or right equivalent invariants have the same rank. Transitivity holds in both cases. For instance let aLx (ax=a, xa=x) and xLb (bx=b, xb=x) then aLb, since ab= ax.b= a.xb= ax= a, and similarly ba=b. Also right equivalence is transitive.

Now if aLc and cRb with c different from a and b, then a,b are not directly left- or right equivalent, but they are still called (indirectly) equivalent, denoted aLRb. Here LR is an equivalence relation since it is easily seen to be reflexive, symmetric and transitive. It follows that if a and b are indirectly equivalent via other invariants, then by transitivity they have all the same rank.

(b) In the *direct* equivalent cases aLb and aRb, the elements ab and ba are not different from a and b, forming L2 and R2 respectively.

For *indirect* equivalence of invariants a and b, the only other intermediate elements are ab and ba, which are invariants, while the four elements {a,b,ab,ba} form an invariant subsemigroup of direct product structure L2xR2. This can be seen as follows.

Assume first that aba=a, bab=b. Composition of aba=a on the right by b yields abab= ab, while bab=b implies baba=ba, sothat ab and ba are invariants. All strings over invariants {a,b}: aa=a, bb=b yield in first instance strings of alternating a and b of forms : a, b, ab, ba, $a(ba)^+$, $b(ab)^+$, where $^+$ denotes repetition at least once. By repetitive application of aba=a and bab=b, all such strings of length>2 are seen to

be further reducible to strings of length≤2, which are just a,b,ab,ba. Hence these four invariants form a closure, with the following equivalences (using aba=a, bab=b):
a R ab since a.ab=aa.b=ab and ab.a=a, ab L b since ab.b=a.bb=ab and b.ab=b,
b R ba since b.ba=bb.a=ba and ba.b=b, ba L a since ba.a=b.aa=ba and a.ba=a.
These relations are depicted in a rectangular form in the figure 3.

Conversely, invariance of ab and ba implies aba=a and bab=b, shown as follows. For indirect equivalence, invariants a,b must have equal rank: $|Qa| = |Qb|$ (lemma 3a), hence exact equality holds in (Qa)b⊆Qb, so Qa.b=Qb (*) and similarly Qb.a=Qa (**). Composing both sides of (*) on the right by .a and applying (**) yields Qa.ba= Qba= Qa. So sequence .ba permutes Qa → Qa. Since ba is invariant, this is the identity permutation, hence (qa)ba=qa for all q, meaning aba=a. Similarly, invariance of ab implies bab=b.

The direct product structure L2xR2 follows from two complementary congruences (preserved partitions), illustrated by figure 3. Denote ab=c and ba=d, then {a=c,b=d} with image L2, and {a=d,b=c} with image R2. The direct product code can be implemented by two independent components x=[x1,x2] of which the first composes as L2 and the second as R2. ◇

The left- and right equivalences can be plotted pairwise in the plane as shown in fig. 3, which also gives the composition tables of L2, R2 and L2xR2= {a,b,ab,ba}. From this rectangular display follows the term **diagonal equivalence** for two indirectly equivalent invariants, since this is the only other form of equivalence. It is denoted by xDy where x and y are obtained by commutation: x=ab and y=ba for some a and b, which themselves are also diagonal equivalent aDb, with a=aba= abba=xy and b=bab= baab=yx. Diagonal equivalence occurs in pairs: if aDb then ab D ba and vice versa.

L2	a	b
a	a	a
b	b	b

R2	a	b
a	a	b
b	a	b

L2xR2	a	b	c	d
a	a	c	c	a
b	d	b	b	d
ab= c	a	c	c	a
ba= d	d	b	b	d
e	a	b	c	d

initial state for unique repr.

Fig.3a

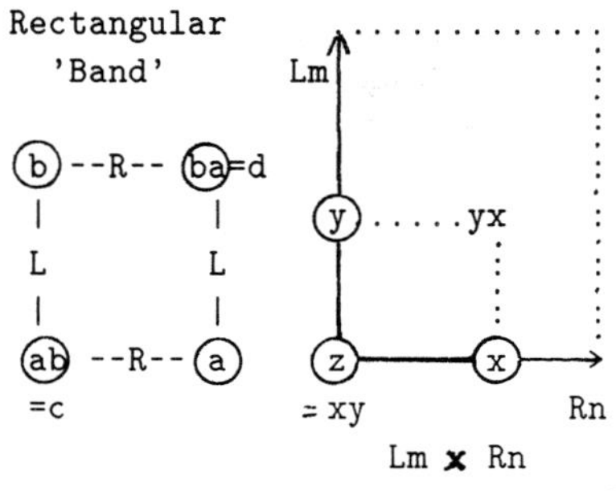

xyx=x , yxy=y

Fig.3b

Image = S/Congruence : L2 = S/{a=c,b=d} ; R2 = S/{a=d,b=c}

Code L2xR2 : x=[x1,x2] , y=[y1,y2] --> xy=[x1,y2] , yx=[y1,x2].

The table of L2xR2, viewed as state machine, has two pairs of equal columns (inputs a=d and b=c), sothat one extra initial state 'e' is required for a unique representation by state transforms.

Lemma 4: (a) In any **invariant** semigroup S: $a \geq aba$, and
(b) a>aba for some a,b only if S is not of constant rank, so:
a=aba for all a,b iff S is invariant of constant rank.

Proof:
(a) We need to show that 'a' commutes with aba, and is left- and right identity for aba. Both follow directly from aa=a and a.aba= aba =aba.a.
(b) If S is not of constant rank, then the minimum rank invariants form a proper ideal Z of S (lemma 1d), and there is an ordered and commuting pair of invariants. Consider invariants 'a' in S–Z and b in Z, then invariant aba is also in Z and has the same (minimal) rank as b, so rank(a)>rank(aba)=rank(b). Hence strict ordering a>aba holds. ◇

The rectangle of equivalent pairs of invariants generalizes to Lm x Rn, with $m,n \geq 2$. The m.n invariants form an m x n matrix, where L- [R-] equivalence holds between elements in the same column [row]. This is the general structure of a constant rank invariant semigroup, also called a rectangular 'band' :

Theorem 1: the following conditions on a finite semigroup S are equivalent:
(a) S is *anti commutative* (no two elements commute: ab=ba implies a=b).
(b) S is *invariant* and of *constant rank.*
(c) aba=a for all a,b in S.
(d) Each pair a,b of invariants in S is equivalent: either directly, forming L2 or R2, or indirectly (diagonal) via ab and ba forming L2xR2.
(e) S is a direct product Lm x Rn of a left- and a right copy semigroup ($m,n \geq 1$).

Notice that rather general conditions (a)(b) imply a very regular structure (e), which is due to the strong properties of *finite* (rank) *associative* (semigroup) algebra.

Proof:
(a) ⇒ (b) : an anti- commutative semigroup S is invariant, because any iteration class x^+ is a commutative subsemigroup, so $|x^+|=1$ for all x, hence each element of S is invariant. Moreover, S is of constant rank because otherwise some pair of invariants a,b would be properly ordered (lemma 4b) and thus commute, contradicting S being anti-commutative.
(b) ⇒ (c) : lemma 4b.
(c) ⇒ (d) : aba=a for all a,b → Pairwise L-, R- or D- equivalent (lemma 3b).
(d) ⇒ (e) : Pairwise equivalence in S implies the direct product structure Lm x Rn with $m,n \geq 1$ as follows. If S contains only left- equivalent invariants then S=Lm where $m=|S|$ and n=1. The other trivial case occurs when S contains n right equivalent invariants, and no left equivalence holds: S=Rn with m=1 and $n=|S|$.

If both left- and right equivalences occur, the Lm x Rn rectangular structure (fig.3b) is seen as follows. Take any invariant z and form two subsets: Lz with all elements y that are left equivalent yLz to z, and Rz containing all x with xRz: right equivalent to z. They intersect only in z, because if w is left- and right equivalent to z, then it cannot differ from z: w=wz=z. Lz and Rz are left- and right copy subsemigroups of S. Let the orders be respectively $|Lz|$=m and $|Rz|$=n. Pairwise equivalence implies n copies of Lz forming a congruence λ of S with image S/λ= Rn. Similarly, congruence ρ

consists of m copies of Rz, yielding image S/ρ= Lm. Since no pair of invariants can be both left- and right equivalent, congruences λ and ρ are orthogonal: S= Lm x Rn.
(e) $\Rightarrow$ (a) : semigroup S= Lm x Rn consists of pairwise equivalent invariants. Then it is anti- commutative which means that no pair commutes. For assume that one pair of distinct invariants a,b commutes: ab=ba, then they are either ordered a<b or a>b (in case ab is a or b), or their product is a third invariant c=ab=ba, their **meet**, that is ordered c<a and c<b. Either case contradicts pairwise equivalence. ◇

3.3 Maximal Subgroups: Periodic G

Lemma 5: of the **iterations** a^i of a semigroup element 'a' with increasing i :
- the tail elements (if any) reduce strictly in rank, and
- the cycle elements (at least one: the invariant of a) have constant minimum rank.

Proof: Consider the successive ranges Qa^i which, due to range-lemma 1a, form a reducing inclusion chain of subsets of Q. Each range is contained properly in the previous one until the cycle is reached at i=t+1. Because as soon as two successive ranges are equal, then so are all next ranges: $Qa^i=Qa^{i+1}$ ==> $Qa^{i+1}=Qa^{i+2}$, etc. (compose left and right by a). So as soon as the cycle is reached, the minimum rank is obtained: the initial tail ranks are strictly decreasing, and from the first periodic element in the cycle all ranks are constant and minimal. ◇

Corollary 3: In a simple semigroup S every element is periodic (has no tail).

This follows directly from the previous lemma and lemma 1d, because if an element of S had a tail, then its iterations would have different ranks, which contradicts the constant rank property of a simple semigroup. To show that a simple semigroup is a disjoint union of isomorphic groups, we first need:

Lemma 6: Maximal subgroups. Let S be a semigroup, then:
(a) Periodic elements generating the same invariant 'e' form a maximal subgroup of S, called the group G_e of e.
(b) Equivalent invariants a~b have isomorphic groups $G_a \cong G_b$:
if aLb via isomorphism a.G_b=G_a, mapping x in G_b to ax in G_a;
if aRb via isomorphism G_b.a=G_a, mapping x in G_b to xa in G_a;
if aDb via isomorphism a.G_b.a=G_a; mapping x in G_b to axa in G_a.

Proof:
(a) Let periodic element x generate invariant e with period p, so x^p=e. Then clearly the inverse of x with respect to e is x^{p-1}. Define x^0=e for consistency in case p=1 (x=e), and denote the inverse of x by x^{-1}. If y is another periodic element generating e, with inverse y^{-1}, then xy has inverse $(xy)^{-1} = y^{-1}.x^{-1}$ since $xy.(xy)^{-1} = x.y.y^{-1}.x^{-1} = x.e.x^{-1} = x.x^{-1} = e$, and similarly $(xy)^{-1}.xy = e$. It follows that xy generates the same invariant as x and y, so closure holds. Inverses are unique, because if x has two inverses x1 and x2 then x1= x1.e= x1.(x.x2)= (x1.x).x2= e.x2= x2. So all periodic elements generating the same invariant form a group.

(b) Let a and b be two right equivalent invariants aRb so ab=b and ba=a, then right composition of G_a with b is a morphism from G_a onto G_b, meaning G_b is an image of G_a, denoted $G_b| G_a$ (divisor relation). This follows, because a is identity for each y in G_a: ay=ya=y, while for each x and y in G_a: xb.yb= xb.ayb= x.ba.yb= x.a.yb= xy.b (*), where we used ba=a. In other words: the image of a composition of elements is the composition of their images.

We need ab=b to show that xb is in G_b, in fact xb generates b upon iteration. This is seen by replacing y in (*) with x, then $(xb)^2 = (x^2).b$, and in general $(xb)^i = (x^i).b$. Let p be the period of x in G_a: x^p=a, then $(xb)^p$= (x^p).b= ab= b in G_b.

So if ab=b and ba=a: a and b are right-copiers for each other, forming right equivalent invariants aRb, then right composition of G_a with b yields image G_b. Similarly, right composition of G_b with *a* yields image G_a. Consequently right equivalent invariants aRb have mutually ordered groups $G_b|G_a$ and $G_a|G_b$, so they are isomorphic $G_a \cong G_b$.

Using left composition by a and b respectively, it follows that also left equivalent invariants have isomorphic groups. And finally, by transitivity, diagonal equivalent invariants have isomorphic groups as well. In that case aDb with (fig.3b) aLba, baLb. Then, in short, a.G_b.a= a.G_{ba}= G_a. In fact the diagonal case covers the other two cases of direct equivalence. ◇

4 Conclusion

Combining all results, we have finally:

Theorem 2: The following conditions on a finite semigroup S are equivalent:
(a) S is simple (has no proper ideal).
(b) S is of constant rank.
(c) S is a disjoint union of isomorphic groups. Invariants (group-identities) form LxR.
(d) for invariants a,b in S: G_a=aSa and aG_ba=G_a
(e) S is a direct product LxRxG of a left- and a right-copy semigroup with a group.

Proof: (a) ⇒ (b) : Corollary 1 (lemma 1d).
(b) ⇒ (c) : Each element x of a constant rank semigroup S is periodic (cor 3). Hence S is a union of as many maximal subgroups as there are invariants, being the subgroup identities (lemma 6a). The subgroups are disjoint because no element can generate two invariants. Constant rank implies that no two invariants are ordered (cor 2), hence they are pairwise equivalent and form a direct product LxR (theoerem 1).

(c) ⇒ (d) : Consider an invariant ‘a’ and elements of form aSa = {axa, x in S}. Let the invariant generated by axa be c=$(axa)^p$ with period p. Since c begins and ends with invariant a, we have ac=ca=c, meaning a≥c, and in fact a=c, since no strict ordering occurs in a constant rank semigroup. Hence $(axa)^p$=a, in other words axa generates invariant ‘a’ for each x, and is thus in G_a. So for each x in constant rank semigroup S, axa is in the max-subgroup containing ‘a’ , denoted as aSa=G_a.

If a,b are two equivalent invariants, with maximal subgroups G_a and G_b, then the group isomorphism is aG_ba=G_a with axa=y, independent of whether it is a left-, a right- or a diagonal equivalence (lemma 6b); the last case covers the first two.

(d) ⇒ (e) : Constant rank semigroup S contains as many disjoint isomorphic groups G as there are invariants, which are the group identities forming a direct product LxR subsemigroup (c). The direct product structure LxRxG follows because there are two orthogonal congruences α = { x≡y if x and y are in the same max-subgroup} and

γ = {x≡y if axa=y for some invariant a} with images S/α=LxR and S/γ=G.

(e) ⇒ (a) : Without going into detail here, in general the direct product of simple semigroups is also a simple semigroup [2, p83, ex8]. Since L, R and G are simple, so is their direct product. ◇

Any set A of state transforms that generate a constant rank closure, is a constant rank state machine M(A,Q). As shown above, in general the closure S= A^+/Q = LxRxG. It is easily verified that Lm has m genetators and m+1 states (see L2, fig.1) with the function of an m-branch; Rn has n generators and n states with an n-reset function, while group G has a permutation machine as generator with $k \leq |G|$ states. Then M is represented over m+1 + n + k states since L,R,G are 'relative prime' (have pairwise no common image, not proven here), and we have:

Corollary 4: A general **constant rank state machine** M has closure LxRxG ; it is the parallel composition of machines with closures L, R, G respectively: a **branch** machine, a **reset** machine and a **permutation** machine.

Further Research

The decreasing-rank basic types of machines (fig.1): monotone iterative type Z, and combinational logic type H (for instance embedding a lower semi-lattice in a boolean lattice), still need to be included, in order to obtain a general structure theory of State Machines. Of course, input and output logic functions should be taken into consideration as well [3] to yield an efficient overall logic design.

In essence, *associative algebra* and the *theory of finite semigroups* [2] need to be translated to *state machine language*, and applied to sequential logic synthesis, similar to the application of boolean algebra to the design of combinational logic circuits. This has been tried before, but with little practical impact, for the following reasons.

Using semigroup theory, Krohn-Rhodes [4] [5] derived, in the sixties, a prime decomposition theorem using only permutation and reset components, restricted further to cascade coupling. Clearly this is not a sufficient level of detail for practical purposes.

▶ All five basic component types [1] should be used, for a natural decomposition.

▶ Also **loop coupling** of components is necessary, to decompose the *'prime' permutation machines* that are not casacade decomposable (with a 'simple group' closure). They can be very complex [1] and are not useful as practical network components.

References.

1. N.F.Benschop: "On State Machine Decomposition and the Five Primitives of Sequential Logic", Internat. Workshop on Logic Synthesis, MCNC, USA, 1987.
2. A.H.Clifford, G.B.Preston: "The Algebraic Theory of Semigroups" Vol I, Mathematical Survey no.7 (AMS) 1961.
3. N.F.Benschop: "Min-Cut Algorithm for State Coding", Internat. Workshop on Logic Synthesis, MCNC, Research Triangle Park, NC, USA, 1989.
4. K.B.Krohn, J.L.Rhodes: "Algebraic Theory of Machines" part I, Tr.Am.Math.Soc. Vol 116, pp 450-464 (1965).
5. A.Ginzburg: "Algebraic Theory of Automata" Academic Press, New York 1968.

Logic and Architecture Synthesis
P. Michel and G. Saucier (Editors)
Elsevier Science Publishers B.V. (North-Holland)

STATE REDUCTION OF INCOMPLETELY SPECIFIED FINITE SEQUENTIAL MACHINES

M. J. Avedillo, J.M. Quintana and J.L. Huertas

Dpto. de Electrónica y Electromagnetismo, Universidad de Sevilla
Dpto. de Diseño Análogico, Centro Nacional de Microeléctronica.
41012 Sevilla, Spain.

ABSTRACT

We present ARNES, an algorithm for the state reduction of incompletely specified sequential machines. The new heuristic method, unlike other reported approaches, does not need to generate any complete set of compatibility classes. Starting from the set of internal states in the given symbolic description of the FSM, the application of a sequence of transformations results in a description with a smaller number of states.

1. INTRODUCTION

Our contribution focuses on the synthesis of PLA-based Finite State Machines. Mainly it is concerned with the initial processing of state tables, a step which is previous to the state assignment. The convenience of this step in optimal design of FSMs has been suggested in [Rude85], [Perk89] in order to obtain more efficient implementations. In particular, a transformation of the initial description aiming at reducing the number of states may result in area savings when starting with behavioral descriptions. Another area where state reduction applies is the test generation for sequential machines. Extensions to the classical D-algorithm [Kubo68] or approximations based on random techniques [Schn75] are ineffective when the number of states in the circuit is large and the tests demand long input sequences [Deva89].

In particular, the problem of optimum FSM realizations can be stated:

Find the FSM's description and state assignment which lead to an optimum FSM implementation.

A great effort is required to deal with this task and simpler options have been taken:

1) Optimum assignment of given FSM.
2) Optimum assignment of minimum FSM. The given FSM is minimized in order to obtain a representation with a minimum number of states.

In general, none of these two approaches leads to optimum FSM implementations. Moreover, both problems are computationally hard and to date no method (other than exhaustive search) is known that solves them. Therefore, heuristic strategies are used.

We have developed and programmed an algorithm, ARNES, which heuristically reduces the number of states in symbolic descriptions of FSMs leading to near-minimal FSMs which are then assigned using an optimal state assignment program. It should be considered an intermediate step towards the concurrent treatment of state minimization and state assignment which will solve the original problem.

2. PREVIOUS HEURISTIC APPROACHES TO STATE REDUCTION

The classic heuristic method for state reduction is [Benn72]. Following the traditional structure of the problem solution, the author proposes a maximal compatible (MC) based approach, although there is no assurance of the existence of a minimum cardinality closed cover set composed uniquely of them. The use of this subset of the prime compatible set reduces (in some cases drastically) the number of compatible candidates. The procedure consists of selecting one of the essential (or quasi-essential) MCs and attempting to satisfy its closure requirements (generating one of the smallest sets of MCs that satisfies the violated closure requirements for the MC selected). The result will be a closed set of MCs that may or may not provide a full cover of the initial set of states. The procedure is repeated until a full cover is performed.

REDUCES [Aved90], like Bennets' algorithm, starts with the set of MCs as candidates to belong to a closed cover. The compatibles are selected according to a set of heuristic rules. Using this set we are able to select that M_i satisfying at least one of the most restrictive constraints (closure constraints are more restrictive than cover ones, and among each type those satisfied by a smaller number of MCs are more restrictive), and maximizing C, a weighted sum of:

a) the number of violated constraints that are satisfied by selecting M_i (positive contribution).

b) the number of constraints that are violated as a consequence of selecting M_i (negative contribution).

The procedure finishes when the selected set of MCs is a closed cover for the given FSM.

We can conclude that a common strategy underlies in both approaches we have briefly described. It consists in generating sets of compatibles (which can be computationally expensive and even prohibitive) and building, incrementally and heuristically, a subset of the previous collection. This subset is a closed cover of the FSM.

3. DESCRIPTION OF ARNES

The state reduction algorithm we propose differs in concept from those approaches described in previous Section. The constructive heuristics above are abandoned in favor of an iterative improvement heuristic. The strategy is similar to that in ESPRESSO-IIC [Bray84] for the heuristic minimization of combinational functions. This is, a set of basic operations is defined which transforms a symbolic description of the FSM in another one with a smaller number of states. The algorithm may be described, from a high level point of view, as a sequence of transformations (functions) which, starting with the initial description of the FSM, results in a succession of intermediate descriptions of such a machine with a decreasing number of states. The process finishes when the application of the functions implemented by the algorithm does no longer reduce the number of states. Its main goal is that it does not need to generate any complete set of compatibles as in classical state reduction algorithms.

Primary objective in ARNES is reducing the number of states in the symbolic description of a FSM being used as an input to other phases of the design process of sequential circuits. Moreover, ARNES, once it obtains a solution with a reduced number of states, operates on it in order to maximize the number of *don't cares* in excitation functions.

In Figure 1 the algorithm control block is described using Pidgin – C.

```
ARNES(T)
{  C = init(T);
   Φ* = Φ = cost(C);
   (Φ,C)=expand(C,T);
   (Φ,C)=reduction(C,T);
   T' = table(C,T);
   if(Φ <Φ*)
   {  ARNES(T');
   }
}
```

Figure 1: Description Pidgin – C of control block of the algorithm

Where:

- T = symbolic description of a FSM using state tables.
- $\mathbb{C}$ = closed cover for table T. This is, closed set of compatibles which covers all internal states in T.
- Φ = cost function. Number of compatibles in $\mathbb{C}$. Number of internal states in the description of the FSM.

Procedure **init**(T) initializes $\mathbb{C}$ to the set of internal states in the initial description, T, which actually is a closed cover of the FSM. Function **cost**($\mathbb{C}$) evaluates the cost of a solution, $\mathbb{C}$. There are two main functions in ARNES: **expand** and **reduction** which we will describe in detail. Once $\mathbb{C}$ has been transformed by the application of basic previous functions, procedure **table**($\mathbb{C}$,T) builds a symbolic description, T', for the FSM defining an internal state for each compatible in $\mathbb{C}$. If the cost of this new description is smaller than that of the initial one, the full process is repeated for T'.

Now, we will explain the main procedures of ARNES.

Procedure **expand** adds states to each compatible C_i in $\mathbb{C}$, includes those compatibles needed to fulfill closure requirements of C_i^+ (expanded compatible) and eliminates those in $\mathbb{C}$, that are now covered. Given C_i to be expanded, an optimum expanded compatible C_i^{opt} is defined such that

$$Card(W(C_i^{opt})) - Card(DIMP(C_i^{opt})) \quad is\ maximum \qquad (1)$$

$$W(C_i^{opt}) = W_1(C_i^{opt}) \cup W_2(C_i^{opt}) \qquad (2)$$

where:

DIMP(C_i^{opt}) : is the closure set, E_i^{opt}, of C_i^{opt} eliminating those compatibles which are members of $\mathbb{C}$.

$W_1(C_i^{opt})$: is the set of compatibles which can be eliminated from $\mathbb{C}$, when substituting C_i by C_i^{opt}, because they are included in C_i^{opt}.

$W_2(C_i^{opt})$: is the set of compatibles which can be eliminated from $\mathbb{C}$, when substituting C_i by C_i^{opt}, because they are included in one compatible in *DIMP*(C_i^{opt}).

If we could enumerate all compatibles $C \supset C_i$, we would select one satisfying (1). Instead, since this operation is computationally too expensive, the basic operation **expand1** obtains C_i^+ that is a good approximation to (1). In Figures 2 and 3 procedures **expand** and **expand1** are described using Pidgin – C.

```
expand(ℂ,T)
/* given a closed cover of T, returns a closed cover composed by larger
   compatibles (eventually prime compatibles) without having generated
   them previously */
{  Φ* = Φ = cost(ℂ);
   for(i = 1; i≦|ℂ|; i++)
   {(W, DIMP, Ci+)=expand1(Ci);
      ℂ=(ℂ ∪ Ci+ ∪ DIMP(Ci+))−Ci−W;
      Φ = cost(ℂ);
      /* Ci+ is a compatible which contains Ci */
      /* DIMP(Ci+) is the closure set, Ei+, of Ci+ minus those
         compatibles in ℂ */
      /* W is the set of compatibles in ℂ contained in Ci+ or in any
         compatible in DIMP*/
   }
   if (Φ<Φ*)
   {expand(ℂ,T);
   }
}
```

Figure 2: Description Pidgin – C of procedure **expand**

Procedure **Expand1** begins by computing the set of states that are compatible to all states in C_i and storing them in *comp*. Then, for each potential expanded compatible $C_i^{+j} = C_i \cup S_j$, where S_j is a state in set *comp*, expression (1) is evaluated. Operations are repeated for one of the expanded compatibles which maximizes (1). The expansion of one compatible finishes when there are not compatible states to the current expanded compatible or (1) is negative for all potential expanded compatibles.

```
expand1(C_i)
/*given C_i, compatible in ℂ, returns C_i^+, which is a good approximation to
(1) */
{  flag=1;
   while( flag == 1)
   {  comp = compatibles(C_i); /* stores in "comp" those states which
         are compatible to all states in C_i */
      if(comp != {∅})
      {  for (each s_j ∈ comp)
            C_i^{+j} = C_i ∪ s_j;
         {  select C_i^{+j} which maximizes expression (1), and store in C_i; }
         if(expression (1) < 0 for all C_i^{+j})
            flag=0; /* C_i expansion is not continued because expanding
            means increasing the number of compatibles in ℂ */
      }
      else
         flag = 0; /* C_i expansion is not continued because these are not
         compatible states to C_i */
   }
}
```

Figure 3: Description Pidgin – C of procedure **expand1**

We remark that after expanding a compatible, $\mathbb{C}$ is still a closed cover for T. Moreover, expanding is only allowed if it does not mean an increment in the cardinality of $\mathbb{C}$.

Procedure **reduction** transforms $\mathbb{C}$ in a new $\mathbb{C}^*$ where each C_i is sequentially substituted by a reduced compatible $C_i^- \subseteq C_i$, such that

$$\{\mathbb{C} - C_i\} \cup C_i^- \quad \textit{is a closed cover}$$

For reducing each C_i, first the set of states in C_i covered by unless another compatible in $\mathbb{C}$, H, is derived. All states in H could be removed from C_i and $\mathbb{C}$ still will be a cover of T, but may be not a closed set of compatibles. Procedure **MCEE** selects the largest subset of H that can be eliminated from C_i, such that $\mathbb{C}$ is still a closed cover of T.

Procedure **reduction** allows to move away from one solution to another of less cost, as the application of the whole procedure to state table T', might lead to less-cost solutions. Moreover, **reduction** eliminates states that are covered by more than one compatible. The minimization of such number of states leads to the maximization of *don't cares* in excitation functions [Perk85]. In Figure 4 we describe the procedure. Figure 5 contains a description of the basic function in **reduction** (procedure **MCEE**). **MCEE** complexity increases exponentially with the number of states in set H. From our experience we conclude this is not a problem even for large machines.

```
reduction(ℂ,T)
  /* returns ℂ* where each compatible Ci is substituted sequentially by Ci⁻,
  minimum compatible which when substituting Ci, verifies that ℂ* is still a
  closed cover of T */
  {
     for (i=1; i<|ℂ|; i++)
     {  H={∅};
        for (j=1; j<|ℂ|; j++)
        {  if (i != j)
           {H=H ∪ (Cj ∩ Ci);
           }
        }
        /* H is the set of states in Ci covered by at least another compatible in ℂ
        */
        if (H != ∅)
        {  Ci⁻ = Ci − MCEE(H); /* largest set of states that may be
           eliminated*/
           ℂ = (ℂ − Ci) ∪ Ci⁻;
        }
     }
  }
```

Figure 4: Description Pidgin – C of procedure **reduction**

4. PRACTICAL RESULTS

The new algorithm has been coded in C and a large set of FSMs used to test it. We have tried many examples of machines from Switching Circuit textbooks and journal papers about minimization (all of them referenced in [Reus86]). In many of them, the gains were significant. Table I shows the number of states, the number of product terms and the size of the PLAs implementing the combinational part of the machines after state assignment and logic minimization, for four different initial descriptions: (I) state tables as they appear in the literature, (II) state tables reduced by the application of an implementation of a classical state reduction algorithm (Bennets' algorithm [Benn72]), (III) state tables reduced by the application of REDUCES and (IV) state tables obtained by ARNES. Reported times include in cases II, III and IV the time invested in processing the initial descriptions. In all cases, the optimal state assignment algorithm we use is the "*i-hybrid complement*" algorithm in NOVA [Vila88].

From Table I we can see that the new algorithm obtains minimum cardinality solutions in 14 of 15 cases, including the well-known (and difficult) 22 state machine (FSM6) in [Hous70]. Also, from this table, it should be clear that the size of the final circuit when ARNES is applied is always lower than the size when starting with original FSMs. In 9 machines ARNES gave less area realizations than solutions supplied by the well-known Bennets' algorithm. Only for FSM1, the implementation of Bennets' algorithm achieved better size results than ARNES. Summarizing the comparison between REDUCES and ARNES, in 8 machines ARNES gave area savings while REDUCES obtained better results for 2 machines.

```
MCEE(H)
/* given H, a set of states in Ci covered by at least another compatible in C,
returns the largest subset of H that can be eliminated from Ci such that
(C -(Ci)) ∪ (Ci - MCEE(H)) is still a closed cover of T */
{  if(all states in H can be eliminated)
   { returns(H);
   }
   else
   {  sj=select-state(H); /* selects the first state in H still not selected
      */
      H- =MCEE(Hj-); /* tries a solution without eliminating sj from
      Ci*/
      H+ =MCEE(Hj+); /* tries a solution eliminating sj from Ci*/
      H'= merge(H-, H+); /* selects the set with the largest
      cardinality*/
      returns(H');
   }
}
```

Figure 5: Description Pidgin – C of procedure **MCEE**

We have also applied ARNES to the machines in the MCNC benchmark[Lisa89]. In table II we summarize the results after state assignment for those machines for which we found descriptions with a reduced number of states. We do not include the machines which have identical primary outputs for any present state and input conditions. Again, the reported time includes, in case II, both state reduction and state assignment phases. Table III shows the time which ARNES required to check whether state reduction was possible for the machines which could not be reduced.

From Table II we can see that almost for half of the machines area savings of more than 50% are achieved when the reduced machine obtained with ARNES is used as input for the state assignment phase. The time to obtain such solutions is less than the one resulting from applying NOVA to the original machine for 9 of them.

5. CONCLUSIONS

We have developed and programmed an algorithm, ARNES, which heuristically reduces the number of states in symbolic descriptions of FSMs leading to near-minimal FSMs. The new iterative improvement method is fast and superior to previous constructive heuristic approaches, particularly, for those machines with a great number of maximal compatibles and closure constraints. Nevertheless, it is well established than the minimum cardinality description is not necessarily the appropriate starting point when the minimum total cost realization is to be found. In fact, from table I, a 11-state description of FSM6 results in a PLA implementing the combinational part of the machine, which is less than 50% of the resulting PLA when same state assignment program is applied to a 9-state description of previous FSM. No information on how state

	(I) Original FSM				(II) Bennets' FSM				(III) REDUCES' FSM				(IV) ARNES' FSM				ns_{min}
	ns	tp	size	time	ns'_B	tp	size	time	ns'_R	tp	size	time	ns'_A	tp	size	time	
FSM1	5	7	98	2.7	4	4	44	1.4	4	4	44	1.2	3	5	55	0.8	3
FSM2	9	10	170	12.1	6	10	140	3.7	5	10	140	3.1	4	7	77	1.8	4
FSM3	7	9	126	4.8	4	7	77	1.5	4	7	77	1.2	4	7	77	1.8	4
FSM4	6	7	98	3.1	4	4	44	1.0	3	4	44	0.8	3	4	44	1.2	3
FSM5	6	8	112	4.3	4	6	66	1.2	4	6	66	1.2	3	5	55	1.2	3
FSM6	22	25	500	261.2	12	16	272	180.3	11	6	102	9.1	9	14	238	55.9	9
FSM7	9	14	238	16.5	5	10	140	2.6	5	10	140	2.7	4	8	88	1.8	4
FSM8	6	8	112	5.0	4	6	66	1.4	4	6	66	1.3	3	6	66	1.2	3
FSM9	9	15	285	35.2	4	9	117	3.0	4	11	143	2.7	3	5	65	1.7	3
FSM10	8	14	224	16.4	5	13	208	4.8	5	13	208	4.7	5	13	208	7.5	4
FSM11	6	9	126	6.1	5	9	126	3.5	5	9	126	3.2	4	7	77	1.5	4
FSM12	9	11	187	15.3	4	7	77	1.7	4	7	77	1.4	3	6	66	1.3	3
FSM13	6	11	176	6.6	3	5	65	1.3	3	5	65	1.4	3	4	52	1.2	3
FSM14	6	7	98	2.8	2	3	24	0.6	2	3	24	0.6	2	3	24	0.4	2
FSM15	10	10	190	23.6	4	6	96	1.9	4	6	96	1.9	4	5	65	1.5	4

ns: # *of states in descriptions from papers;*
ns'_B: # *of states in descriptions obtained by Bennets' algorithm;*
ns'_R: # *of states in descriptions obtained by REDUCES;*
ns'_A: # *of states in descriptions obtained by new algorithm;*
ns_{min}: minimun number of states ;
tp: # *of product terms;*
size: PLA size evaluated by NOVA;
time: time in seconds in a SUN workstation 3/260;

Table I

	(I) Original FSM						(II) ARNES' FSM			
	ni	no	ns	tp	size	time	ns'_A	tp	size	time
bbara	4	2	10	25	550	17,1	7	20	380	14,7
bbsse	7	7	16	29	957	47,0	13	29	957	81,6
beecount	3	4	7	10	190	7,8	4	9	144	4,7
ex1	9	19	20	50	2600	350,3	18	43	2236	1027,0
ex2	2	2	19	36	756	95,0	7	13	195	11,3
ex3	2	2	10	18	324	14,5	5	9	135	6,4
ex5	2	2	9	18	324	11,6	3	6	72	3,0
ex7	2	2	10	17	306	17,3	4	9	108	3,7
lion9	2	1	9	9	153	8,7	4	7	77	3,2
mark1	5	16	15	17	646	51,5	12	16	608	60,0
opus	5	6	10	16	448	11,6	9	16	448	14,45
scf	27	56	121	146	19126	9676,6	97	128	16768	11217
sse	7	7	16	29	957	49,4	13	29	957	81,7
tbk	6	3	32	170	5100	7840	16	52	1404	1458,8
train11	2	1	11	12	204	14,5	4	6	66	3,0

Table II

FSM	time	FSM	time
bbtas	0,2	kirkman	99,0
cse	8,3	lion	0,1
dk14	0,3	mc	0,2
dk15	0,2	planet	77,5
dk16	0,6	planet1	78,7
dk17	0,2	s1	19,5
dk27	0,2	sand	430,8
dk512	0,3	shiftreg	0,1
ex4	1,7	styr	83,9
ex6	0,8	tav	0,3
keyb	10,9	train4	0,1

Table III

reduction will affect the assignment and logic minimization steps is available if state reduction and assignment are performed sequentially. This is why this work should be considered an intermediate step towards the concurrent treatment of both tasks, which will solve the optimal FSM realization problem.

REFERENCES

[Aved90] M. J. Avedillo, J.M. Quintana and J.L. Huertas, "A New Method for the State Reduction of Incompletely Specified Finite Sequential Machines", *Proceedings of EDAC'90*, pp. 552-556, Glasgow, March 1990.

[Benn72] R. G. Bennetts, J. L. Washington and D. W. Lewin, "A Computer Algorithm for State Table Reductions," *IERE Radio and Electron. Eng.*, Vol. 42, pp. 513-520, Nov. 1972.

[Bray84] R. K. Brayton, G.D. Hatchel, C. McMullen, and A.L. Sangiovanni, "Logic Minimization Algorithms for VLSI Synthesis". Hingham, MA, Kluwer Academic Pub., 1984.

[Deva89] S. Devadas, H-K. Tony Ma, A. Richard Newton and A. Sangiovanni-Vincentelli, "A Synthesis and Optimization Procedure for Fully and Easily Testable Sequential Machines", *IEEE Trans. on CAD, Vol-8,* pp. 1100-1107, October 1989.

[Kubo68] H. Kubo, "A Procedure for Generating Test Sequences to Detect Sequential Circuit Failures", *NEC Res. Dev.* 12: 69-78, 1968.

[Hous70] S. House, "A New Rule for Reducing CC Tables," *IEEE Trans. on Computers* (Short Notes), November 1970.

[Lisa89 R. Lisanke, "Introduction of Synthesis benchmark", Int. Workshop on Logic Synthesis, North Carolina, 1989.

[Perk85] M. A. Perkowski and N. Nguyen, "Minimization of Finite State Machines in SuperPeg", Proceeding of the Midwest Conference on Circuits and Systems, Lusville, Kentucky, pp 139-147, August 1985.

[Perk89] M. A. Perkowski, J. Liu, "Generation and Optimization of Finite State Machines from Parallel Program Graphs", *DIADES Research Group Report 10/89*, Dept. EE PSU."

[Reus86] B. Reusch and W. Merzenich, "Minimal Coverings for Incompletely Specified Sequential Machines", *Acta Informatica*, No. 22, pp. 663-678, 1986.

[Rude85] R. Rudell, A. Sangiovanni-Vincentelli and G. De Micheli, "A Finite-State Machine Synthesis System", *Proceedings of ISCAS 1985*, pp.647-650, May 1985.

[Schn75] H.D. Schnurmann, E. Lindbloom and R.G. Carpenter, "The Weighted Random Test-Pattern Generator", *IEEE Trans. on Computers*, July 1975.

[Vila88] T. Vila,"NOVA", in *Octtools User's Manual*, UCB, March, 1988.

Logic and Architecture Synthesis
P. Michel and G. Saucier (Editors)
Elsevier Science Publishers B.V. (North-Holland)

Determining the Optimal Cycle Time in Controller Synthesis

Rumi Zahir and Wolfgang Fichtner
Integrated Systems Laboratory, ETH Zürich
8092 Zürich, Switzerland

Abstract

This paper presents an algorithm to determine the optimal clock cycle time of an arbitrary data path by analyzing the timing constraints on the system interface and on the internal system components. By employing a clock cycle time that is shorter than the delay of the slowest functional unit, time-consuming units are optimally scheduled across multiple cycles. In this way controller size and complexity can be traded against execution speed.

1 Introduction

In most behavioral synthesis systems [3, 12, 2, 8, 4] the clock cycle time t_{cy} is an a priori, user-defined quantity. In others [6, 10, 11] t_{cy} is determined by the slowest functional unit in the data path. This type of clocking scheme synthesis leads to suboptimal solutions whenever the operation delays of the functional units have a large variance because fast functional units are underutilized. One possibility to enhance performance is to shorten the clock cycle time and schedule long operations in multiple cycles. Some synthesis systems support chained and/or multi-cycle operations. However, all systems (except [10, 11]) base their scheduling on a fixed clock cycle time t_{cy}. This paper presents a method for automatically determining the optimal clock cycle time for a given data path configuration.

Practically no synthesis systems take timing constraints on the external system interface or on the internal components into consideration during synthesis. Their underlying assumption is that all functional units are purely combinational circuits, i.e. they do not require any clocking. In real designs, however, clocked components such as register files, ROMs, dynamic PLAs and RAMs are frequently used.

The methodology described in this paper determines the optimal clock cycle time of an arbitrary data path by analyzing the timing constraints on the *system interface* and on the *individual functional units* of a data path. This typically leads to better solutions because fast combinational operations are naturally chained together and slow functional units are scheduled across multiple cycles.

The next section briefly reviews related work. Section 3 outlines the *Gestalt* system. Section 4 presents an example and discusses the clock cycle optimization algorithm. Finally section 5 gives an outlook on future work.

2 Related Research

2.1 Clocking Scheme Synthesis

Based on a user-defined initial clock cycle, the SPAID system [6] determines which operations are multi-cycle operations. The actual clock cycle time is the delay required by the slowest remaining single-cycle functional unit. From this SPAID computes the required number of control steps and the total execution time. During synthesis the clock cycle time is fixed.

In the Chippe design system [2] an expert system controls the clock cycle time used by the Slicer scheduler [8]. Slicer supports multi-cycle operation and chaining. After scheduling the controller is generated by the Cogent sub-system. However, the expert system in Chippe lacks rules to optimize between controller cost and clock cycle time [2].

The synthesis tools of the USC ADAM system Sehwa [10] and Maha [11] base their clock cycle definition on CSSP [9]. CSSP determines the optimal partitioning and clock cycle time for a pipelined implementation of a given data flow graph. CSSP enumerates all possible cycle times starting at the delay of the slowest functional unit and then chooses the cycle time that results in the minimal execution time. This method promotes chaining but precludes multi-cycle scheduling of operations.

Most synthesis systems concentrate their optimization efforts on data path generation. They make trade-offs between data path area and speed, while keeping the number of necessary control steps limited in order to keep the control logic small. The clock cycle time is either determined a priori by the user or is set to the propagation delay of the slowest functional unit. This paper shows that performance can be gained by employing a *shorter* clock cycle time, thus trading controller size and complexity against execution speed.

2.2 Interface Timing Constraints

Digital systems tend to have rather complex timing characteristics. Most synthesis systems do not consider interface timing constraints during synthesis and produce purely self-synchronous designs. Consequently it can take an effort equal to the synthesis effort to embed a synthesized system into its final surroundings. Two different approaches have addressed this problem.

The *Janus* system [1] takes the timing descriptions of two interfaces and synthesizes interface hardware to match the two systems timing constraints. Interface synthesis as a post-processing step is practical as long as the interface timing constraints are not critical to system performance. A cleaner approach is the one taken in the *System's Architects Workbench* [12]. The control step scheduling is performed by *CSTEP* [7]. *CSTEP* uses interface specifications to guide the scheduler. However, scheduling is based on a user-supplied clock cycle time.

The ability to synthesize systems that adhere to timing specifications is a significant step in order for synthesis to become practical. Timing constraints do not only occur at the periphery of a system but also at internal component boundaries. Therefore component as well as interface timing constraints have to be used to guide controller synthesis.

3 The Gestalt System

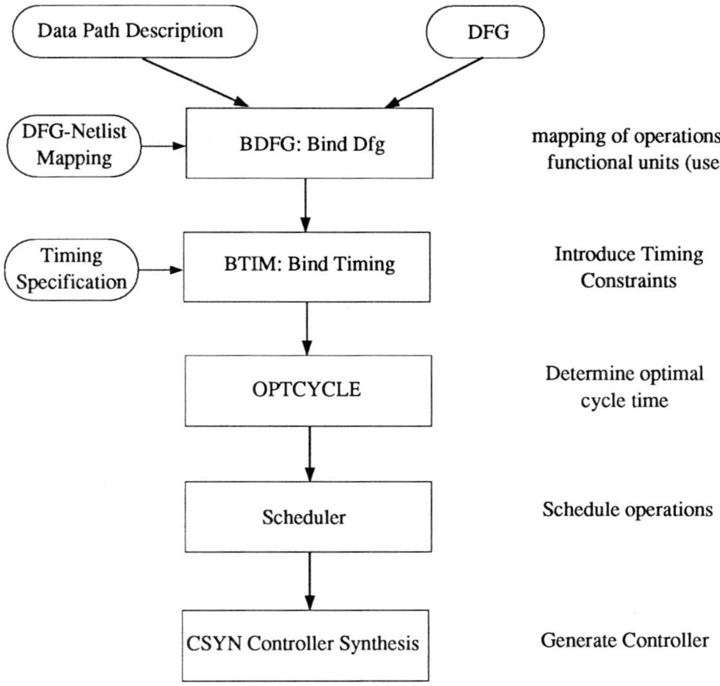

Figure 1: *Gestalt* System Overview

The *Gestalt* controller synthesis system allows the designer to get a quick performance and cost assessment of the control logic that will operate his data path. As shown in figure 1 the designer provides *Gestalt* with four input descriptions: a behavioral specification in form of a *data flow graph (DFG)*, the data path architecture in form of a *netlist*, a *DFG-Netlist mapping* and finally a *timing description* specifying the interface timing characteristics for each functional unit used in the data path.

Gestalt uses the functional unit mapping to determine the feasibility of the desired behavior on the given netlist. It does *not* perform data path synthesis. By analyzing the timing constraints of the functional units in the data path and on the system boundary, *Gestalt* determines the optimal clock cycle time for the controller. Once the clock cycle time has been established, a list scheduling algorithm is used to bind individual operations to control steps. The schedule has to comply with the resource and timing constraints imposed by the architecture mapping and the functional unit timing requirements. Finally, *Gestalt* produces a full controller specification. This paper, however, is concerned only with the clock cycle time optimization (OPTCYCLE box in figure 1).

4 Clock Cycle Time Optimization

The next section introduces an example that is used throughout this paper to illustrate the clock cycle optimization method. The following sections briefly explain the specification of the data path and of the timing constraints. Section 4.4 describes how the optimal clock cycle time is calculated and finally section 4.5 presents the results.

4.1 An Example

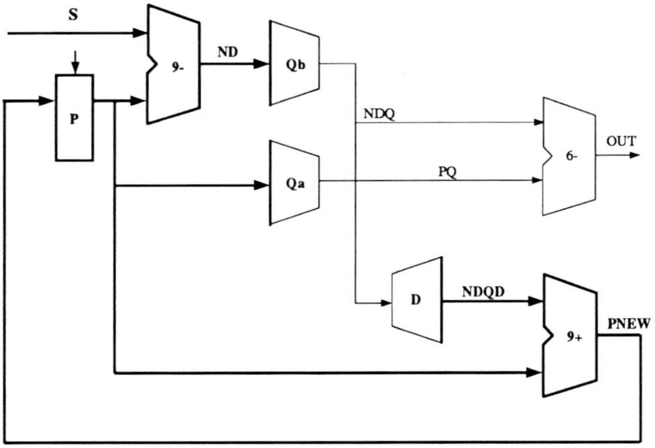

Figure 2: Data Flow Graph of the Video Coder Example

The data flow graph shown in figure 2 is part of a video data compression system. It compresses 9-bit video data (thick lines) into 6-bit values (fine lines) using a hybrid delta PCM compression algorithm with a 9-bit prediction value (P). Data compression proceeds as follows: first the difference ND between signal and prediction value is calculated. The compressed output value OUT is the difference between the quantized difference ND and the quantized prediction value PQ. Finally the new prediction value is computed by adding the quantization error NDQD to the old prediction value. The required operators in the data path are: addition (9+,6+), subtraction (9-), quantization (Qa, Qb) and dequantization (D).

4.2 Applying the DFG-Netlist Mapping

The first step in figure 1 (BDFG) binds the data flow graph to the data path netlist using the *DFG-Netlist mapping* provided by the designer. Consistency checks verify that all necessary registers and connections for the execution of the data flow graph exist in the netlist. The result is a bound data flow graph in which each node has been assigned a functional unit in the data path.

The netlist for the Video Coder contains 4 registers (P, PQ, S and OUT), a Q unit, a D unit, a 9-bit ALU and a 6-bit ALU. Table 1 indicates the mapping, and figure 3 shows

Mappings		Timing Specification	
DFG Element	functional unit	T_{setup} [ns]	$T_{propagation}$ [ns]
P, PQ, S, OUT	Register	5	10
-	MUX	-	5
Qa, Qb	Q	3	15
D	D	3	16
9+, 9−	ALU9	-	40
6+	ALU6	-	30

Table 1: Mapping and Timing Specifications in Video Coder Data Path

the resulting bound data flow graph. As there is only one Q unit, the two quantization operations Qa and Qb have to be sequenced. The same holds for the operations 9+ and 9− that share the 9-bit ALU. These resource constraint arcs are drawn in thick lines. Functional units that require a clock are marked by an extra vertical arrow. The numbers on the arcs in figure 3 are not part of the bound data flow graph. They are added by the next step that introduces the timing constraints.

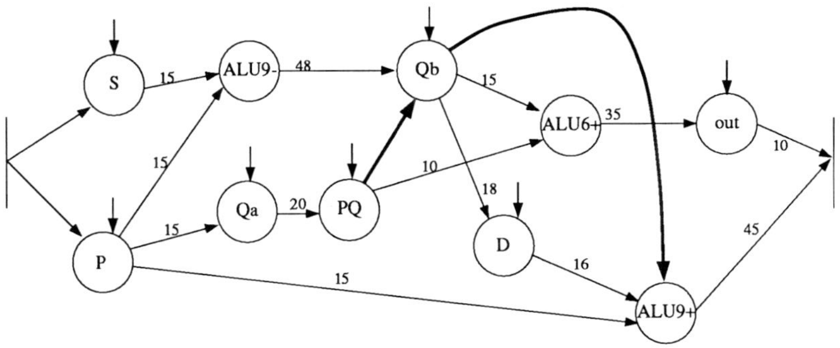

Figure 3: Bound Data Flow Graph

4.3 Introducing Timing Constraints

The second step in figure 1 (BTIM) uses the timing specifications to transform the bound data flow graph into a timing constraint graph. The data flow graph defines data dependencies and operation sequence. The timing specifications of the functional units determine the timing constraints between successive operations. Figure 4 shows the timing constraint specifications for a register, a clocked functional unit (a latched, dynamic PLA) and a combinational functional unit (a ripple-carry adder).

Applying the timing specifications from table 1 to the Video Coder example produces the delay values that annotate the arcs in figure 3. Hold time constraints are not shown.

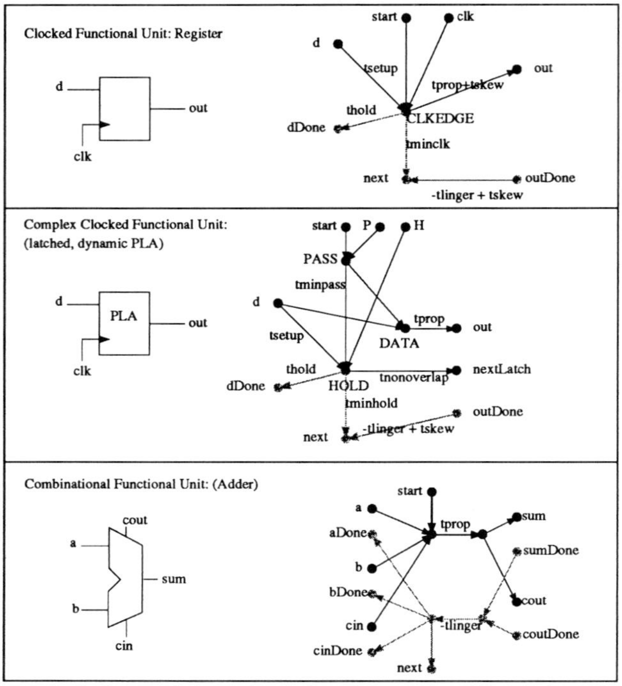

Figure 4: Timing Constraint Specifications

4.4 Finding the Optimal Clock Cycle

To determine the optimal cycle time the timing constraint graph is traversed and the nodes of all combinational units are merged into the preceding clocked node forming a so-called *clock-edge node*. The timing constraints are updated at the new clock-edge node as is shown in figure 5. This step is repeated until the modified graph contains clocked clock-edge nodes only. All nodes in this *clock-edge graph* require a clock edge,

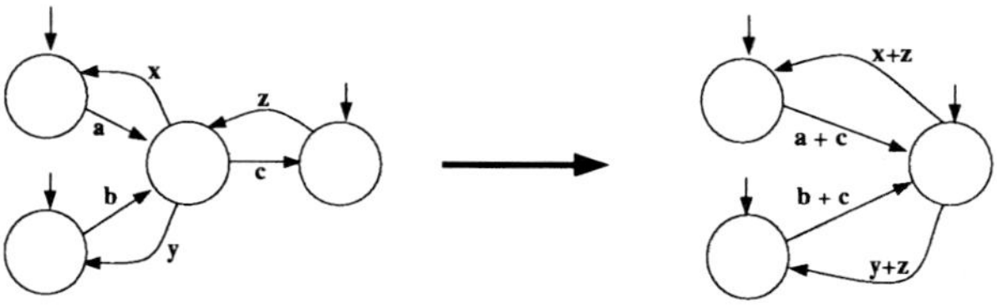

Figure 5: Collapsing Timing Constraints in Clock-Edge Nodes

i.e. they required at least one clock cycle and cannot be chained into a single cycle. The number of nodes on the critical path in the clock-edge graph is a lower bound on the number of required cycles.

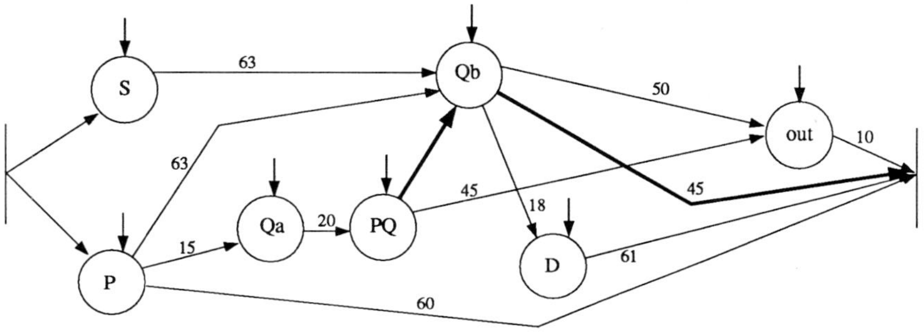

Figure 6: Clock-Edge Graph

The clock-edge graph for the Video Coder example is shown in figure 6. On this graph the optimal cycle time, t_{cy}, is calculated in a two step procedure:

1. Find lower and upper bound for t_{cy}: t_{cymin} and t_{cymax}
2. Scan the range of feasible cycle times for an optimal solution that minimizes the cost function: $\text{Cost}(t_{cy}) = T_{exec}(t_{cy}) * n_{cy}$

Definition of Variables

d_i	=	propagation delay of the ith arc of the clock-edge graph
k	=	number of arcs in the clock-edge graph
d_{max}	=	$\max(d_1, d_2, \ldots, d_k)$
d_{min}	=	$\min(d_1, d_2, \ldots, d_k)$
t_{cy}	=	cycle time
$t_{cyabsmin}$	=	absolutely minimal cycle time (technology dependent parameter)
t_{cymin}	=	lower bound for t_{cy}
t_{cymax}	=	upper bound for t_{cy}
n_{cy}	=	number of cycles
$n^i_{cy}(t_{cy})$	=	$\lceil d_i/t_{cy} \rceil$, number of cycles of the ith arc in the clock-edge graph
$N_{crit}(t_{cy})$	=	number of cycles required for the critical path [1]
$T_{exec}(t_{cy})$	=	$N_{crit}(t_{cy}) * t_{cy}$, total execution time
$\text{Cost}(t_{cy})$	=	$T_{exec}(t_{cy}) * N_{crit}(t_{cy})$

Calculation of lower and upper bounds of t_{cy}

The natural upper bound for t_{cy} is the maximum delay d_{max}. Increasing t_{cy} beyond d_{max} does not enhance performance. Thus:
$t_{cymax} = d_{max}$.

[1]Note that this is also the number of states in the controller. The critical path length can depend on the choice of t_{cy}.

When making the choice of the lower bound for t_{cy} two points have to be considered. One is the absolute lower bound at which a controller can be implemented. This value $t_{cyabsmin}$ is technology dependent. In this paper we assume a *fixed* value for $t_{cyabsmin}$, although really it is a function of the number of states in the controller $N_{crit}(t_{cy})$. On the other hand there is no point in choosing a clock cycle time that is shorter than the propagation delay of the fastest functional unit. Frequently this is the propagation delay of the registers. Here we take t_{cymin} to be half of the minimal functional unit delay d_{min} or the technological minimum if that is bigger:

$$t_{cymin} = \begin{cases} \lceil d_{min}/2 \rceil & \text{if } (\ 2 * d_{min} > t_{cyabsmin}) \\ t_{cyabsmin} & \text{otherwise} \end{cases}$$

Scanning the t_{cy} interval

As the cost function is not continuous, the whole clock cycle time range has to be scanned to find the optimal solution. The step size Δt_{cy} at which the t_{cy} interval is traversed is determined by:

$$\Delta t_{cy} = \begin{cases} t_{cyabsmin} & \text{if } (2 * d_{min} > t_{cyabsmin}) \\ \lceil d_{min}/2 \rceil & \text{otherwise} \end{cases}$$

The clock cycle time range is scanned as follows:

for t_{cy} = t_{cymin} to t_{cymax} step Δt_{cy} do

1. Calculate the number of cycles $n^i_{cy}(t_{cy})$ for all clock-edge graph arcs
2. Determine critical path length $N_{crit}(t_{cy})$
3. Compute execution time $T_{exec}(t_{cy})$ and associated cost. Update minimal cost value and remember the corresponding t_{cy} value.

The critical path in a graph without a circuit can be found in linear time to the number of arcs [5]. Since the number of loop iterations is independent of the number of arcs in the graph, the complexity of this algorithm is $O(k)$.

4.5 Results

Iterating across the cycle time range of the clock-edge graph of the Video Coder (figure 6) produces the graphs reproduced in figure 7. The graph on the left shows $\text{Cost}(t_{cy}), T_{exec}(t_{cy})$ and $N_{crit}(t_{cy})$ plotted against the cycle time t_{cy}. The graph on the right shows $\text{Cost}(N_{crit}), T_{exec}(N_{crit})$ and $t_{cy}(N_{crit})$ plotted in function of N_{crit}[2]. Figure 7 shows that although a 23 ns clock cycle time results in minimal execution time, it requires a higher controller cost than the optimal cost solution with a clock cycle time of 35 ns. The corresponding values for the number of cycles in the critical path can easily be identified in the right hand graph in figure 7. The two best solutions for the Video Coder clock cycle time are summarized in table 2. The optimal clocking scheme is 5 cycles with a clock cycle time of 35 ns.

[2]In figure 7 N_{crit} is labelled as Ncy.

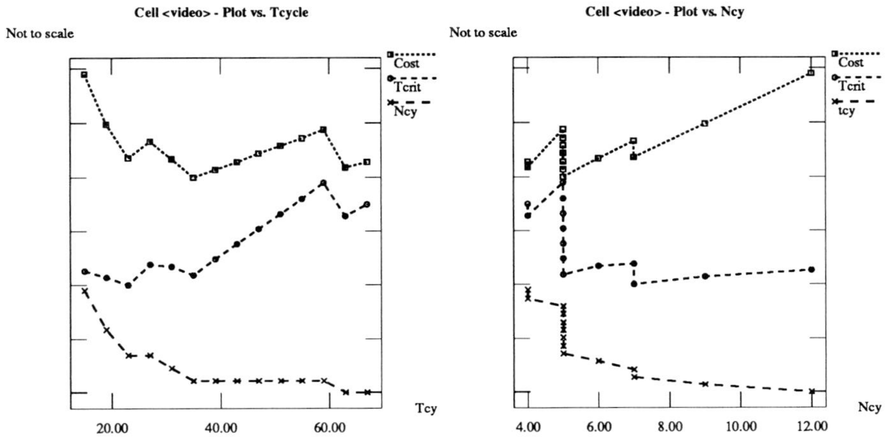

Figure 7: Clock Cycle Times for Video Coder Data Path

t_{cy} [ns]	$N_{crit}(t_{cy})$	T_{exec} [ns]	Cost
35	5	175	875
63	4	252	1008

Table 2: Best Clock Cycle Times for the Video Coder Data Path

5 Summary and Future Work

The *Gestalt* controller synthesis system is being developed at the Integrated Systems Laboratory (IIS) at ETH Zürich. Data path specification such as netlists and functional unit parameters are extracted from the VLSI[3] standard cell design system used at IIS. The clock cycle optimization algorithm described in section 4 has been implemented in C++ and was used to generate the graphs shown in figure 7. The run times for the current implementation are in the order of seconds. There are several improvements to be made to the algorithm:

1. The determination of the lower bound for t_{cy} (t_{cymin}) and the step Δt_{cy} are not satisfactory.

2. The scanning of the t_{cy} interval should use a more sophisticated iteration method.

3. The cost function should take actual controller cost into account. The current implementation assumes a linear relationship between number of required cycles (states) and controller size.

[3]VLSI is a trademark of VLSI Technology Inc.

Future work is going to take a closer look at the specification of timing constraints, the handling of control flow and at various design styles for controller synthesis.

References

[1] Gaetano Boriello and Randy H. Katz. Synthesis and Optimization of Interface Transducer Logic. In *Proc. of the IEEE Int. Conf. on Computer Aided Design*, pages 274–277, 1987.

[2] Forrest D. Brewer and Daniel D. Gajski. Knowledge Based Control in Micro-Architecture Design. In *Proc. of the Design Automation Conference*, pages 203–209, 1987.

[3] Raul Camposano and Wolfgang Rosenstiel. Synthesizing Circuits from Behavioral Descriptions. *IEEE Transactions on Computer-Aided Design*, 8(2):171– 180, Feb 1989.

[4] E.F. Girczyc and J.P. Knight. An Ada to Standard Cell Hardware Compiler Based on Graph Grammars and Scheduling. In *Proc. of the IEEE Int. Conf. on Computer Design*, pages 726–731, 1984.

[5] M. Gondran and M. Minoux. *Graphs and Algorithms.* John Wiley & Sons, 1986.

[6] Baher S. Haroun and Mohamed I. Elmasry. Architectural Synthesis for DSP Silicon Compilers. *IEEE Transactions on Computer-Aided Design*, 8(4):431–447, April 1989.

[7] J.A. Nestor and D.E. Thomas. Behavioral Synthesis with Interfaces. In *Proc. of the IEEE Int. Conf. on Computer Aided Design*, pages 112–115, 1986.

[8] Barry M. Pangrle and Daniel D. Gajski. State Synthesis and Connectivity Binding for Microarchitecture Compilation. In *Proc. of the IEEE Int. Conf. on Computer Aided Design*, pages 210–213, 1986.

[9] Nohbyung Park and Alice C. Parker. Synthesis of Optimal Clocking Schemes. In *Proc. of the Design Automation Conference*, pages 489–495, 1985.

[10] Nohbyung Park and Alice C. Parker. Sehwa: A Software Package for Synthesis of Pipelines from Behavioural Specifications. *IEEE Transactions on Computer-Aided Design*, 7(3):356–370, March 1988.

[11] Alice C. Parker and Jorge T. Pizarro and Mitch Mlinar. Maha: A Program for Datapath Synthesis. In *Proc. of the Design Automation Conference*, pages 461–466, 1986.

[12] D.E. Thomas and E.M. Dirkes and R.A. Walker and J.V. Rajan and J.A. Nestor and R.L. Blackburn. The System Architect's Workbench. In *Proc. of the Design Automation Conference*, pages 337–343, 1988.

Logic and Architecture Synthesis
P. Michel and G. Saucier (Editors)
Elsevier Science Publishers B.V. (North-Holland)

A Formal Approach to Control-Unit Synthesis

M. Mahmood †, F. Mavaddat ‡, and M.I. Elmasry †

† Department of Electrical and Computer Engineering
‡ Department of Computer Science
University of Waterloo, Waterloo, Ontario, Canada N2L 3G1
Fax: (519) 746-5195

The first phase of control-unit synthesis is to translate a high-level behavioral specification into a logical description, such as a PLA profile, or a state-table. In this paper, we propose a new approach to automating this phase. The method is described in two parts. First, we use the DT0L category of L systems to model the data-path of a register-transfer design. Given a data-path, and a set of register-transfer operations referred to as the goal-set, we then use the DT0L model to verify that the goal-set can be computed with that data-path. If the computation is possible, we also generate the local microcode that computes the goal-set on the data-path. Second, given a data-path and an algorithmic state machine specification of the expected behavior, we transform the specification into a variation of Mishchenko's microprogram model, and generate a state-table using the local microcode generation technique. The method is efficient, uniform, and correct-by-construction.

1. Introduction

A register-transfer-type digital system consists of a data-path and a control-unit. The functionality of the data-path depends on the control-unit. By designing the appropriate control-units, we can map different computations onto the same data-path or map the same computation onto different data-paths. If a library of pre-designed data-paths is available, a control synthesis tool should be able to select a suitable data-path from the library and map a given behavioral specification onto the data-path. Such a tool should also be able to explore the architectural design space by generating several control-unit designs for a given data-path, and selecting a suitable design that satisfies the design constraints [1].

Control-unit synthesis can be split into two phases [1]:

1. The translation of a high-level behavioral specification into a logical description, such as a PLA profile [1, 2], a microcode [3, 4], or a state-table.

2. The generation of the layout from the logical description [5].

In the past, most control-unit synthesis tools [2, 3, 4] have ignored the microcode generation subtask. While the work by Evangelisti et al. [6] has addressed microcode generation, the control generated by their method could be incorrect, and requires a post-verification step.

Here, we propose a new method for automating the first phase of control-unit synthesis. We start with a description of a data-path and a high-level specification of the expected behavior; we then translate the specification into a state-table representation of a finite state machine (FSM) that controls the data-path to realize the behavior.

We use the DT0L category of L systems [7] to model the register-transfer behavior of data-paths [8]. Unlike other formal language models of data-path, DT0L systems capture both parallelism and synchronous time. This helps to generate 'correct-by-construction' microcode, and eliminates the post-verification procedures needed by others [6]. An *algorithmic state machine (ASM)* type language is used to describe the behavior of a digital system [9].

In Section 2, we describe the DT0L model of data-paths. In Section 3, we discuss the microcode generation algorithm. In Section 4, we describe a method of translating an ASM behavioral specification into a control-unit, and give an example of control-unit generation. We summarize our results, in Section 5.

2. DT0L Model of Data-Paths

A DT0L system is a triple, $G = (\Sigma, H, \omega)$, where Σ is an alphabet, $\omega \in \Sigma^*$ is referred to as the *axiom,* and H is a finite set of homomorphisms from Σ^* into Σ^*. The *language* of G, defined by $\mathbf{L}(G) = \{x \in \Sigma^* | x = \omega \text{ or } x = h_1 \cdots h_k(\omega), \text{ where } h_1, \ldots, h_k \in H \text{ and } k \geq 1\}$, is referred to as a *DT0L language*.

A data-path is mapped into a corresponding DT0L system by applying a number of transformation steps. First, we derive a *closed data-path* by adding a suitable amount of hardware to the data-path to model the i/o environment. Figure 1 shows the closed data-path created by the Splicer system [10]. This data-path is subsequently used to demonstrate our control-unit synthesis method. Next, we transform the closed data-path to a directed acyclic graph (DAG), by cutting all the unit-delays of the closed data-path, thereby separating the master and slave parts. Finally, we map the DAG into a corresponding DT0L system $G_{dp} = (\Sigma, H, \omega)$, where:

- $\Sigma = \{x | x$ is a function or a unit-delay identifier for each node of the DAG $\}$.
- $H = \{h | h$ is derived from each selector-less graph $\}$; selector-less graphs are obtained by partially evaluating the DAG with respect to each microinstruction $b \in \{0,1\}^n$, where n is the number of selector elements in the data-path. We write $h(a) = b$ as $a \rightarrow b$, and refer to this as a rewrite-rule.
- $\omega \in \Sigma_d^*$ is a concatenation of the data-path's output unit-delay identifiers.

See [8] for a detailed discussion of the graph transformation steps.

3. Local Microcode Generation

A *local microprogram* is a restricted form of program in which there are no branches into or out of the program, and is specified by a set of register-transfer operations

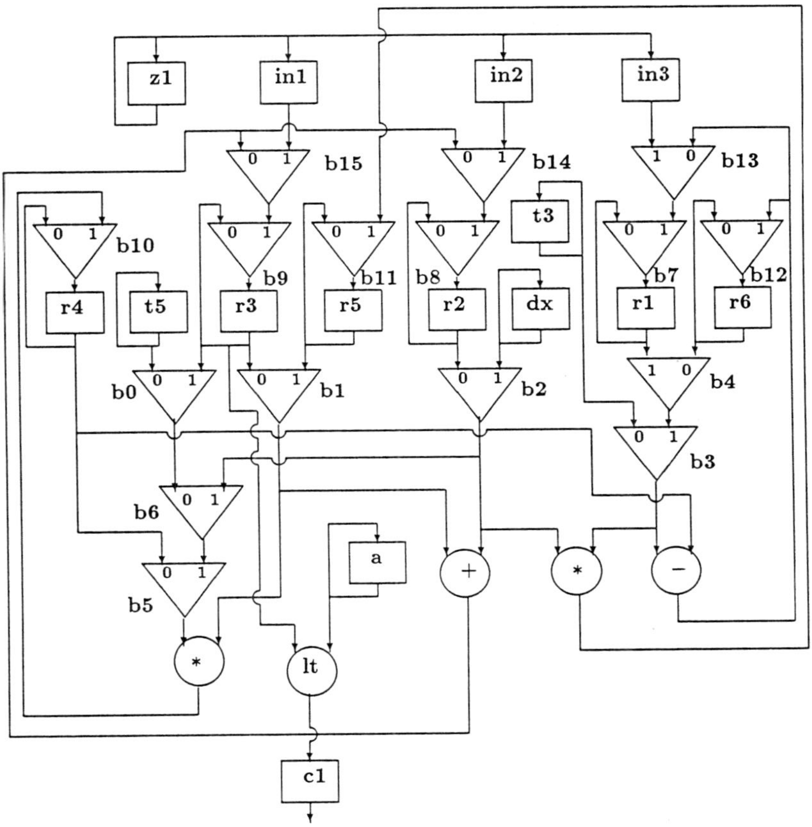

FIGURE 1

A closed data-path created by the Splicer system, in our notation. There are simple modifications to the original design to fit our notation. Rectangles, circles, and triangles denote unit-delays, functionals, and selectors, respectively. *b*0 through *b*15 are microinstruction (selector) inputs. The data-path computes the specification for solving a differential equation.

referred to as a *query* or a *goal-set* (see Figure 2). In [8], we formulated the problem of whether a given query can be computed on a given data-path as the membership problem of the query in the corresponding DT0L language. We also showed that if the query is a member, the local microcode for computing the query can be derived from the parse tree. Although there exists a DT0L-membership algorithm with polynomial time complexity [11], it does not construct the parse tree, and thus can not be used to solve the above problem.

We have developed a heuristic backtracking algorithm for solving our version of the local microcode generation problem. The algorithm consists of an algebraic transformer and a DT0L parser. The algebraic transformer generates algebraically equivalent goal-sets by applying one or more transformations to the given query. The parser tries to generate the corresponding local microcode by using one of the generated goal-sets.

The parser uses an 'inverted representation' of H, defined as a set of $(r, \bar{p})$ pairs, where r is a rewrite-rule and $\bar{p} \in \{0,1,x\}^n$ is the corresponding *partial-instruction*. At step i, the parser selects a non-conflicting set of rewrite-rules, θ_i, to rewrite a goal-set, $Goal_i$, as the new goal-set, $Goal_{i+1}$, and generates the corresponding microinstruction by merging the set of partial-instructions corresponding to θ_i. The parser terminates when all goal-set parts rewrite to themselves. The local microcode is a sequence of microinstructions. Figure 2 shows an example of local microcode generation.

For typical data-paths, the unpruned search space traversed by the algorithm is very large. The user sets bounds on the depth of the search space of both the algebraic transformer and the parser. The algebraic transformer saves some intermediate results to avoid redundant work during backtracking. We have also developed several heuristics to improve the performance of the algorithm. To search promising paths first, the algorithm can choose to rewrite the subgoals with the lesser number of rewrites, or it can choose the largest subgoals first; it also applies more promising rewrite-rules first. To prune the search space, the algorithm uses several techniques, such as branch and bound search, and lookahead functions. For a detailed discussion of the algorithm and the heuristics, see [12, 13].

We have generated local microcode for several data-paths created by other high-level synthesis tools, including the data-path of Figure 1. The heuristics improved the search time by a factor ranging from 1.4 to 3600. The parser generated an optimal first derivation for about 90% of the queries. The parser can backtrack and generate other derivations. In some cases, the parser found new optimal derivations which use less hardware than the results reported in the literature, such as the derivation of Figure 2. For a detailed discussion of the experimental results, see [12, 13].

4. Control-Unit Synthesis

4.1. Behavioral Specification

We use an ASM-type specification language to describe the behavior of a digital system. While the proposed language is similar to many other FSM-type languages [14], the syntax of our language is tailored to the translation steps, discussed later.

Let Ψ be the set of state names of a specification. A specification is a collection of *specification blocks*, one corresponding to each state in Ψ. Each specification block is a (S, I_s) pair, where S is in Ψ and I_s is a conditional statement defined as follows:

Step	Goal-set	Rewrite-Rule	Time
1	$\{(r3 \leftarrow +(r3, dx)),$ $(r1 \leftarrow -(-(r1, *(*(t5, r3), *(dx, r1))),$ $*(dx, *(r2, t3)))),$ $(r2 \leftarrow +(*(dx, r1), r2))\}$	$r3 \rightarrow\ r3$ $r1 \rightarrow\ -(r1, r4)$ $r2 \rightarrow\ r2$	4
2	$\{(r3 \leftarrow +(r3, dx)),$ $(r1 \leftarrow -(r1, *(*(t5, r3), *(dx, r1)))),$ $(r4 \leftarrow *(dx, *(r2, t3))),$ $(r2 \leftarrow +(*(dx, r1), r2))\}$	$r3 \rightarrow\ r3$ $r1 \rightarrow\ -(r1, r4)$ $r4 \rightarrow\ *(dx, r5)$ $r2 \rightarrow\ r2$	3
3	$\{(r3 \leftarrow +(r3, dx)),$ $(r1 \leftarrow r1),$ $(r4 \leftarrow *(*(t5, r3), *(dx, r1))),$ $(r5 \leftarrow *(r2, t3)),$ $(r2 \leftarrow +(*(dx, r1), r2))\}$	$r3 \rightarrow\ r3$ $r1 \rightarrow\ r1$ $r4 \rightarrow\ *(r4, r5)$ $r5 \rightarrow\ *(r2, t3)$ $r2 \rightarrow\ +(r5, r2)$	2
4	$\{(r3 \leftarrow +(r3, dx)),$ $(r1 \leftarrow r1),$ $(r4 \leftarrow *(t5, r3)),$ $(r5 \leftarrow *(dx, r1)),$ $(r2 \leftarrow r2)\}$	$r3 \rightarrow\ +(r3, dx)$ $r1 \rightarrow\ r1$ $r4 \rightarrow\ *(t5, r3)$ $r5 \rightarrow\ *(dx, r1)$ $r2 \rightarrow\ r2$	1

	Sequence of Microinstructions															
Time	$b0$	$b1$	$b2$	$b3$	$b4$	$b5$	$b6$	$b7$	$b8$	$b9$	$b10$	$b11$	$b12$	$b13$	$b14$	$b15$
1	0	0	1	1	1	1	0	0	0	1	1	1	x	x	x	0
2	x	1	0	0	x	0	x	0	1	0	1	1	x	x	0	x
3	x	1	1	1	1	1	1	1	0	0	1	x	x	0	x	x
4	x	x	x	1	1	x	x	1	0	0	x	x	x	0	x	x

FIGURE 2

Local microcode derivation for the data-path of Figure 1 and the goal-set of Step 1. The first table shows the rewriting steps. The rewritings of unit-delay elements $t3$, dx and a are not shown. The third column shows the corresponding scheduling and register-binding. The second table shows the corresponding sequence of microinstructions. The derivation does not use the unit-delay $r6$. Therefore, the data-path of Figure 1, originally proposed in [10], can be optimized by removing $r6$ and its associated hardware from the design.

- (Z, N) is a conditional statement (or *consequent*), if N is in Ψ and Z is a goal-set. To interpret (Z, N), the elements of the goal-set Z are evaluated and assigned, and the specification block (N, I_n) is selected for next interpretation.

- If P is a *proposition*, and C_1 and C_2 are conditional statements, then $if(P, C_1, C_2)$ is also a conditional statement, where a *proposition* is a boolean expression in the sum-of-product form, whose literals are the data-path status expressions and/or control-inputs. To interpret this form, P is evaluated. The control-unit interprets C_1 if P is true, and C_2 otherwise.

A specification block is interpreted in one time unit. We define an order among the consequents of a specification block, and use (Z_j, N_j) to refer to the j-th consequent. Figure 3 shows a specification that computes a "differential equation". It consists of two specification blocks corresponding to states S_1 and S_2. This example has been used in several data-path synthesis tools [10, 15], and may be regarded as a benchmark.

% Mapping $g = \{$(x, r3), (y, r2), (u, r1), (a,a), (dx, dx), (3, t3), (5, t5) $\}$

$$
\begin{array}{lll}
S_1 : if(run, & ([x \leftarrow in1, y \leftarrow in2, u \leftarrow in3], & S_2), \\
 & ([\,], & S_1)); \\
 & & \\
S_2 : if(lt(x,a), & ([\ (u \leftarrow\ -(-(u, *(*(u,dx), *(5,x))), *(dx, *(3,y)))), & \\
 & (x \leftarrow\ +(x,dx)), (y \leftarrow\ +(y, *(u,dx)))], & S_2), \\
 & (\ [(done \leftarrow true)], & S_1));
\end{array}
$$

FIGURE 3

Behavioral specification for solving a differential equation. Variables $in1$, $in2$, $in3$, and $z1$ model the environment. The signals run and $done$ are control-input and module-status, respectively. $lt(x, a)$, $(done \leftarrow true)$, and $(x \leftarrow\ +(x, dx))$ are a status expression, a output signal, and a register-transfer operation, respectively.

4.2. Intermediate Representation

We use a variation of Mishchenko's model, proposed in 1967, for specifying micro-programs [16]. The model has also been used for synthesizing finite state automata [17], and control-units [1, 18, 19].

Let Φ be the set of state names of a finite state machine. A Mishchenko model is a collection of *transition-blocks*, one corresponding to each state in Φ. Each *transition-block*, TB_i, is an explicit representation of the transition and output mappings associated with the state $S_i \in \Phi$, and is expressed as:

$$
\begin{array}{llll}
S_i & f_{i1}(\bar{c}) & Z_{i1} & S_{i1} \\
 & f_{i2}(\bar{c}) & Z_{i2} & S_{i2} \\
 & \cdots & \cdots & \cdots \\
 & f_{ir}(\bar{c}) & Z_{ir} & S_{ir},
\end{array}
\qquad (1)
$$

where, for $j = 1, \cdots, r$, (Z_{ij}, S_{ij}) is a consequent, $\bar{\mathbf{c}}$ is a vector of status and/or control-input variables, and $f_{i1}, f_{i2}, \ldots, f_{ir}$ are boolean expressions in the sum-of-product form such that

$$f_{ij} \cdot f_{ik} = 0 \ \text{ for } \ j \neq k \qquad \text{and} \qquad \bigvee_{j=1}^{r} f_{ij} = 1.$$

The elements of the set $\{f_{i1}, f_{i2}, \ldots, f_{ir}\}$ are called the *antecedents* of TB_i. We use the notation '$S_i^{(j)}$:' to refer to the j-th row of TB_i; namely,

$$S_i^{(j)} : \ f_{ij}(\bar{\mathbf{c}}) \ \ Z_{ij} \ \ S_{ij}. \tag{2}$$

The interpretation of a Mishchenko's model is as follows. Let S_i denote the present state of the finite state machine. The antecedents associated with the transition block TB_i are evaluated. Let $f_{ik}(\bar{\mathbf{c}}) = 1$ for some $k, 1 \leq k \leq r$. Then (Z_{ik}, S_{ik}) is interpreted as explained in Section 4.1.

The following example of a transition-block pair computes the greatest common divisor of two integers:

$$\begin{array}{llll}
S_1 & run & \{(a \leftarrow in2), (b \leftarrow in1)\} & S_2 \\
 & \neg run & \{(a \leftarrow a), (b \leftarrow b)\} & S_1 \\
S_2 & c1 & \{(a \leftarrow a), (b \leftarrow b)\} & S_1 \\
 & \neg c1 \wedge c2 & \{(a \leftarrow b), (b \leftarrow -(a,b))\} & S_2 \\
 & \neg c1 \wedge \neg c2 & \{(a \leftarrow a), (b \leftarrow -(b,a))\} & S_2,
\end{array}$$

where run is a control-input variable, and $c1$ and $c2$ are the status variables representing the status expressions $eq(b,a)$ and $lt(b,a)$, respectively.

4.3. Expansion Techniques

We now present two transformations for expanding transition-blocks. We make extensive use of the transformations when we translate an intermediate representation into a FSM.

First, the interpretation of (2), on a given closed data-path, may require $m \geq 2$ time steps. This is accomplished with new auxiliary states, and can be written as (3), where $S_i.u_{2j}, \ldots, S_i.u_{mj}$ are the names of the auxiliary states, and for $k = 1, \ldots, m$, Z_{ij}^k is the goal-set corresponding to the k-th time step in the computation of Z_{ij}, derived using the local microcode generation algorithm of Section 3.

$$\begin{array}{llll}
S_i^{(j)} : & f_{ij}(\bar{\mathbf{c}}) & Z_{ij}^1 & S_i.u_{2j} \\
S_i.u_{2j} & true & Z_{ij}^2 & S_i.u_{3j} \\
\cdots & \cdots & \cdots & \cdots \\
S_i.u_{mj} & true & Z_{ij}^m & S_{ij}
\end{array} \tag{3}$$

State expansion.

$$\begin{array}{llll}
S_i^{(j1)} : & p_1 & Z_{ij} & S_{ij} \\
S_i^{(j2)} : & p_2 & Z_{ij} & S_{ij} \\
\cdots & \cdots & \cdots & \cdots \\
S_i^{(jn)} : & p_n & Z_{ij} & S_{ij}
\end{array} \tag{4}$$

Product expansion.

Second, let $f_{ij}(\bar{\mathbf{c}}) = p_1 \vee p_2 \vee \cdots \vee p_n$, where p_k is a product term, $1 \leq k \leq n$. We rewrite (2) as a product expanded form (4).

4.4. Translating Specifications into State-Tables

The control-unit synthesis method transforms a behavioral specification, namely, a collection of specification blocks, into a state-table representation of the control-unit. The translation is done in three steps:

Step 1: We extend the data-path to include its i/o environment. This leads to a closed data-path, as shown in Figure 1. The unit-delay, $c1$, captures the data-path status. Such unit-delays are referred to as status variables. The first two components of the corresponding DT0L system, namely Σ and H, are derived using the graph transformation techniques discussed in Section 2.

Step 2: Given a specification block, $B_i = (S_i, I_{s_i})$, a closed data-path, and a mapping g from the specification variables to the closed data-path's unit-delays, we transform B_i into the following transition-block pair:

$$\begin{array}{cccc} S_i & true & Z_i^t & S_i^e \\ S_i^e & f_{i1}(\bar{\mathbf{c}}) & Z_{i1} & S_{i1} \\ & f_{i2}(\bar{\mathbf{c}}) & Z_{i2} & S_{i2} \\ & \cdots & \cdots & \cdots \\ & f_{ir}(\bar{\mathbf{c}}) & Z_{ir} & S_{ir}, \end{array} \qquad (5)$$

where

- Z_i^t is the goal-set corresponding to B_i's test-phase, using the mapping function g. Each status expression of B_i leads to a subgoal in Z_i^t. We assign each status expression of B_i to one of the status variables of the data-path. All other storage variables are assigned to themselves to preserve the specification semantics.

- For $j = 1, \ldots, r$, Z_{ij} is the goal-set corresponding to the j-th consequent of B_i, using the mapping function g. Each register-transfer operation of the j-th consequent of B_i leads to the generation of a subgoal in Z_{ij}. All other storage variables are assigned to themselves to preserve the specification semantics.

- For $j = 1, \ldots, r$, S_{ij} is the state label corresponding to the j-th consequent of B_i.

- For $j = 1, \ldots, r$, $f_{ij}(\bar{\mathbf{c}})$ is a boolean expression in sum-of-product form, whose literals are the control-input and/or data-path status variables. $f_{ij}(\bar{\mathbf{c}})$ is derived by traversing B_i's decision tree to the j-th consequent. See [2, 9] for a more complete description of the boolean expression generation method.

After applying Step 2 to the specification block corresponding to state S_2 of Figure 3, we derive:

$$\begin{array}{cccc} S_2 & true & Z_2^t & S_2^e \\ S_2^e & c1 & Z_{21} & S_2 \\ & \neg c1 & Z_{22} & S_1, \end{array}$$

where

$$\begin{aligned} Z_2^t &= \{(c1 \leftarrow lt(r3,a)), (r1 \leftarrow r1), (r2 \leftarrow r2), (r3 \leftarrow r3), (a \leftarrow a), (dx \leftarrow dx), \\ &\quad (t3 \leftarrow t3), (t5 \leftarrow t5)\}, \\ Z_{21} &= \{(r1 \leftarrow -(-(r1, *(*(r1,dx), *(t5,r3))), *(dx, *(t3,r2)))), (r3 \leftarrow +(r3,dx)), \\ &\quad (r2 \leftarrow +(r2, *(r1,dx))), (t5 \leftarrow t5), (a \leftarrow a), (dx \leftarrow dx), (t3 \leftarrow t3)\}, \\ Z_{22} &= \{(r1 \leftarrow r1), (r2 \leftarrow r2), (r3 \leftarrow r3), (a \leftarrow a), (dx \leftarrow dx), (t3 \leftarrow t3), (t5 \leftarrow t5)\}. \end{aligned}$$

Step 3: This step deals with the application of state and product expansion methods to the results of Step 2, if needed. The application of the state expansion technique to (5) is given by the following two expansions:

$$\begin{array}{llll} S_i & true & {Z_i}^1 & S_i.u_2^t \\ S_i.u_2^t & true & {Z_i}^2 & S_i.u_3^t \\ \cdots & \cdots & \cdots & \cdots \\ S_i.u_{m-1}^t & true & {Z_i}^{m-1} & S_i^e \end{array} \qquad (6)$$

Test-phase expansion.

$$\begin{array}{llll} S_i^{e(j)}: & f_{ij}(\bar{\mathbf{c}}) & Z_{ij}^1 & S_i^e.u_{2j} \\ S_i^e.u_{2j} & true & Z_{ij}^2 & S_i^e.u_{3j} \\ \cdots & \cdots & \cdots & \cdots \\ S_i^e.u_{m_jj} & true & Z_{ij}^{m_j} & S_{ij} \end{array} \qquad (7)$$

Execution-phase expansion (j-th row).

Next, the product expansion technique is applied to the first row of (7). Finally, the state-table is generated by replacing all goal-sets by their corresponding micro-instruction. The implementation of the results of Step 3 transformations using PLA-type structures is straightforward [1, 5].

Figure 4 shows the state-table corresponding to the specification of Figure 3 generated by the above transformations. The behavioral specification of Figure 3 has two states, while the corresponding state-table of Figure 4 has five states. The three auxiliary states are introduced by the state expansion technique.

5. Conclusion

This paper presented a new method for control-unit synthesis. The register-transfer behavior of data-paths is modeled by a DT0L system. First, we discussed the local microcode generation algorithm which is based on DT0L parsing. Next, we described the control-unit synthesis method. The algorithmic behavior of a digital system is specified by an ASM-type language. The ASM specifications are transformed into a collection of transition-blocks. The transition-blocks are then transformed into a state-table by applying state and product expansion techniques. We demonstrated the method by deriving the state-table for a known data-path and a behavioral specification.

Present State	Inputs $c1$	run	Next State	b0	b1	b2	b3	b4	b5	b6	b7	b8	b9	b10	b11	b12	b13	b14	b15	*done*
S_1	x	0	S_1	x	x	x	x	x	x	x	0	0	0	x	x	x	x	x	x	0
S_1	x	1	S_2	x	x	x	x	x	x	x	1	1	1	x	x	x	1	1	1	0
S_2	0	x	S_1	x	x	x	x	x	x	x	0	0	0	x	x	x	x	x	x	1
S_2	1	x	$S_2.u_{21}$	0	0	1	1	1	1	0	0	0	1	1	1	x	x	x	0	0
$S_2.u_{21}$	x	x	$S_2.u_{31}$	x	1	0	0	x	0	x	0	1	0	1	1	x	x	0	x	0
$S_2.u_{31}$	x	x	$S_2.u_{41}$	x	1	1	1	1	1	1	1	0	0	1	x	x	0	x	x	0
$S_2.u_{41}$	x	x	S_2	x	x	x	1	1	x	x	1	0	0	x	x	x	0	x	x	0

FIGURE 4

The synthesized state-table corresponding to the specification of Figure 3 and the data-path of Figure 1. The status expression, $lt(r3, a)$, is assigned to the data-path status variable, $c1$. The last four microinstructions are the local microcode for computing the goal-set Z_{21}.

References

[1] A. Ginetti and C. Trullemans, A Control Unit Generator, IEEE Comp Euro 88, 1988.

[2] C.A. Papachristou and D. Cornett, Generation and Implementation of State Machine Controllers: A VLSI Approach, *Micrprocessiong and Microprogramming*, vol. 16(2-3), pp. 73–82, Sept.-Oct. 1985.

[3] A.W. Nagle, R. Cloutier, and A.C. Parker, Synthesis of Hardware for the Control of Digital Systems, *IEEE Trans. on Computer-Aided Design of Integrated Circuits and Systems*, vol. CAD-1(4), pp. 201–212, Oct. 1982.

[4] L.-F. Sun, J.-M. Liaw, and T.-M. Parng, Automated Synthesis of Microprogrammed Controllers in Digital Systems, *IEE Proceedings∼E Computers and Digital Techniques*, vol. 135(4), pp. 231–240, July 1988.

[5] A.L. Sangiovanni-Vincentelli, An Overview of Synthesis Systems, in: *Proc. IEEE 1985 Custom Integrated Circuits Conference*, pp. 221–225, May 1985.

[6] C.J. Evangelisti, G. Goertzel, and H. Ofek, Using the Dataflow Analyzer on LCD Descriptions of Machines to Generate Control, in: *Proc. 4th International Symposium on Computer Hardware Description Languages*, pp. 109–115, 1979.

[7] G. Rozenberg and A. Salomaa, *The Mathematical Theory of L Systems*, Academic Press, 1980.

[8] M. Mahmood, F. Mavaddat, M.I. Elmasry, and M.H.M. Cheng, A Formal Language Model of Local Microcode Synthesis, in: L.J.M. Claesen (ed.), *Formal VLSI Specification and Synthesis: VLSI Design Methods-I*, pp. 23–41, North-Holland, 1990.

[9] C.R. Clare, *Designing Logic Systems Using State Machines*, McGraw-Hill, 1973.

[10] B.M. Pangrle and D.D. Gajski, Design Tools for Intelligent Silicon Compilation, *IEEE Trans. on Computer-Aided Design of Integrated Circuits and Systems*, vol. CAD-6(6), pp. 1098–1112, Nov. 1987.

[11] N.D. Jones and S. Skyum, Recognition of Deterministic ET0L Languages in Logarithmic Space, *Information and Control*, vol. 35, pp. 177–181, 1977.

[12] M. Mahmood, F. Mavaddat, and M.I. Elmasry, Experiments with an Efficient Heuristic Algorithm for Local Microcode Generation, *IEEE International Conference on Computer Design: VLSI in Computers and Processors*, Sept. 1990.

[13] M. Mahmood, *A Formal Approach to VLSI Control-Unit and Microcode Synthesis*, Ph.D. thesis, University of Waterloo, Canada (in preparation).

[14] J.R. Southard, MacPitts: An Approach to Silicon Compilation, *IEEE Computer*, vol. 16(12), pp. 74–82, Dec. 1983.

[15] P.G. Paulin, J.P. Knight, and E.F. Girczyc, HAL: A Multi-Paradigm Approach to Automatic Data Path Synthesis, in: *Proc. 23rd Design Automation Conference*, pp. 263–270, June 1986.

[16] A.T. Mishchenko, Transformation of Microprograms, *Kibernetika (Cybernetics)*, vol. 3(6), pp. 7–13, 1967.

[17] A.T. Mishchenko, The Formal Synthesis of an Automaton by a Microprogram, Part I, *Kibernetika (Cybernetics)*, vol. 4(3), pp. 24–31, 1968, Part II, *Kibernetika (Cybernetics)*, vol. 4(5), pp. 21–27, 1968.

[18] M. Davio, J.-P. Deschamps, and A. Thayse, *Digital Systems with algorithm implementation*, John Wiley & Sons, 1983.

[19] L.D. Coraor, P.T. Hulina, and O.A. Morean, A General Model for Memory-Based Finite-State Machines, *IEEE Trans. on Computers*, vol. C-36(2), pp. 175–184, Feb. 1987.

Low Level Synthesis I

Logic and Architecture Synthesis
P. Michel and G. Saucier (Editors)
Elsevier Science Publishers B.V. (North-Holland)

Efficient Computation of Exact and Simplified Observability Don't Care Sets for Multiple-Level Combinational Networks

Maurizio Damiani   Giovanni De Micheli

Center for Integrated Systems
Stanford University
Stanford, CA 94305

Abstract

This work presents new algorithms for the extraction of exact and approximate Observability Don't Care sets (ODC sets) in a combinational multiple-level logic network. The proposed algorithms are efficient because they use local information, *i.e.* the computation of the ODC for a vertex in the network requires only the knowledge of the *don't care* sets at the adjacent vertices.

Two approaches are proposed. The former computes the exact ODC sets. The latter computes ODC subsets and can be used when partial information on the *don't care* sets is available. The algorithms for the simplified computation finds the largest subsets of the actual ODC sets, given the information available at the adjacent vertices.

1 Introduction

Over the past few years, the problem of computing efficiently and correctly the Observability Don't Care (ODC) sets has emerged as a central one in the synthesis of combinational networks [1], [2], [3]. The knowledge of the ODC sets is important in several respects, namely: 1) local minimization of functions in a Boolean network, 2) synthesis of 100% testable networks, 3) test pattern generation [4].

Bartlett proposed in [1] a computation of the ODC sets requiring the representation of the primary output expression in terms of the network intermediate variables. Such a representation may be subject to the explosion in size of the representation. Muroga proposed in [3] exhaustive simulations of the circuit for determining the observability *don't care* sets of vertices with reconvergent fanout.

Other authors pointed out that it would be desirable and computationally much more efficient to derive the ODC set of a vertex of a Boolean network from the ODC sets of its direct fanout vertices [5], [7]. They showed, however, that a straightforward application of this idea could lead to erroneous results, because of the effect of reconvergent fanout. Therefore only approximate solutions have been proposed to compute the ODC sets [7], [8].

In this paper we show how an exact computation of ODC sets is indeed possible by using only local information. We propose an algorithm for such a computation that traverses the network backward from the primary outputs to the primary inputs processing each vertex only once.

Unfortunately this method, leading to an exact computation of the ODC sets, involves implicitly Boolean complementations and it is consequently prone to the well-known phenomenon of "combinational explosion". We therefore propose two other algorithms for the

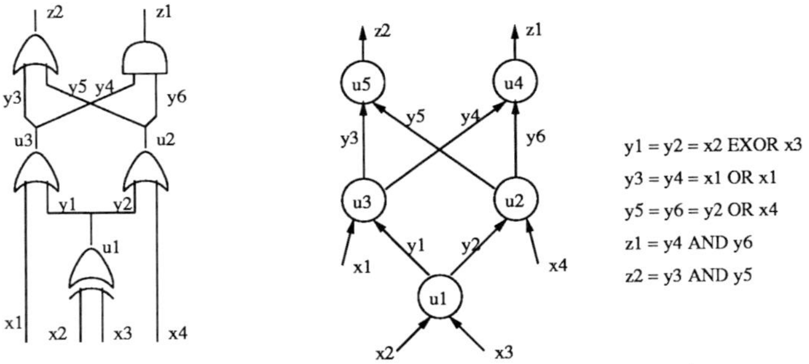

Figure 1: A multiple-level combinational circuit and its associated Boolean network representation.

computation of subsets of the actual ODC sets, still based on local information. The first algorithm computes an ODC subset at each vertex of the Boolean network from the ODC subsets of its direct fanout vertices. The second one computes *both* subsets of the actual *care* and *don't care* sets from those of its fanout. It will be shown that the subsets computed by the second algorithm include those computed by the first one, thus providing a better approximation.

2 Definitions and Notations

2.1 Combinational Boolean Networks.

In this paper we model multiple-output combinational circuits by Boolean networks [1]. A Boolean network N with n input vertices and m output vertices realizes a function $\underline{\mathcal{F}} : B^n \to B^m$ [1], where B is the Boolean set $\{0, 1\}$. Underlining is used for denoting vector quantities throughout this paper.

The network is specified by its acyclic network graph $G = (V, E)$. The elements of the vertex set $V = V^I \cup V^G \cup V^O = \{v\}$ are in one-to-one correspondence with primary inputs, logic gates, and primary outputs, respectively. There is a directed edge e from a vertex μ to a vertex ν if the output of the gate in μ is connected to an input of the gate in ν. In contrast to [1], and similarly to [3], we associate a network variable y_i to each edge $e_i \in E$. An example of a circuit and of its associated Boolean network is shown in Fig. (1).

A network variable y_i is said to be a fanout (fanin) variable of a vertex $\nu \in V$ if e_i is an edge whose tail (head) end-point is ν. We associate to each vertex ν an expression $f_\nu(y_1, y_2, \cdots, y_n)$ of its fanin edge variables. The expression f_ν, describing the functionality of the gate in ν, specifies all the fanout variables of ν in terms of the fanin variables.

2.2 Observability *don't care* sets.

By cutting an edge e_i and by considering y_i as a primary input variable, the new network N^c realizes a function $\underline{\mathcal{F}}^c(\underline{x}, y) : B^{n+1} \to B^m$. Given a primary input assignment $\underline{x}_0$, the variable y_i of N is not observable if the *vector* equality [6]

$$\underline{\mathcal{F}}^c(\underline{x}_0, 0) \;=\; \underline{\mathcal{F}}^c(\underline{x}_0, 1). \tag{1}$$

is satisfied. Recall that the cofactor $f|_{y_i}$ of a function f with respect to y_i is the function obtained by assuming $y_i = 1$. Similarly, $f|_{\overline{y}_i}$ is the function obtained by setting $y_i = 0$. The *vector* function:

$$\underline{ODC}_{y_i} \;=\; \underline{\mathcal{F}}^c|_{\overline{y}_i} \,\overline{\oplus}\, \underline{\mathcal{F}}^c|_{y_i} \stackrel{\text{def}}{=} \overline{\left(\frac{\partial \underline{\mathcal{F}}}{\partial y_i}\right)}. \tag{2}$$

therefore describes the observability of y_i. In particular, y_i will be observable at the kth network output ($k = 1, \cdots, m$) if the kth component of $\underline{ODC}_{y_i}$ is 0. The quantity $\partial \underline{\mathcal{F}} / \partial y_i$ is usually termed ***Boolean difference*** [6] of $\underline{\mathcal{F}}$ with respect to the (possibly internal) variable y_i.

The observability *care* set of y_i ($\underline{OC}_{y_i}$) is defined by the complement of $\underline{ODC}_{y_i}$, *i.e.* by $\overline{\underline{ODC}_{y_i}}$. Its kth component describes the network configurations that make the variable y_i observable at the kth output.

Given a network, it is possible in principle to compute $\underline{ODC}_y$ for any internal variable y by flattening the network N^c and applying Eq. (2). We show here that it is possible to avoid the flattening operation on the network and compute exact and approximate versions of the functions $\underline{ODC}_y$ with a single traversal of the network.

For example, if the *don't care* set of a vertex ν is known, then it is easy to obtain an expression of $\underline{ODC}_y$ for any fanin variable y of ν from [1]:

$$\underline{ODC}_y = \underline{ODC}_\nu + (1, 1, \cdots, 1)^T \overline{\left(\frac{\partial f_\nu}{\partial y} \right)}. \tag{3}$$

The vector $(1, 1, \cdots, 1)^T$ is used to add $\partial f_\nu / \partial y$ to all the components of the vector $\underline{ODC}_y$.

If the network has a tree structure, then it is possible to obtain all the ODC sets by traversing backwards the network and applying iteratively Eq. (3). If a vertex has reconvergent fanout, however, the observability conditions of the vertex do not coincide with those of its fanout variables. We present here how the observability *don't care* of a vertex can be derived from those of its fanout variables.

2.3 Observability *don't care* sets in presence of reconvergent fanout

Consider a vertex ν with reconvergent fanout, and suppose that all the edges $e_1, e_2, \cdots, e_n$, whose tail-end point is ν, are cut. Let $\underline{\mathcal{F}}^c(\underline{x}, y_1, y_2, \cdots, y_n)$ be the function realized by the new network N^c, obtained by adding the variables $y_1, \cdots, y_n$, corresponding to the cut edges, to the primary inputs. Then, the observability ***don't care*** set of ν is described by the function

$$\underline{ODC}_\nu = \underline{\mathcal{F}}^c(\underline{x}, 0, \cdots, 0) \overline{\oplus} \underline{\mathcal{F}}^c(\underline{x}, 1, \cdots, 1) \tag{4}$$

or, equivalently, by

$$\underline{ODC}_\nu = \underline{\mathcal{F}}^c|_{\overline{y}_1, \overline{y}_2, \cdots, \overline{y}_n} \overline{\oplus} \underline{\mathcal{F}}^c|_{y_1, y_2, \cdots, y_n}. \tag{5}$$

For the sake of simplicity, we describe first the case in which $n = 2$, so that there are only two fanout variables, y_1 and y_2. We will generalize the result down below.

The observability *don't care* set is described by the function

$$\underline{ODC}_\nu = \underline{\mathcal{F}}^c|_{\overline{y}_1, \overline{y}_2} \overline{\oplus} \underline{\mathcal{F}}^c|_{y_1, y_2}. \tag{6}$$

By manipulating Eq. (6), $\underline{ODC}_\nu$ can be rewritten as

$$\underline{ODC}_\nu = \left(\underline{\mathcal{F}}^c|_{\overline{y}_1, \overline{y}_2} \overline{\oplus} \underline{\mathcal{F}}^c|_{y_1, \overline{y}_2} \right) \overline{\oplus} \left(\underline{\mathcal{F}}^c|_{y_1, \overline{y}_2} \overline{\oplus} \underline{\mathcal{F}}^c|_{y_1, y_2} \right) \tag{7}$$

where the term $\underline{\mathcal{F}}^c|_{y_1, \overline{y}_2}$ has been "added and subtracted" in Eq. (6).

From Eq.(2), the first term within parentheses is $\underline{ODC}_{y_1}|_{\overline{y}_2}$, while the second parentheses describe $\underline{ODC}_{y_2}|_{y_1}$. It then follows that

$$\underline{ODC}_\nu = \underline{ODC}_{y_1}|_{\overline{y}_2} \overline{\oplus} \underline{ODC}_{y_2}|_{y_1}. \tag{8}$$

[1]Notice that Eq. (3) may contain internal network variables. These, however, may be resolved by back-substitution to obtain an expression of primary input variables only, so that there is no real contrast with the definition (2).

Notice that in Eq. (5), $\underline{ODC}_{\nu}$ can be also rewritten as

$$\underline{ODC}_{\nu} = \left(\underline{\mathcal{F}}^c|_{\overline{y}_1,\overline{y}_2} \overline{\oplus} \underline{\mathcal{F}}^c|_{\overline{y}_1,y_2}\right) \overline{\oplus} \left(\underline{\mathcal{F}}^c|_{\overline{y}_1,y_2} \overline{\oplus} \underline{\mathcal{F}}^c|_{y_1,y_2}\right) = \underline{ODC}_{y_2}|_{\overline{y}_1} \overline{\oplus} \underline{ODC}_{y_1}|_{y_2} \tag{9}$$

from which we obtain the identity

$$\underline{ODC}_{\nu} = \underline{ODC}_{y_1}|_{\overline{y}_2} \overline{\oplus} \underline{ODC}_{y_2}|_{y_1} = \underline{ODC}_{y_1}|_{y_2} \overline{\oplus} \underline{ODC}_{y_2}|_{\overline{y}_1}. \tag{10a}$$

Similarly, for the *care* set we obtain

$$\underline{OC}_{\nu} = \underline{OC}_{y_1}|_{\overline{y}_2} \oplus \underline{OC}_{y_2}|_{y_1} = \underline{OC}_{y_1}|_{y_2} \oplus \underline{OC}_{y_2}|_{\overline{y}_1}. \tag{10b}$$

These identities will be used in Sect. (3), when deriving approximations to $\underline{ODC}_{\nu}$.

The algebraic manipulation above can be extended to the general case of n fanout variables as follows. It can be easily verified that for $n \geq 2$ the following identity can be derived from Eq. (5):

$$\underline{ODC}_{y} = \left(\underline{\mathcal{F}}^c|_{\overline{y}_1,\overline{y}_2,\cdots,\overline{y}_n} \overline{\oplus} \underline{\mathcal{F}}^c|_{y_1,\overline{y}_2,\cdots,\overline{y}_n}\right) \overline{\oplus}$$

$$\left(\underline{\mathcal{F}}^c|_{y_1,\overline{y}_2,\cdots,\overline{y}_n} \overline{\oplus} \underline{\mathcal{F}}^c|_{y_1,y_2,\cdots,\overline{y}_n}\right) \overline{\oplus} \cdots \overline{\oplus} \left(\underline{\mathcal{F}}^c|_{y_1,y_2,\cdots,y_{n-1},\overline{y}_n} \overline{\oplus} \underline{\mathcal{F}}^c|_{y_1,y_2,\cdots,y_{n-1},y_n}\right) \tag{11}$$

This can be rewritten as:

$$\underline{ODC}_{\nu} = \overline{\bigoplus}_{i=1}^{n} \underline{\mathcal{F}}^c|_{y_1,\cdots,y_{i-1},\overline{y}_i,\cdots,\overline{y}_n} \overline{\oplus} \underline{\mathcal{F}}^c|_{y_1,\cdots,y_i,\overline{y}_{i+1},\cdots,\overline{y}_n} \tag{12}$$

From Eq. (2), each term of the sum in Eq.(11) is recognized to be $\underline{ODC}_{y_i}|_{y_1,\cdots,y_{i-1},\overline{y}_{i+1},\cdots,\overline{y}_n}$.

Eq. (12) can be thus rewritten as

$$\underline{ODC}_{\nu} = \overline{\bigoplus}_{i=1}^{n} \underline{ODC}_{y_i}|_{y_1,\cdots,y_{i-1},\overline{y}_{i+1},\cdots,\overline{y}_n}. \tag{13}$$

For the *care* set, we have:

$$\underline{OC}_{\nu} = \bigoplus_{i=1}^{n} \underline{OC}_{y_i}|_{y_1,\cdots,y_{i-1},\overline{y}_{i+1},\cdots,\overline{y}_n}. \tag{14}$$

Similarly to the case of 2 fanout variables, changing the order in which the fanout variables are complemented results in different expressions of $\underline{OC}_{\nu}$. For n fanout variables, there are $n!$ possible orderings, hence $n!$ different possible expressions for $\underline{ODC}_{\nu}$. In particular, let $(i_1, i_2, \cdots, i_n)$ denote a permutation of $(1, 2, \cdots, n)$. Then, the following identities hold:

$$\underline{ODC}_{\nu} = \overline{\bigoplus}_{j=1}^{n} \underline{ODC}_{y_{i_j}}|_{y_{i_1},\cdots,y_{i_{j-1}},\overline{y}_{i_{j+1}},\cdots,\overline{y}_{i_n}}. \tag{15}$$

$$\underline{OC}_{\nu} = \bigoplus_{j=1}^{n} \underline{OC}_{y_{i_j}}|_{y_{i_1},\cdots,y_{i_{j-1}},\overline{y}_{i_{j+1}},\cdots,\overline{y}_{i_n}}. \tag{16}$$

2.4 An exact algorithm for the ODC sets computation

Given the ODC set of a vertex ν, it is possible to compute the ODC set of all its fanin edges by means of Eq. (3). In turn, Eq. (13) allows us to compute a vertex ODC set, given those of its fanout variables. It is thus now possible to visit the Boolean network backwards from the primary outputs to its inputs and to determine the ODC sets of each vertex also in presence of reconvergent fanout.

The following algorithm performs the computation of the ODC sets. It uses the subset S of the vertices whose ODC set is known. Initially S is the set of primary output vertices with empty fanout set.

OBSERVABILITY(G);
$S :=$ {*primary output vertices with empty fanout set*};
while ($S \neq V$) {
select $\nu \in \{V - S\}$ such that $FO(\nu) \subseteq S$;
foreach fanout variable y_i of ν {
$\mu = head\ vertex\ of\ edge\ e_i$
$\underline{ODC}_{y_i} = \underline{ODC}_{\nu} + \overline{(\partial f_\mu / \partial y_i)}$ /* compute ODC_y by Eq. (3) */
}
$\underline{ODC}_{\nu} = \overline{\oplus_{i=1}}^{n}\ \underline{ODC}_{y_i}|_{y_1,\cdots y_{i-1},\overline{y}_{i+1}\cdots\overline{y}_n}$ /* compute $\underline{ODC}_{\nu}$ by Eq. (13) */
$S := S \cup \{\nu\}$;
}

We illustrate here the operation of the algorithm on the circuit shown in Fig. (1). At the beginning, $S = \{u_4, u_5\}$, and

$$\underline{ODC}_{u_4} = \begin{pmatrix} 0 \\ 1 \end{pmatrix} ; \underline{ODC}_{u_5} = \begin{pmatrix} 1 \\ 0 \end{pmatrix}$$

The first and second component of the vectors $\underline{ODC}$ describe the observability with respect to z_1 and z_2, respectively.

First the vertices u_2, u_3 are considered. By applying Eqs. (3) and (13)

$$\underline{ODC}_{u_3} = \underline{ODC}_{y_3}|_{\overline{y}_4} \overline{\oplus} \underline{ODC}_{y_4}|_{y_3} = \begin{pmatrix} 1 \\ y_5 \end{pmatrix} \overline{\oplus} \begin{pmatrix} \overline{y}_6 \\ 1 \end{pmatrix} = \begin{pmatrix} \overline{y}_6 \\ y_5 \end{pmatrix}$$

$$\underline{ODC}_{u_2} = \underline{ODC}_{y_6}|_{\overline{y}_5} \overline{\oplus} \underline{ODC}_{y_5}|_{y_6} = \begin{pmatrix} \overline{y}_4 \\ 1 \end{pmatrix} \overline{\oplus} \begin{pmatrix} 1 \\ y_3 \end{pmatrix} = \begin{pmatrix} \overline{y}_4 \\ y_3 \end{pmatrix}$$

Then $S = \{u_2, u_3, u_4, u_5\}$ and u_1 is selected. Its fanout variables are y_1, y_2, and according to Eq. (3) :

$$\underline{ODC}_{y_1} = \begin{pmatrix} \overline{y}_6 + x_1 \\ y_5 + x_1 \end{pmatrix} = \begin{pmatrix} \overline{y}_2 \overline{x}_4 + x_1 \\ y_2 + x_4 + x_1 \end{pmatrix}$$

$$\underline{ODC}_{y_2} = \begin{pmatrix} \overline{y}_4 + x_4 \\ y_3 + x_4 \end{pmatrix} = \begin{pmatrix} \overline{y}_1 \overline{x}_1 + x_4 \\ y_1 + x_1 + x_4 \end{pmatrix}$$

so that, using Eq. (13),

$$\underline{ODC}_{u_1} = \underline{ODC}_{y_1}|_{\overline{y}_2} \overline{\oplus} \underline{ODC}_{y_2}|_{y_1} = \begin{pmatrix} \overline{x}_4 + x_1 \\ x_4 + x_1 \end{pmatrix} \overline{\oplus} \begin{pmatrix} x_4 \\ 1 \end{pmatrix} = \begin{pmatrix} x_1 x_4 \\ x_1 + x_4 \end{pmatrix}$$

which is the correct result.

Finally the algorithm computes the ODC sets of the primary inputs, that may be used as external ODC sets for the minimization of the logic feeding the circuit.

The product of all the components of $\underline{ODC}_{\nu}$ gives the conditions for which the gate output at vertex ν is not observable at *any* output. For example, in the case of the gate in u_1, the product of the components of $\underline{ODC}_{u_1}$ yields $x_1 x_4$.

3 Computation of approximate ODC sets

It is of practical interest to consider the case in which the ODC sets are approximated by subsets, because of their size. Note that excess approximations of the ODC sets are of no practical value for Boolean minimization. Therefore it is useful to derive ODC subsets at a vertex from the ODC sets of its fanout vertices. Unfortunately, Eq. (7) may not yield an ODC subset from subsets of the actual ODC sets of each edge variable.

For example, in the circuit of Fig. (1), if we assume

$$\underline{ODC}_{y_1} = \begin{pmatrix} x_1 \\ x_1 \end{pmatrix}$$

(actually a subset of the true $\underline{ODC}_{y_1}$), we find for $\underline{ODC}_{u_1}$ the estimate:

$$\underline{ODC}_{u_1} = \begin{pmatrix} x_1 x_4 + \overline{x}_1 \overline{x}_4 \\ x_1 \end{pmatrix}$$

which is no longer a subset of the true $\underline{ODC}_{u_1}$.

Two different approaches for the computation of ODC subsets are considered here. In particular we show two formulae that can replace Eq. (13) in the previous algorithm. The first formula computes *both* subsets of the actual *care* and *don't care* sets of a vertex from those of its fanout variables. The second one, simpler but less accurate, computes ODC subsets only.

We give first formulae for computing ODC_ν in the case of two fanout variables. We then present their generalization to larger fanout.

3.1 Propagation of *care* and *don't care* subsets

Let y_1 and y_2 denote the variables corresponding to the edges in the fanout of a vertex ν. Assume that only subsets $\underline{\widetilde{ODC}}_{y_i}$, $\underline{\widetilde{OC}}_{y_i}$ of $\underline{ODC}_{y_i}, \underline{OC}_{y_i}$ are available. Then, $\underline{\widetilde{ODC}}_\nu$ and $\underline{\widetilde{OC}}_\nu$ computed by:

$$\underline{\widetilde{ODC}}_\nu = \underline{\widetilde{ODC}}_{y_1}|_{\overline{y}_2} \underline{\widetilde{ODC}}_{y_2}|_{y_1} + \underline{\widetilde{ODC}}_{y_1}|_{y_2} \underline{\widetilde{ODC}}_{y_2}|_{\overline{y}_1} +$$

$$+ \underline{\widetilde{OC}}_{y_1}|_{\overline{y}_2} \underline{\widetilde{OC}}_{y_2}|_{y_1} + \underline{\widetilde{OC}}_{y_1}|_{y_2} \underline{\widetilde{OC}}_{y_2}|_{\overline{y}_1} \tag{17a}$$

and by

$$\underline{\widetilde{OC}}_\nu = \underline{\widetilde{OC}}_{y_1}|_{\overline{y}_2} \underline{\widetilde{ODC}}_{y_2}|_{y_1} + \underline{\widetilde{OC}}_{y_1}|_{y_2} \underline{\widetilde{ODC}}_{y_2}|_{\overline{y}_1} +$$

$$+ \underline{\widetilde{ODC}}_{y_1}|_{\overline{y}_2} \underline{\widetilde{OC}}_{y_2}|_{y_1} + \underline{\widetilde{ODC}}_{y_1}|_{y_2} \underline{\widetilde{OC}}_{y_1}|_{\overline{y}_1} \tag{17b}$$

are subsets of the true $\underline{ODC}_\nu$ and $\underline{OC}_\nu$, respectively. To verify Eq. (17), it is sufficient to observe that, by expanding Eq. (10a) in sum of products form,

$$\underline{ODC}_\nu = \underline{ODC}_{y_1}|_{\overline{y}_2} \underline{ODC}_{y_2}|_{y_1} + \underline{OC}_{y_1}|_{\overline{y}_2} \underline{OC}_{y_2}|_{y_1} + \underline{ODC}_{y_2}|_{\overline{y}_1} \underline{ODC}_{y_1}|_{y_2} + \underline{OC}_{y_2}|_{\overline{y}_1} \underline{OC}_{y_1}|_{y_2} \tag{18a}$$

Similarly, it can be verified from Eq. (10b) that

$$\underline{OC}_\nu = \underline{OC}_{y_1}|_{\overline{y}_2} \underline{ODC}_{y_2}|_{y_1} + \underline{ODC}_{y_1}|_{\overline{y}_2} \underline{OC}_{y_2}|_{y_1} + \underline{OC}_{y_2}|_{\overline{y}_1} \underline{ODC}_{y_1}|_{y_2} + \underline{ODC}_{y_2}|_{\overline{y}_1} \underline{OC}_{y_1}|_{y_2}. \tag{18b}$$

Since we assumed $\underline{\widetilde{ODC}}_{y_i} \subseteq \underline{ODC}_{y_i}$; $\underline{\widetilde{OC}}_{y_i} \subseteq \underline{OC}_{y_i}$, $i = 1, 2$, the relations $\underline{\widetilde{ODC}}_\nu \subseteq \underline{ODC}_\nu$ and $\underline{\widetilde{OC}}_\nu \subseteq \underline{OC}_\nu$ are immediately verified.

3.2 Computation of *don't care* subsets only

If only subset of *don't care* conditions $\underline{\widetilde{ODC}}_{y_i}$ are available, then, $\underline{\widetilde{ODC}}_\nu$ can be computed from:

$$\underline{\widetilde{ODC}}_\nu = \underline{\widetilde{ODC}}_{y_1}|_{\overline{y}_2} \underline{\widetilde{ODC}}_{y_2}|_{y_1} + \underline{\widetilde{ODC}}_{y_1}|_{y_2} \underline{\widetilde{ODC}}_{y_2}|_{\overline{y}_1} \tag{19}$$

To verify Eq. (19) it is sufficient to observe that its right member matches the first elements in the sum (17a). Therefore $\underline{\widetilde{ODC}}_\nu$, as computed by Eq. (19), is also contained in $\underline{ODC}_\nu$. Note that the ODC sets computed by Eq. (17a) always include those derived from Eq. (19).

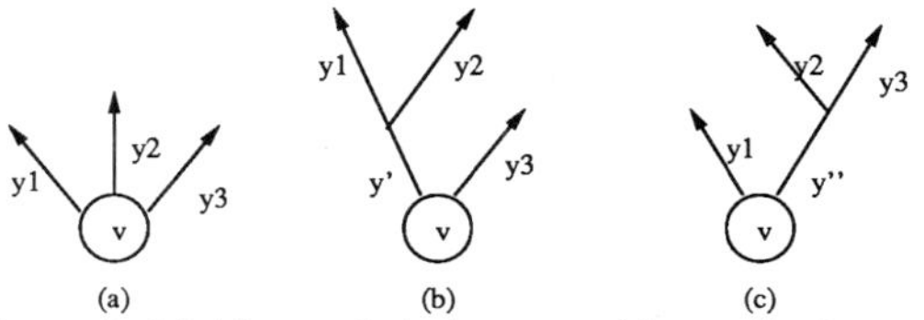

Figure 2: Multi-way fork decompositions of a fanout tree.

Therefore, the propagation of both *care* and *don't care* sets provide a better approximation to the exact ODC sets.

The ODC subsets computed by Eq. (17a), (17b) and (19) have a local maximality property, in the sense that no other cube of the cofactors $\underline{\widetilde{ODC}}_{y_i}|_{y_j}$, $\underline{\widetilde{OC}}_{y_i}|_{y_j}$ appearing in those equations can be proved to be always contained in $\underline{ODC}_\nu$.

3.3 Generalization to larger fanout

A vertex ν with fanout $n > 2$ can always be replaced by a multi-way fork of interconnections, as shown in Fig. (2). Eqs. (17) (or, more simply, Eq. (19)) could be applied iteratively to this fork. In this way a subset of ODC_ν can be obtained in an $O(n)$ number of applications of Eqs. (17) or (19).

The result, however, depends on the selected tree. Consider, for example, the case of three fanout variables, shown in Fig. (2), and suppose, for simplicity, to apply Eq. (19) only. If the fanout tree of ν is decomposed as in Fig. (2b), the resulting expression of $\underline{\widetilde{ODC}}_\nu$ is

$$\underline{\widetilde{ODC}}_\nu = \underline{\widetilde{ODC}}_{y_1}|_{\bar{y}_2\bar{y}_3}\underline{\widetilde{ODC}}_{y_2}|_{y_1\bar{y}_3}\underline{\widetilde{ODC}}_{y_3}|_{y_1y_2} + \underline{\widetilde{ODC}}_{y_2}|_{\bar{y}_1\bar{y}_3}\underline{\widetilde{ODC}}_{y_1}|_{y_2\bar{y}_3}\underline{\widetilde{ODC}}_{y_3}|_{y_1y_2} +$$

$$\underline{\widetilde{ODC}}_{y_1}|_{\bar{y}_2y_3}\underline{\widetilde{ODC}}_{y_2}|_{y_1y_3}\underline{\widetilde{ODC}}_{y_3}|_{\bar{y}_1\bar{y}_2} + \underline{\widetilde{ODC}}_{y_2}|_{\bar{y}_1\bar{y}_3}\underline{\widetilde{ODC}}_{y_1}|_{y_2\bar{y}_3}\underline{\widetilde{ODC}}_{y_3}|_{\bar{y}_1\bar{y}_2} \tag{20}$$

If it is decomposed as in Fig. (2c), we instead obtain the following expression, different from the previous one:

$$\underline{\widetilde{ODC}}_\nu = \underline{\widetilde{ODC}}_{y_2}|_{\bar{y}_3\bar{y}_1}\underline{\widetilde{ODC}}_{y_3}|_{y_2\bar{y}_1}\underline{\widetilde{ODC}}_{y_1}|_{y_2y_3} + \underline{\widetilde{ODC}}_{y_3}|_{\bar{y}_2\bar{y}_1}\underline{\widetilde{ODC}}_{y_2}|_{y_3\bar{y}_1}\underline{\widetilde{ODC}}_{y_1}|_{y_2y_3} +$$

$$\underline{\widetilde{ODC}}_{y_2}|_{\bar{y}_3y_1}\underline{\widetilde{ODC}}_{y_3}|_{y_2y_1}\underline{\widetilde{ODC}}_{y_1}|_{\bar{y}_2\bar{y}_3} + \underline{\widetilde{ODC}}_{y_3}|_{\bar{y}_2\bar{y}_1}\underline{\widetilde{ODC}}_{y_2}|_{y_3\bar{y}_1}\underline{\widetilde{ODC}}_{y_1}|_{\bar{y}_2\bar{y}_3} \tag{21}$$

The largest expression that can be obtained, starting from the variable *don't care* subsets, is

$$\underline{\widetilde{ODC}}_\nu = \underline{\widetilde{ODC}}_{y_1}|_{\bar{y}_2\bar{y}_3}\underline{\widetilde{ODC}}_{y_2}|_{y_1\bar{y}_3}\underline{\widetilde{ODC}}_{y_3}|_{y_1y_2} + \underline{\widetilde{ODC}}_{y_1}|_{\bar{y}_3\bar{y}_2}\underline{\widetilde{ODC}}_{y_3}|_{y_1\bar{y}_2}\underline{\widetilde{ODC}}_{y_2}|_{y_1y_3} +$$

$$\underline{\widetilde{ODC}}_{y_2}|_{\bar{y}_1\bar{y}_3}\underline{\widetilde{ODC}}_{y_1}|_{y_2\bar{y}_3}\underline{\widetilde{ODC}}_{y_3}|_{y_2y_1} + \underline{\widetilde{ODC}}_{y_2}|_{\bar{y}_3\bar{y}_1}\underline{\widetilde{ODC}}_{y_3}|_{y_2\bar{y}_1}\underline{\widetilde{ODC}}_{y_1}|_{y_2y_3} +$$

$$\underline{\widetilde{ODC}}_{y_3}|_{\bar{y}_1\bar{y}_2}\underline{\widetilde{ODC}}_{y_1}|_{y_3\bar{y}_2}\underline{\widetilde{ODC}}_{y_2}|_{y_3y_1} + \underline{\widetilde{ODC}}_{y_3}|_{\bar{y}_2\bar{y}_1}\underline{\widetilde{ODC}}_{y_2}|_{y_3\bar{y}_1}\underline{\widetilde{ODC}}_{y_1}|_{y_3y_2} \tag{22}$$

larger than (or equal to) both estimates (20) and (21).

In the general case, by expanding Eq. (13) in sum-of-products form, it is immediate to verify that the product

$$\underline{\widetilde{ODC}}_{y_1}|_{\bar{y}_2,\cdots,\bar{y}_n}\underline{\widetilde{ODC}}_{y_2}|_{y_1,\bar{y}_3,\cdots,\bar{y}_{n-1},\bar{y}_n}\cdots\underline{\widetilde{ODC}}_{y_n}|_{y_1,y_2,\cdots,y_{n-1}} \tag{23}$$

is certainly contained in $\underline{ODC}_\nu$. Let $(i_1, i_2, \cdots, i_n)$ denote a permutation of $(1, 2, \cdots, n)$. From the expression (15) of the exact $\underline{ODC}_\nu$ associated to that permutation it follows that

$$\underline{\widetilde{ODC}}_{y_{i_1}}|_{\overline{y}_{i_2},\cdots,\overline{y}_{i_n}} \underline{\widetilde{ODC}}_{y_{i_2}}|_{y_{i_1},\overline{y}_{i_3},\cdots,\overline{y}_{i_n}} \cdots \underline{\widetilde{ODC}}_{y_{i_n}}|_{y_{i_1},y_{i_2},\cdots,y_{i_{n-1}}} \tag{24}$$

is also always contained in $\underline{ODC}_\nu$. Let π_n denote the set of all possible permutations $\underline{i} = (i_1, i_2, \cdots, i_n)$ of $(1, 2, \cdots, n)$. The sum

$$\sum_{\underline{i}\in\pi_n} \underline{\widetilde{ODC}}_{y_{i_1}}|_{\overline{y}_{i_2},\cdots,\overline{y}_{i_n}} \underline{\widetilde{ODC}}_{y_{i_2}}|_{y_{i_1},\overline{y}_{i_3},\cdots,\overline{y}_{i_n}} \cdots \underline{\widetilde{ODC}}_{y_{i_n}}|_{y_{i_1},y_{i_2},\cdots,y_{i_{n-1}}} \tag{25}$$

is therefore a subset of $\underline{ODC}_\nu$.

Similarly to the case of two fanout variables, it can be shown that no other product of the variables appearing in Eq. (25) can be added, while maintaining the first member of Eq. (25) an ODC subset. From the expansion of Eq. (13), all candidate products involve in fact the complement of at least one fanout variable ODC subset, *i.e.* an OC *superset*. If such a product is added, it is no longer possible to guarantee $\underline{\widetilde{ODC}}_\nu \subseteq \underline{ODC}_\nu$. Eq. (25) therefore represents the largest possible estimate.

Unfortunately, its computation requires a factorial amount of time, and is therefore not applicable to large fanouts. Note, however, that for most circuits the number of fanout stems for a node is typically small.

When subsets of the true $\underline{OC}_{y_i}$ sets are also available, then the largest possible subsets of $\underline{ODC}_\nu, \underline{OC}_\nu$ can be computed as follows. A sum-of products expression of $\underline{ODC}_\nu$, $\underline{OC}_\nu$ can be obtained by expanding the expressions (13) and (14). By substituting each $\overline{\underline{ODC}_{y_i}}$ and $\overline{\underline{OC}_{y_i}}$ with $\underline{\widetilde{OC}}_{y_i}$ and $\underline{\widetilde{ODC}}_{y_i}$, respectively, an expression of ODC and OC subsets is obtained. Different subsets are obtained by repeating the same procedure using a permutation of the variables $y_1, \cdots, y_n$. The largest possible subset is the sum of these subsets.

4 Relation to previous works.

Methods for computing observability *don't care* sets have been proposed in [8], [3] and [7]. None of these methods maintains separate components for the observability at different outputs, and therefore must either make some form of approximations in case of reconvergent fanout, or abandon the network traversal method.

The most "conservative" approach is taken in the program MIS [8]. There, the observability *don't care* of a gate in a vertex ν is computed by assuming the complete observability of its fanout gates. The observability $ODC_{y_i}^{MIS}$ of each fanout variable y_i of ν is thus obtained from Eq. (3) by neglecting the first term of the sum. MIS computes the observability *don't care* set of ν as

$$ODC_\nu^{MIS} = \prod_{y_i} \overline{\frac{\partial f_{\mu_i}}{\partial y_i}} \tag{26}$$

In the above equation, μ_i is the head vertex of the edge e_i. The technique presented in [3] considers NOR networks only. It consists in particular of applying Eq. (3) to gates with a single fanout stem. If z is the output of a NOR gate with input variables $y_1, y_2, \cdots, y_n$,

$$ODC_{y_i} = ODC_z + \frac{\partial z}{\partial y_i} = ODC_z + \sum_{j=1, j\neq i}^{n} y_j \tag{27}$$

which corresponds (with a different formalism) to Eq. (3.1) in [3]. The cases of not-reconvergent and reconvergent fanout are treated separately. In the case of not reconvergent fanout, the observability of the gate output ν is computed (cfr. Eq. (3.4) in [3]) from those of the fanout variables y_i as

$$ODC_\nu = \prod_{i=1}^{n} ODC_{y_i} \tag{28}$$

In the case of reconvergent fanout, since the different components of the observability are not kept separate, the network traversal paradigm must be abandoned. Muroga proposes an approach based upon *two exahustive simulations* of the circuit, with the reconvergent gate output stuck-at 0 and stuck-at 1, respectively. The observability *don't care* set for the gate is then computed as the set of primary input configurations that produce the same output in the two simulations. In other words, Eq. (1) is solved by "brute force".

In [7] ODC subsets are computed as follows. Variables are associated to gate outputs rather than edges. Eq. (3) is used to determine the observability of fanin variables for each vertex. In presence of reconvergent fanout lines, the observability *don't care* sets associated to each variable are considered as estimates of the true ODC of the vertex. Note that, because of the different definition of variables, in presence of reconvergent fanout the observability of a vertex seems to depend on itself. Consider, for example, the circuit of Fig. 1, and assume that the gate in u_4 does not exist. The method finds two estimates of the observability of u_1 as

$$ODC'_{u_1} = u_1 + x_1; \quad ODC''_{u_1} = u_1 + x_4. \tag{29}$$

This apparent dependency is eliminated by invoking the RESTRICT operation on the expressions ODC'_{u_1}, ODC''_{u_1}. which drops all the cubes containing u_1 or $\overline{u}_1$. The RESTRICT operation on an expression $expr$ with respect to a variable var can be defined as follows:

$$RESTRICT(expr, var) \stackrel{\text{def}}{=} expr|_{var}\, expr|_{\overline{var}} \tag{30}$$

In the example of Fig. 1,

$$RESTRICT(ODC'_{u_1}, u_1) = ODC'_{u_1}|_{u_1} ODC'_{u_1}|_{\overline{u_1}} = x_1;$$

$$RESTRICT(ODC''_{u_1}, u_1) = ODC''_{u_1}|_{u_1} ODC''_{u_1}|_{\overline{u_1}} = x_4. \tag{31}$$

RESTRICT is therefore similar to the cofactor operation of Eqs. (8) and (17). Eq. (17) shows, however, that only part of the cubes (containing either u_1 or $\overline{u}_1$) actually needs to be dropped. The estimates obtained after the application of RESTRICT are finally ANDed. In the above example, $ODC_{u_1} = x_1 x_4$. It has been proven in [7] that the final result is a subset of the true ODC set. The exact result is, in the above example, $ODC_{u_1} = x_1 + x_4 \supseteq x_1 x_4$.

4.1 Implementation and Experimental Results

The algorithm OBSERVABILITY with exact and simplified formulae has been implemen ted in program SPY. SPY is written in C and consists of about 3000 lines of code. We have successfully tested SPY on several benchmark circuits. We report here on ly on some examples, including critical ones proposed by Brglez and Fujiwara. Table I shows the experimental results obtained by running SPY on some of these circuits. The number of literals refers to the average size of unminimized expressions of the ODC sets, as generated by the algorithm. Final literal counts are typically smaller, but we report here on the size of the uncompacted ODC sets because they are the direct implementation of the formulae shown in the paper. The run times are in s econds, over a DEC 3100 workstation.

5 Summary

In this paper we have presented a novel algorithm for the extraction of exact and approximate Observability Don't Care sets (ODC sets) in a combinational multilevel logic network. The proposed algorithms are efficient because they use local information, i.e. the computation of the ODC for a vertex requires only the knowledge of the *don't cares* at the vertex immediate fanout. Two approaches have been proposed. The former computes the exact ODC sets. The latter computes ODC subsets and can be used when partial information on the *don't cares* is available.

Circuit	Exact formula (13)		Simplified formula (25)	
	CPU time	# literals	CPU time	# literals
C17	0	2	0	2
C432	105	1990	68	2017
C499	12	75	6	88
C880	11	70	6	60
C1908	*	*	175	273
C6288	*	*	4295	689
apex6	40	38	28	43
apex7	26	9	18	11
CM138	0	2	0	2

Table 1: CPU time and average number of literals for the ODC sets of some benchm ark circuits

6 Acknowledgement

This research was supported in part by a fellowship of the Rotary Foundation and by AT & T and DEC, jointly with NSF, under a PYI award program. We also acknowledge support from NSF under contract # MIP 8719546.

The helpful discussions with Frederic Mailhot and Michiel Ligthart are here gratefully acknowledged.

References

[1] K. A. Bartlett, R. K. Brayton, G. D. Hachtel, R. M. Jacoby, R. Rudell, A. Sangiovanni-Vincentelli, and A. Wang, " Multilevel Logic Minimization Using Implicit Don't Cares", *IEEE Transactions on CAD/ICAS*, vol. CAD-7, No. 6, pp. 723-739, June 1988.

[2] G. D. Hachtel and M. R. Lightner, " Top-Down Synthesis of Multilevel Logic Networks" *Proc. ICCAD 1987*, pp. 316-319, S. Clara, Nov. 1987.

[3] S. Muroga, Y.Kambayashi, H.Lai and J.Culliney, "The Transduction Method - Design of Logic Networks Based on Permissible Functions ", *IEEE Trans. Comp.*, vol. 38, No. 10, pp. 1404-1424, 1989.

[4] D. Bostick, G. D. Hachtel, R. M. Jacoby, M. R. Lightner, P. Moceyunas, C. R. Morrison, and D. Ravenscroft, " The Boulder Optimal Logic Design System", *Proc. ICCAD 1987*, pp. 62-65, S. Clara, Nov. 1987.

[5] R. K. Brayton and F. Somenzi, "Boolean Relations and the Incomplete Specification of Logic Networks", *Proc. VLSI '89*, pp. 231-240, Munich, August 1989.

[6] H. Fujiwara, *Logic Design and Design for Testability*, MIT Press, Cambridge, 1985.

[7] G. D. Hachtel, R. M. Jacoby, P. H. Moceyunas, " On Computing and Approximating the Observability Don't Care Set ", *Proceedings on the International Workshop on Logic Synthesis*, Research Triangle Park, May 1989.

[8] R. Brayton, R. Rudell, A. Sangiovanni-Vincentelli, A. Wang , "MIS: A Multiple-Level Logic Optimization System", *IEEE Transactions on CAD/ICAS*, Vol. CAD-6, No. 6, pp. 1062-1081, November 1987.

Logic and Architecture Synthesis
P. Michel and G. Saucier (Editors)
Elsevier Science Publishers B.V. (North-Holland)
IFIP, 1991

Lexicographical Factorization Minimizing the Critical Path and the Routing Factor of Multilevel Logic

P. Abouzeid, G. Saucier, F. Poirot

Institut National Polytechnique de Grenoble/CSI
46, Avenue Félix VIALLET
38031 Grenoble Cedex FRANCE

A method is proposed to decrease the routing factor and critical path of standard cell logic in a multilevel synthesis environment. It is mainly based on lexicographical boolean expressions allowing to simplify input dependency. A first set of results shows that the goal is achieved.

INTRODUCTION

Multilevel synthesis of a set of boolean functions consists of a factorization step followed by a technology mapping phase. For the factorization step, the optimization objective is usually the minimization of the number of literals of the factorized expressions. Algebraic division and more sophisticated Boolean division techniques are commonly used [2, 3]. The number of literal occurences is drastically reduced. Once the second step which is the technology mapping is performed, the global optimization result can be analyzed in terms of area and performances (critical path). Two experimental results can be easily outpointed. First, if the factorization is really pushed forward, thus reaching really a quasi-minimal number of literals, an important increase of the channel congestion appears as well as a poor critical path. The channel area may be 6 or 7 times the area of the gates for large logic blocks. Rather than improving the result afterwards by modification rules, we propose here to control immediately the factorization step by reducing in an adequate manner the factorizations performed. The wiring consists of input connections and of intercell connections. We suggest to control input dependency by creating structured logic cones [1] and intercell connection by controlling severely the fanout of common logic blocks.
A first section recalls briefly the basic idea. A second section defines how the input ordering is chosen. A third section gives how further minimization and factorization compatible with this input ordering are performed and finally shows the first results.

1. BASIC IDEA

There exists a huge number of ways to factorize a boolean expression. If we just restrict ourselves to factorizations based on algebraic division, the number of kernels of a boolean function is enormous. Therefore, if we want to analyze kernel and cokernel intersections as well as product term intersections between functions more or less in an exhaustive approach, the problem will not be tractable for complex circuits. Therefore rather than to restrict the explored solution space in an arbitrary way, ASYL [4] restricts the target architecture achieving simultaneously 2 goals: the algebraic problem is more tractable, the logic network will be more wireable exhibiting a reduced routing factor. The target logical architecture aims at constructing the different functions as logic cones in which the inputs enter in an ordered way (the same for each function) while controlling the intersection between these cones (figure 1). The restriction of input ordering ensures an optimized input wiring, the cone intersection control ensures a controlled intercell wiring especially in term of fanout control.

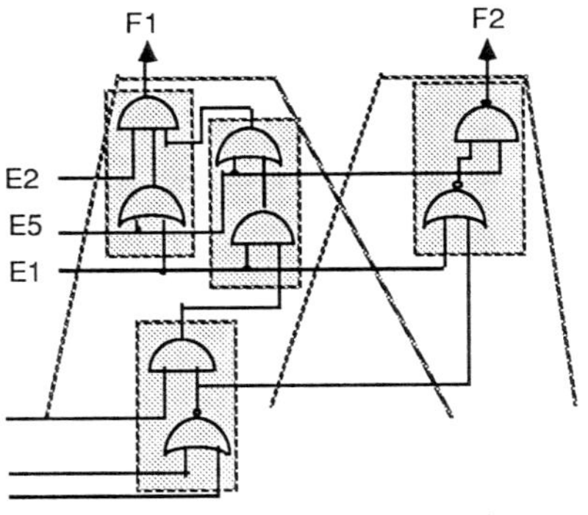

Figure 1: virtual logic cones with the input ordering E4, E6, E3, E1...

During the technology mapping [5, 7, 9], the mapping will be performed layer by layer in these cones and a classical place and route tool will be helped by this logical structuring.
The general strategy will be the following:

Step 1: choice of factorization (kernels) compatible with an input ordering and definition of this input ordering.
Step 2: algebraic division.
Step 3: further minimization compatible with the input order.
Step 4: creation of subfunctions compatible with the input order.

2. CHOICE OF FACTORIZATIONS COMPATIBLE WITH AN INPUT ORDERING; DEFINITION OF AN INPUT ORDERING

At this step the factorizations considered are based on kernels of any level for each function. If a kernel appears to be common, it is considered as a subfunction and may be used at most by k functions (fanout limitation). No kernel nor cokernel nor product term intersection between functions is considered at this step. The problem is to select a set of factorizations compatible with an input ordering. For this, we used an algebraic approach based on lexicographical expression of boolean function [6, 10]

2.1. Preliminary definitions

Definition 0: to a product term written X.Y.Z is attached a precedence relation between its variables starting that X precedes Y precedes Z.

Definition 1: a lexicographical expression of a Boolean function is a sum of product terms such that the precedence relations associated with its product terms are compatible.
The union of these precedence relations defines a partial or total ordering of the variable called the reference ordering.
Examples:

F = acb + ^abd + ^c^b^d + ^def defines a total ordering of the variables: {a c b d e f}
G = ab + abd + ac defines a partial ordering (a precedes b precedes d & a precedes c) and would be compatible with {abdc}, {acbd}...

Definition 2 : a factorized lexicographical expression of a Boolean function is a factorized expression which by flattening (applying the distributivity laws between the sums and the products) leads to a lexicographical expression. One or several reference orders can be attached to such an expression.

Consider two functions : F' = e[^ac + ^f(c + d)] + ^ac^d + ^c(db + ^d^b)
G' = c[^fe + ^a(e + ^d)] + f^e(^c + a)

F' is a lexicographical factorized expression, but G' is not.

Relation with the target architecture

A tree corresponding to a lexicographical factorized expression of a Boolean function is a tree in which inputs enter in an ordered manner on each path of the tree. This ordering is the reference ordering of the lexicographical expression. A lexicographical factorized Boolean expression is therefore the algebraic target of our approach.
As an example figure 2 shows a lexicographical factorized tree corresponding to the preceding function F' = e[^ac + ^f(c + d)] + ^ac^d + ^c(db + ^d^b).

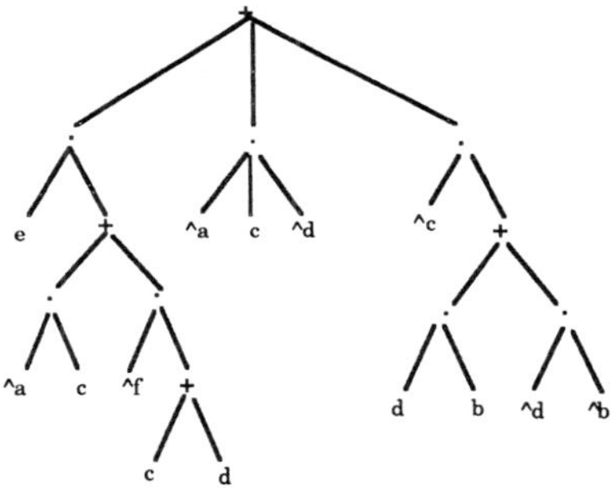

figure 2: a lexicographical factorized tree

The general idea in the proposed synthesis approach is to select elementary factorizations leading to optimized lexicographical expressions of the functions.

Definition 3: an elementary factorization based on a kernel and cokernels extraction induces a precedence relation stating that the literals of the cokernels precede the literals of the kernel.
The factorization ^c(b + ^d^b) induces that (c) precedes (db). We write this down by c ---> (db).
Intuitively, this is related to the logical schematic of figure 2 in which we see that input c enters the logic cone before d, b.

Definition 4 :an elementary factorization is compatible with a reference order if the precedence relation induced between the literals is respected in the reference order.
In the expression F = c^ad + a^c + c^d + ^d^b + b^ad the factorization d^a (c+b) is compatible with the reference orders {d a c b} or {d a b c}.

Definition 5 :two elementary factorizations are compatible if there exists a reference order compatible with both of them.
The general problem is to find an optimal set of compatible factorizations. An optimal set of compatible factorizations is a set of compatible factorizations allowing to save a maximal number of literals. As this compatibility relation is not transitive, a global and classical approach looking for maximal compatibility classes is not feasible. Therefore a greedy algorithm considering the factorizations in a decreasing gain order is adopted.

2.2. A greedy algorithm to choose a set of compatible factorizations

Initially, a kernel extractions phase is performed and ends up with a list of kernels and cokernels ranked in a decreasing gain order. The gain is the maximal gain of literals saved by factorizations (related to all cokernels). With each kernel is associated the precedence relations between input variables. Starting from the highest gain factorization a next compatible factorization will be chosen.

A compatible factorization does not violate any input ordering of the previously chosen factorizations. To find out if a factorization is compatible with the previous ones, a current precedence matrix will update all the precedence relations between variables. If N_i is the number of the input variables, the precedence matrix associated with an elementary precedence relation R: B ---> B' is a matrix N_i x N_i denoted Pr (R_i) in which the element (i,j) is 1 if x_i precedes x_j and 0 if x_j precedes x_i in R. All the other elements contain a don't care. Two matrices are uncompatible if x_i (x_j) precedes x_j (x_i) in one matrix and x_j (x_i) precedes x_i (x_j) in the other one.The current precedence matrix denoted Pr_c contains the information about all updated precedence relations.
The matrix Pr_c is antisymmetric. Suppose that at step i, the current precedence matrix is Pr_c and that a new relation R is chosen. The updating of the precedence relation leads to consider the following new precedence relations expressed in the matrix ΔPr_c:

$$\Delta Pr_c = PR_c^{step\ i} \times Pr(R) + Pr(R) \times Pr_c^{step\ i} \qquad (1)$$

The matrix multiplication is based on the following based rules where (-) represente a don't care value:

$$0 * 0 = 0$$
$$1 * 1 = 1$$
$$0 * (-) = (-)$$
$$1 * (-) = (-)$$

This incremention procedure of the precedence relations when adding a relation B--->B' states that any ancestor of an element of B in the current state inherits new successors which are the elements of B' and similarly the elements of B' inherit new ancestors which are the ancestors of any element of B.

In fact, due to the antisymmetry of this matrix, this computation is redundant and it is sufficient to do the operation :

$$\Delta Pr_c = (Pr_c^{step\ i} \times Pr(R))^* \qquad (2)$$

where (*) means a completion procedure by antisymmetry. That means that as soon as an element (i,j) in the product matrix takes the values 1 or 0, the element (j,i) takes the complementary value.
The current updated precedence matrix becomes :

$$Pr_c^{step\ i+1} = Pr_c^{step\ i} + Pr(R) + \Delta Pr_c \qquad (3)$$

To test if a new relation is compatible with the previous ones, it is sufficient to verify if the new relation would not introduce inconsistencies in the current precedence matrix.

Let us mention that the sum (3) can be suppressed if the matrix Pr_c and Pr(R) are extended to Pr'_c and Pr'(R) where a matrix Pr' is obtained from a matrix Pr by adding the diagonale one, the final relation is then

$$Pr'^{\,step\ i+1}_c = (Pr'^{\,step\ i}_c \times Pr'(R))^* \qquad (4)$$

To illustrate this, consider three precedence relations corresponding to three elementary factorizations ranked in a decreasing gain ordering N_i = 8 and the inputs are {a, b, c, d, e, f, g, l}
(1) ab ---> efg
(2) e ---> fl
(3) cd ---> al

Step 1.The first relation is chosen and the initial precedence matrix is the matrix corresponding to this first relation (figure 3).

	A	B	C	D	E	F	G	L
A					1	1	1	
B					1	1	1	
C								
D								
E	0	0						
F	0	0						
G	0	0						
L								

Figure 3: the matrix of the initial relation

Step 2. The second relation is chosen ; the precedence refinement produced by this new relation is given in figure 4b:

$$\Delta Pr : (Pr_C \text{ x } Pr \text{ (relation 2)})^*$$

	A	B	C	D	E	F	G	L
A					1	1	1	
B					1	1	1	
C								
D								
E	0	0						
F	0	0						
G	0	0						
L								

Pr 0

	A	B	C	D	E	F	G	L
A								
B								
C								
D								
E						1		1
F					0			
G								
L					0			

Pr (relation 2)

	A	B	C	D	E	F	G	L
A						1		1
B						1		1
C								
D								
E								
F	0	0						
G								
L	0	0						

Δ Pr

Figure 4a: updating of the matrix

The current precedence matrix is now: $Pr_C = Pr_O + Pr(\text{relation } 2) + \Delta Pr$

	A	B	C	D	E	F	G	L
A					1	1	1	1
B					1	1	1	1
C								
D								
E	0	0				1		1
F	0	0			0			
G	0	0						
L	0	0			0			

Figure 4b: the updated matrix

Step 3. In a similar way, the influence of relation 3 leads to a further updating of the precedence relation :

$$\Delta Pr_C = (Pr_C \text{ x } Pr(\text{relation } 3))^*$$

	A	B	C	D	E	F	G	L
A			0	0				
B								
C	1							1
D	1							1
E								
F								
G								
L			0	0				

Pr (relation 3)

	A	B	C	D	E	F	G	L
A					1	1	1	1
B					1	1	1	1
C								
D								
E	0	0				1		1
F	0	0			0			
G	0	0						
L	0	0			0			

Pr_C

	A	B	C	D	E	F	G	L
A								
B								
C					1	1	1	1
D					1	1	1	1
E			0	0				
F			0	0				
G			0	0				
L			0	0				

Δ Pr

The current matrix is then : $Pr_C + Pr$ (relation 3) $+ \Delta Pr$

	A	B	C	D	E	F	G	L
A			0	0	1	1	1	1
B					1	1	1	1
C	1				1	1	1	1
D	1				1	1	1	1
E	0	0	0	0		1		1
F	0	0	0	0	0			
G	0	0	0	0				
L	0	0	0	0	0			

Figure 4c: final matrix

As long as the updating of the precedence matrix is feasible, the factorization is considered as accepted and the procedure ends up with the definition of a set of compatible factorizations.

2.2.1.Definition of the input ordering

The input ordering or partial orderings can be deduced from the final precedence matrix.
A preprocessing step may simplify the matrix by mapping the equivalent inputs. Two inputs are equivalent if the corresponding rows (and columns) are identical. This leads to a relabelling of the input variables. The final ordering is then trivial and is as follows :

1. Identify the rows containing only 1's and don't care, they constitute the current group; at least one variable belongs to such set.
2. Cancel the rows and the columns corresponding to these elements. Go back to step 1.

In the previous example we obtain bcd ---> a ---> e ---> fgl
The following flowchart summarizes the approach which is a greedy one:

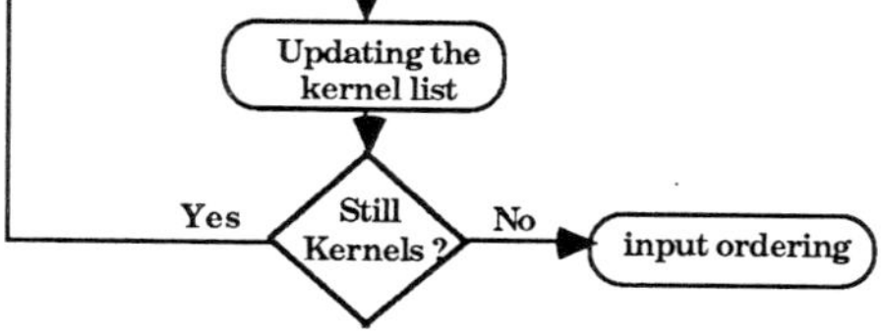

3. ALGEBRAIC DIVISION FOR LEXICOGRAPHICAL EXPRESSIONS AND FURTHER SIMPLIFICATION

3.1. Algebraic division

The algebraic division in this specific environment is much simpler for 2 reasons:
First, it can be shown that 2 factorizations compatible lexicographically (compatible with the same reference order) are also algebraically compatible. Two factorizations are algebraically compatible if they can be performed simultaneously, they do not concern the same product

terms. Due to a lack of space the demonstration connot be given here. This avoids a cumbersome research of compatible kernels.

Secondly, the division procedure itself is simplified.
Due to the lexicographical order, the cokernels are structured by an inclusion tree comparable to a binary decision tree. The kernels appearing at the leaves are the kernels of degree 0. This is true for the initial functions and for the subfunctions created as common subfunctions (set of common product terms between functions for instance).
As an example, consider the function:

F = dbca + dbce + dbkf + dh^k + ^d^bca + ^d^bce + ^d^bkf +
^d^eh + ^db^ca + ^db^ae + ^hk^f
with a reference order d ---> b ---> c ---> aefhk

The set of factorizations is:
1) db(ca + ce + kf)
2) ^d^b(ca + ce + kf)
3) dbc(a + e)
4) ^d^bc(a + e)
5) ^d(^bca + ^bce + ^bkf + ^eh + b^ca + b^ae)
6) d(bca + bce + bkf + h^k)
7) ^db(^ca + ^ae)

and the tree structuring the cokernels is the following

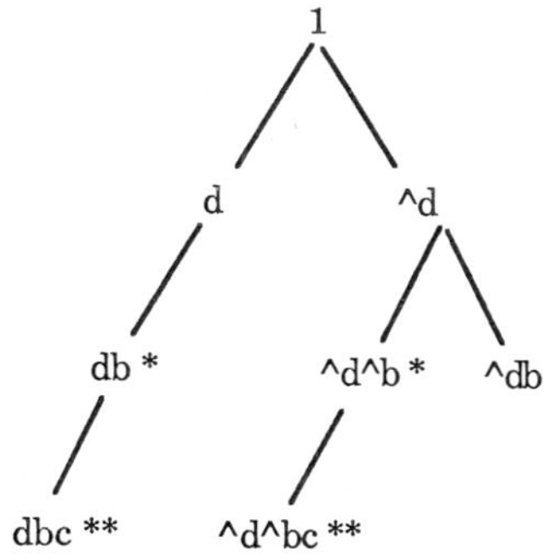

Figure 5: cokernels tree

(*) and (**) means that all the related cokernels have the same kernel.

The result of the division is immediatly obtained starting from the root and embedding the different levels of kernels.

$\Rightarrow$ F = d [b(c(a + e) + fk) + h^k] + ^d [^b(c(a + e) + fk) + ^eh + b(^ca + ^ae)]
+ ^f^hk

3.2. Further minimization

This step aims at looking for further minimizations still compatible with the lexicographical order which are not based on kernel extraction. Suppose for instance that the reference order is {abecfik} and the following kernel exists **abc(ef + ik)**.
Rather than rejecting completely this factorization, a partial compatible factorization is extracted and performed **ab(ecf + cik).**
Therefore, all the partial minimizations compatible with the reference order are processed in this step.

3.3. Subfunction creation

The previous steps end up with a set of factorized lexicographical trees. Starting from these trees, common subtrees are identified starting from the leaves (figure 6). They become subfunctions if their size is above a first parameter p, and are put in common in no more than p' trees (fanout limitation).

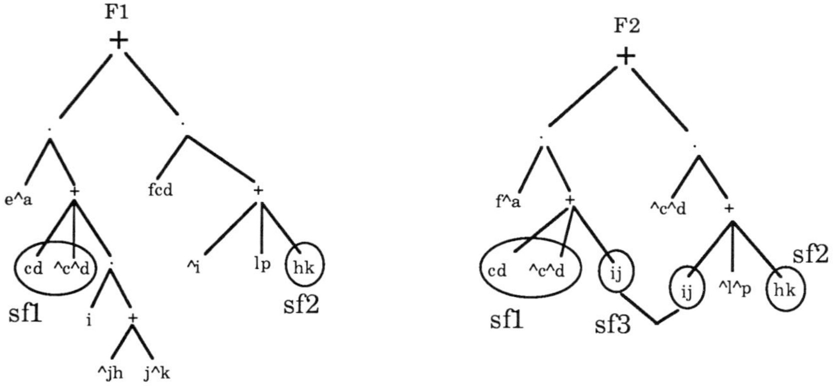

Figure 6: subfunction creation

4.EXAMPLES

The first example aims at showing the importance of the input dependency control and thus the importance of the proposed lexicographical ordering.
Consider the function F having 18 inputs and 1 output:

F = cde + ace^f^i + ceai^l + gh^e^a + ^a^eikl + ^e^aikp^y + ^e^aikp^z + d^ca^g^i + d^cahl^m + d^cah^p^yz + d^ca^phtb + ^d^a^gh^kl + ^a^d^gk^l^m + fg^hi + fgkp^yz + fg^ilm^p + fg^ilmy^zb + fg^ilmy^qs^z + fg^ilmy^qt^z

The function has first been factorized with no restriction by classical kernel and common cubes extraction and then through the lexicographical ordering. It is shown that the lexicographical ordering induces a decrease of the routing factor of 24% and a decrease of the global area of 15%.

Figure 7 shows the two layouts of the standard cells of VLSI Technology in 1.5µ CMOS technology.

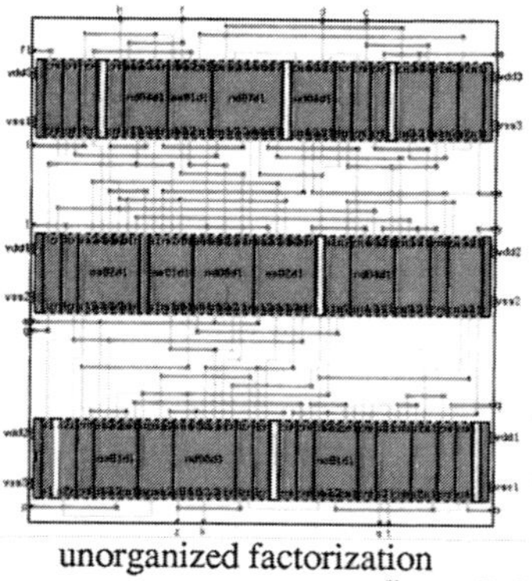

unorganized factorization

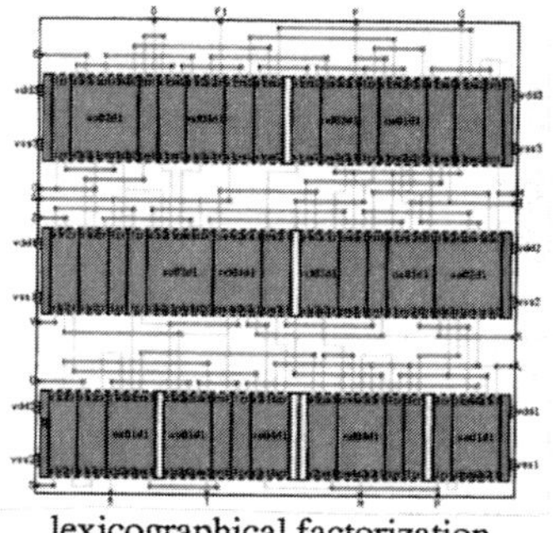

lexicographical factorization

figure 7: Layouts of the F function.

In a second example three levels of filtering during the creation of subfunctions have been applied on the benchmark misex1 [8]. As expected, it is shown that an excessive filtering suppressing completely the common parts of the cones is not suitable. A medium filtering accepting common factors if they are not too small leads to a good solution. Here common factors are accepted until they contains at least 3 literals The "excessive" factorization putting anything possible in common decreases the number of literals but increases the area of the wiring channels. These three solutions, shown in table 1, are all compared to the classical factorization methods and the layout measures are taken within the same environment (VLSI Technology). Notice also that the number of literals is more important in the lexicographical expression (79 instead 69 for the medium filtering). This means that the literal number is not of high importance. For more serious comparisons, the critical path is computed each time to verify if there is no parasitic effect on performance. The critical path stayed equivalent for the three layouts.
A third example (misex2) shows that for small circuits for which the wiring area is not excessive, the decrease of the routing factor may induce an increase of the gate area. Thus the global area is slightly increased. Therefore, further research aims at reaching an adequate control of the techniques proposed to yield a final improvement of the circuit area and performance.
Starting from the previous example the medium filtering has been adopted. The next 2 benchmarks (5xp1, sao2) show the most striking advantage of the lexicographical expression which is the optimization of the critical path.

	misex1				misex2		5xp1		sao2	
	UF	SF	MF	OF	UF	MF	UF	MF	UF	MF
Global area (mm2)	0,1936	0,2107	0,1810	0,1881	0,3157	0,3237	0,5955	0,3505	0,7272	0,5844
Routing factor	1	0,88	0,85	0,93	1,17	0,95	1,49	1,35	1,84	1,52
Routing area (mm2)	0,0968	0,0986	0,0831	0,0906	0,1702	0,1577	0,3563	0,2013	0,4711	0,3525
Critical path (ns)	24,9	21,3	20,2	22,7	30	31,2	23,6	23,8	43,3	32,7
Literal number	69	93	79	69		141		126		184

UF Unorganized Factorization (Non lexicographical factorization)

SF Strongest Filtering
MF Medium Filtering
OF O-Filtering
} Lexicographical factorization

Table 1: comparative results.

CONCLUSION

The multilevel synthesis method proposed here shows how to decrease the routing factor of standard cell logic. As for any optimization problem, it is important to consider the goal in an early stage of the design. The solution proposed here relies on an adequate factorization phase. Rather than to minimize extensively the literal number, not really related to physical layout, the factorization prepares a simple optimized wiring between input pads and the standard cells and simple inter-cell wiring by controlling internal fanouts. The goal is achieved even for small circuits and will be even more exciting for large examples.

References

[1] P. Abouzeid, K. Sakouti, G. Saucier, F. Poirot. "Multilevel synthesis minimizing the routing factor", 27th DAC, 24-27 June 1990.
[2] R. K. Brayton, C. T. McMullen. "The decomposition and factorization of boolean expressions", ISCAS Proceeding pp49-54, Mai 82.
[3] R. Brayton,R. Rudell, A. Wang, and A. Sangiovanni-Vincentelli."MIS: a multiple-level logic optimization system". In IEEE Transactions onCAD, pp 1062-1081,November 1987.
[4] M. Crastes de Paulet, C. Duff, G. Saucier "ASYL: a Logic and Architecture Design Automation System". EURO ASIC 89, pp 183-209, January 1989.
[5] E. Detjeus, G. Gannot. "Technology Mapping in MIS". ICCAD, Santa Clara (USA), November 1987, pp. 116-119.
[6] J. Kuntzmann, "Algèbre de Boole", Edition Dunod, Paris, 1968.
[7] R. Lisanke, F. Berglez, G. Kedem. "McMAP: A Fast Technology Mapping Procedure for Multi-Level Logic Synthesis", ICCD-88, pp 252-256.
[8] Benchmark set, Microelectronic Center of North Carolina, 1989.
[9] K. Sakouti, G. Saucier, "A fast and effective technology mapper on an autodual library of standard cells", IFIP Working Conference on Logic and Architecture Synthesis, Paris, France, 30.05 - 01.06 1990.
[10] G. Saucier, R. Leveugle, P.Abouzeid, "A channelless layout for multilevel synthesis with compiled cells" International Conference on Computer Design (ICCD'89), Cambridge, USA, 2-4 Octobre 1989.

Logic and Architecture Synthesis
P. Michel and G. Saucier (Editors)
Elsevier Science Publishers B.V. (North-Holland)

Corolla Partitions and Don't Cares

Sujit Dey[1] **Franc Brglez**[2] **Gershon Kedem**[1]

[1]**Department of Computer Science**
Duke University, Durham, NC 27706

[2]**Microelectronics Center of North Carolina**
Research Triangle Park, NC 27709

Abstract

This paper introduces a circuit partitioning method based on analysis of reconvergent fanout. We define a corolla as a set of overlapping reconvergent fanout regions. We partition the circuit into a set of non-overlapping corollas and use the corollas to resynthesize the circuit. We show that resynthesis of large benchmark circuits consistently reduces transistor pairs and layout area while improving delay and testability. In the second part of the paper, we analyze the role of don't cares to further minimize the corollas.

1 Introduction

Today, more designs are being specified at high levels of abstraction. This can result in complex Boolean circuits that cannot always be efficiently optimized. The partition hierarchy introduced at high-level can improve the effectiveness of synthesis tools on individual partitions; however, the choice of partitions at high-level can significantly affect the overall design quality. Our work is aimed at partitioning very large Boolean networks in a way that will best match to capabilities of current logic synthesis tools.

There are two basic approaches to multi-level logic synthesis: 1) algebraic/boolean factoring, starting with a PLA and producing a minimized multi-level circuit, and, 2) synthesis of multi-level network of simple gates into another optimized multi-level network. There are a number of effective algorithms for minimizing, decomposing and factoring two-level specifications [1,2]. In contrast, when specifications are in the form of a multi-level Boolean network, there is no single effective optimization technique that would optimize the size of such networks. Some of the early techniques used local transformations [3] and identification and removal of redundancies [4]. More recently, global flow analysis [5], partitioning based on clustering of nodes and operations [6,7], and use of don't cares have shown potential for further minimization [8,9].

We introduce a new technique to partition a large multi-level circuit, based on an analysis of signal reconvergence. The signal reconvergence regions between the fanout points and all other nodes define a unique set of primitives in our partitions, the petals. We define a corolla as a set of overlapping petals, forming a multi-input, multi-output module. We partition the circuit into a set of non-overlapping corollas.

We discuss the application of corolla-based partitioning to improve the synthesis of large multi-level logic circuits. The first part of the paper introduces stem regions, petals and corollas. We summarize the results of corolla partitioning and logic resynthesis on a variety of large circuit benchmarks. Most notably, we find that logic resynthesis based on corolla partitioning consistently reduces recovergent fanout branches, transistor pairs and layout area while improving circuit delay and testability. The details of the partitioning algorithm can be found in [10,11]. The second part of the paper analyzes the resynthesis potential of typical corollas that have been extracted. We explore the use of don't cares to further minimize the corollas in the local as well as global context.

2 Stem Regions, Petals and Corollas

We introduce the key concepts used in our partitioning approach beginning with stem regions. We define a petal as the basic primitive of our partitioning, and define our partition, corolla, as a collection of petals. We model the circuit with an acyclic graph $G(V, E)$, directed from the primary inputs to the primary outputs. Each primary input, gate, and fanout point of the circuit is represented by a distinct node in the graph, and each wire in the circuit is represented by an edge in the graph. For each gate node x in the graph, there is an associated Boolean function f_x, and conversely x is the boolean variable representing f_x. Figure 1a shows a multi-level combinational circuit consisting of simple gates, and Figure 1b shows the corresponding graph. The shaded nodes in the graph represent the fanout stems to distinguish them from gates and inputs which are drawn as unshaded nodes. For our partitioning approach, we introduce a primary stem region as a subgraph of stem region defined in [12].

Definition 1 (Fanout Stem and Fanout Branches) *A node of the circuit graph with an outdegree greater than one is called a* **fanout stem**, *and its outgoing edges are called its* **fanout branches**.

In Figure 1b, traversing from the primary output side to the primary input side, y, u, l, m, n, g, h and a are the fanout stems. The fanout branches of stem m are (m, p) and (m, q).

Definition 2 (Reconvergent fanout stem and Reconvergent node) *Let $s \in V$ be a stem node. If there are more than one disjoint path from s to another node $r \in V$, then s is a* **reconvergent fanout stem** *and r is a* **reconvergent node** *for s. If s has no such reconvergent node, then s is a* **non-reconvergent fanout stem**.

In Figure 1b, u, n, g, h and a are reconvergent fanout stems, while y, l, m are non-reconvergent fanout stems. The reconvergent nodes of stem a are i, o, p, q, r and v.

Definition 3 (Closing reconvergent node) *A* **closing reconvergent node** *of a stem s is a reconvergent node of s that does not drive any other reconvergent node of s.*

In Figure 1b, o, p, q and v are the closing reconvergent nodes of fanout stem a.

Definition 4 (Primary Stem region) *The* **primary stem region** *of a reconvergent fanout stem s consist of all the nodes that are located on all the paths from s to any of its closing reconvergent nodes, and all the output edges of these nodes.*

In the graph of Figure 1b, the primary stem region of a consists of the nodes a, c, d, e, f, g, h, $i, j, k, l, m, n, o, p, q, r, s$ and v, and all their output edges.

Definition 5 (Exit lines) *An edge e in a primary stem region of s is termed an* **exit line** *for s if e is not the ouput of a closing reconvergent node of s, and e drives a node which is not included in the primary stem region of s.*

In Figure 1b, edges $(a, b), (n, w), (n, t)$ are the exit lines of the primary stem region of a.

2.1 Petals and Corollas

A petal is the basic primitive of our partitions, whereas a corolla is a collection of petals forming a partition of the circuit graph. We partition each stem region into a set of disjoint petals, and dynamically cluster petals to partition the circuit graph G(V,E) into a set of disjoint corollas.

Definition 6 (Petal) *Let s be a reconvergent fanout stem. Each petal of $s, P(s)$, is a biconnected component [13] of the primary stem region of s, with directions of edges ignored.*

Let $P_i(s)$ denote the nodes of the i-th petal of a reconvergent stem s. If $q > 1$ petals are identified for s, then s is split into q fanout stem nodes $\{s_1, \ldots, s_q\}$, so that each s_i is the new stem node for the petal $P_i(s)$. We connect s to each s_i by an edge (s, s_i). Notice that in effect we have partitioned the fanout branches of stem node s into q subsets. We exclude s from $P_i(s)$, and include the node s_i in $P_i(s)$. Consequently, we have the following property for the petals of a stem.

Property 1 *The petals of a stem are disjoint, that is, for a reconvergent stem s, if $P_i(s) \neq P_j(s)$, then $P_i(s) \cap P_j(s) = \phi$,*

Definition 7 (Tip and Exit of a Petal) *A* **tip** *of a petal of stem s is a closing reconvergent node of s included in the petal. An* **exit** *is any edge in the petal which drives a node not included in the petal.*

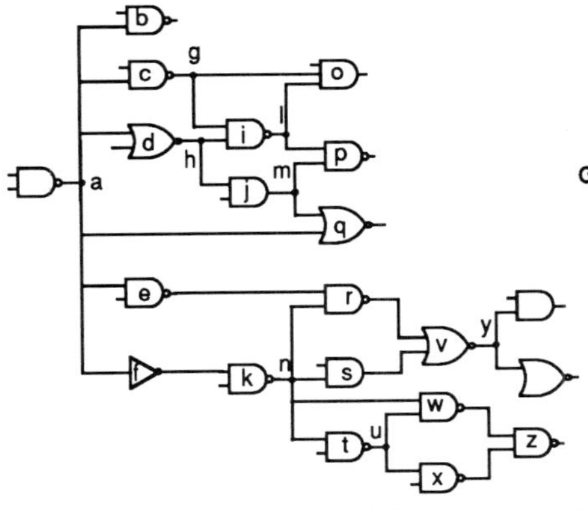

Figure 1a: A gate-level circuit.

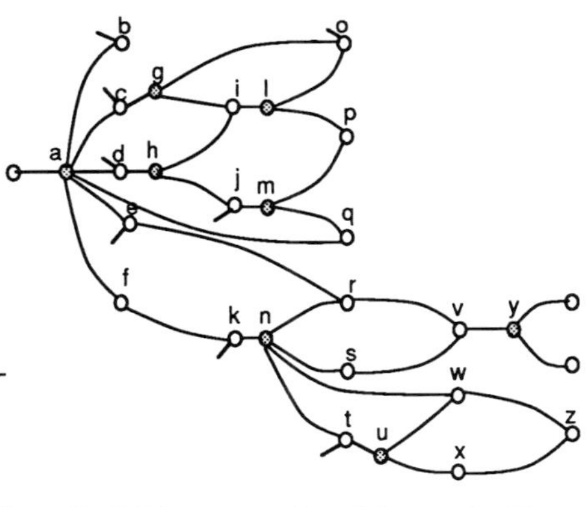

Figure 1b: DAG representation of the circuit in Figure 1a

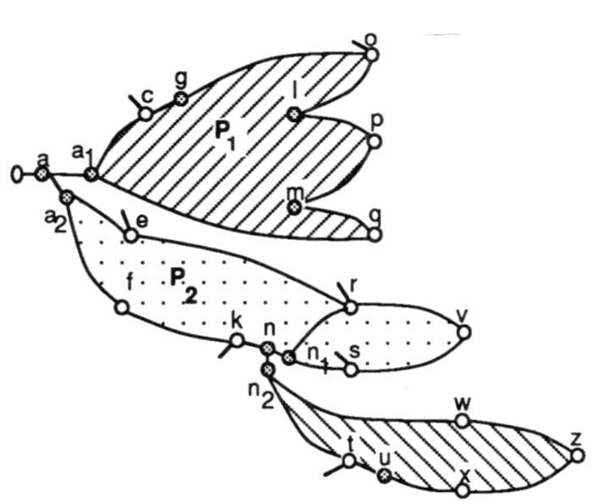

Figure 1c: DAG partitioning into petals

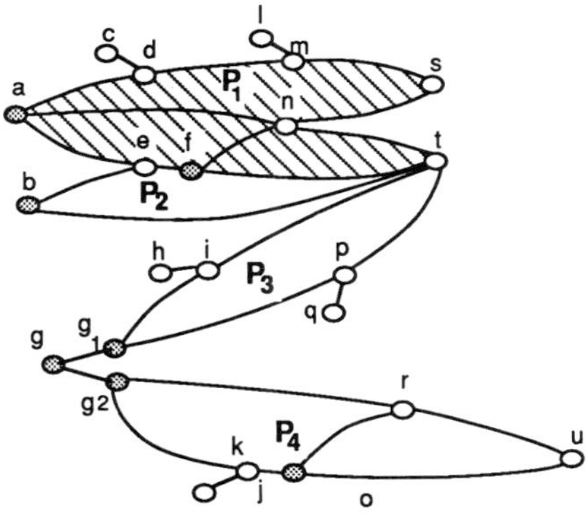

Figure 2: Overlapping petals form corollas

Let us consider the reconvergent fanout stem a in Figure 1b again. We have two biconnected components of the undirected primary stem region of a. Hence, we split the stem node a into two, and introduce the split nodes as shown in Figure 1c. Now we have the petals $P_1 = \{o, p, q, l, m, i, j, g, h, c, d, a_1\}$, and $P_2 = \{v, r, s, n, k, e, f, a_2\}$; a_1 and a_2 are the new stem nodes of P_1 and P_2, and P_1 and P_2 are disjoint. The tips of P_1 are o, p and q; it has no exits. The tip of P_2 is v, and it has an exit, $(n, n1)$. Similarly, the stem region of fanout stem n is partitioned into two petals with new stems n_1 and n_2, and tips v and z respectively as shown in Figure 1c. Notice that the petal rooted at stem n_1 is contained in the petal P_2.

Definition 8 (Overlap of Petals) *We define a natural relation* **overlap** (ρ) *on the set of petals by saying that petal* P_i *overlaps petal* P_k, *denoted by* $P_i \rho P_k$, *if either (i)* $P_i \cap P_k \neq \phi$, *or, (ii)* $P_i \cap P_j \neq \phi$ *AND* $P_j \rho P_k$. This relation, overlap, is an equivalence relation that partitions the petals of G into equivalence classes $C_1, C_2, \ldots, C_k$ such that two distinct petals are in the same class if and only if they overlap.

Definition 9 (Corolla) *A* **maximal corolla** *is an equivalence class formed by the relation overlap. A* **(constrained) corolla** *is any subset of a maximal corolla.*

In partitioning the circuit graph, we make the following requirement:

Property 2 *Any two corollas should be non-overlapping or disjoint.*

The inputs of a corolla C, denoted by $I(C)$, are those nodes which do not have any of their input edges included in C, and which drive at least one node in C. An input node of C which has more than one fanout branch included in C is a **stem input** of C. All the other inputs of C are **line inputs**. The output nodes of a corolla C, denoted by $O(C)$, are the tips and exits of the individual petals constituting the corolla.

The graph of Figure 2 shows some overlapping petals. The petals P_1, P_2 and P_3 overlap with each other. Petal P_4 does not overlap with any other petal in the graph. There are two equivalence classes, and hence, two maximal corollas: $C_1 = \{P_1, P_2, P_3\}$, and $C_2 = \{P_4\}$. The stem inputs of the maximal corolla C_1 are a, b, and g_1, and its line inputs are c, h, l, q; the output nodes of the maximal corolla C_1 are s and t. Since a maximal corolla contains all the overlapping petals, partitioning into maximal corollas is unique. However, forming (constrained) corollas is not unique; in this example, the candidates to form a (constrained) corolla which is a subset of the maximal corolla C_1 are the petals P_1, P_2 and P_3. There are seven different ways of combining these petals to form seven different (constrained) corollas.

Definition 10 (C-Support) *The* **C-Support** *of a node* y *of corolla* C *consists of those input nodes of* C *which have at least one path to* y. *In other words, C-Support(y) is the subset of* $I(C)$ *on which* f_Y *depends.*

In Figure 2, considering the maximal corolla C_1, the C-Support sets for the output nodes s and t are: C-Support(s) $= \{a, b, c, l\}$, C-Support(t) $= \{a, b, g, h, q\}$. As opposed to the C-Support set for a node y, Support(y) is all the primary inputs of the circuit which affect y.

3 Benchmark Results of Partitioning and Resynthesis

We have applied corolla-based partitioning and resynthesis to the ISCAS85 and a subset of ISCAS89 benchmarks. The main objective is to reduce the **Layout Area** of the circuit. While it is not always feasible to flatten large multi-level circuits to a PLA and then resynthesize the PLA using algebraic/boolean factoring [1,2], the individual corollas produced by our partitioning algorithm can be. Once we have partitioned the circuit into non-overlapping corollas, we resynthesize each corolla independently. We flatten the corolla to a two-level PLA, minimize the PLA using ESPRESSO [14], and synthesize the minimized PLA using a logic minimization tool [1,2]. If the cost of the resynthesized corolla is less than the cost of the original corolla, we glue back the resynthesized corolla in place of the original one, preserving the input and output mapping of the original corolla.

Table 1 summarizes the data we have obtained in the experiment before (bef) and after (aft) partitioning and resynthesis, as well as the percent reduction (%Rd). Gate inputs, fanout branches and untestable faults refer to a gate level netlist prior to technology mapping, whereas transistor pairs, average of critical-path rise and fall delays, and layout area refer to a standard cell realization after technology mapping. The technology mapping [15] is applied to both the original circuit (before partitioning) and the reduced circuit (after partitioning and resynthesis). The results shown have been obtained with MCNC-distributed standard cell library and VPNR automatic place and route system [16].

Ckt	Gate Inps			Branches			Trans. Pairs			Layout Area		
	bef	aft	%Rd	bef	aft	%Rd	bef	aft	%Rd	bef	aft	%Rd
t481	110	59	46.4	60	20	66.7	101	53	47.5	108	49	**54.6**
c432	443	424	4.3	295	260	13.5	373	357	4.3	484	444	**8.3**
c499	408	376	7.8	256	256	0.0	702	672	4.3	672	642	**5.5**
c880	681	598	12.2	437	378	13.5	579	566	2.3	805	752	**6.6**
c1355	1032	896	13.2	768	560	27.1	878	704	19.8	1313	692	**47.3**
c1908	1336	1174	12.1	995	901	9.5	1032	933	9.6	1359	1208	**11.1**
c2670	1880	1624	13.6	1224	1047	14.5	1487	1287	13.5	2136	1720	**19.5**
c3540	2716	2461	9.4	1821	1638	10.1	2061	1883	8.7	4530	3587	**20.8**
c5315	4073	3742	8.1	2830	2633	7.0	3365	3112	7.5	5623	5120	**9.0**
c6288	4800	4760	0.8	3840	3777	1.6	4231	4161	1.7	7803	7255	**7.0**
c7552	5610	4920	12.3	3833	3421	10.8	4462	3920	12.5	6367	5231	**17.9**
s444	336	295	12.2	239	204	14.7	311	254	18.3	417	309	**25.9**
s713	591	497	15.9	266	190	28.6	386	292	24.4	505	371	**26.6**
s9234	3585	2886	19.8	3390	2838	16.3	4193	3847	8.3	8884	7504	**15.5**

Ckt	Untested Faults			Avg. Delay		
	bef	aft	%Rd	bef	aft	%Rd
t481	0	0	NA	15.8	10.4	34.2
c432	41	40	2.4	71.7	55.6	22.4
c499	8	0	100	42.8	37.5	12.9
c880	0	0	NA	69.3	36.7	47
c1355	8	0	100	46.8	39.9	14.7
c1908	9	7	22.2	69.3	66.6	3.9
c2670	116	64	44.8	79.7	98.9	-24.1
c3540	133	51	61.7	75.5	76.3	-1.1
c5315	62	26	58.1	71.1	76.5	-7.6
c6288	34	33	2.9	220.8	194.0	12.1
c7552	145	73	49.7	213.4	189.6	11.2
s444	14	3	78.6	20.5	17.6	14.2
s713	48	1	97.9	43.6	30.4	30.3
s9234	471	365	22.5	90.8	89.8	1.1

Table 1. Partitioning and resynthesis improves layout area, testability and delay.

The results in Table 1 show that corolla-based partitioning and resynthesis consistently reduces number of gate inputs, reconvergent branches and number of transistors: all these have contributed to a consistent reduction in the layout area, which is considerable for some of the benchmarks. In addition, we have reduced the number of untestable faults and the average critical-path delay for most of the benchmarks. The gain in resynthesis of the 100% testable benchmark circuit t481, which was generated by algebraic factorization from a PLA, points out that the corolla partitioning and resynthesis technique can be also effectively applied to circuits generated by synthesis tools which use factorization techniques.

4 Corolla Resynthesis without Don't Cares

Partitioning a DAG into non-overlapping corollas and replacing nodes covered by each corolla with a module-node produces a new graph with smaller number of nodes and reduced signal reconvergence. Many, if not all, of the fanout nodes of the original graph are covered by corollas and we now have the option of resynthesizing each corolla separately and replacing it with a new circuit that may have reduced signal reconvergence, transistor count or both. The reasons for gain in resynthesis of a corolla can be due to several factors: 1) factorization and technology mapping produce results that are closer to optimum for corollas than for a circuit as a whole, 2) corollas could have been formed from several individually designed circuit parts that have been merged during flattening of hierarchy, 3) some corollas may have had redundant nodes that were removed during resynthesis. We use two small circuits, extracted as corollas from a larger circuit, to illustrate the gains that we may encounter when resynthesizing a corolla. The corollas before and after resynthesis are shown in Figures 3 and 4. Note that each of the original circuits can be further partitioned into smaller corollas. The circuit in Figure 3 is a simple bit-comparator (greater-or-equal) that was obtained during hierarchical synthesis, using minimal and fully testable "greater"and "equal" modules. The hierarchical synthesis in this case produced not only a circuit that is not minimal in size, it also contains several redundancies. Using global redundancy identification and removal techniques such as discussed in [17], circuits can be optimized by removing one redundancy at a time. However, having identified and resynthesized the circuit in Figure 3 as a single corolla, we remove all redundancies in a single step and simultaneously optimize the circuit.

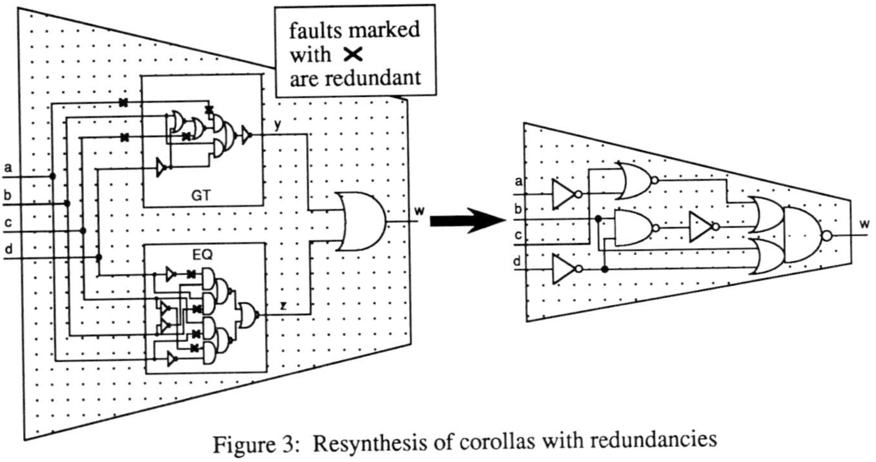

Figure 3: Resynthesis of corollas with redundancies

It can be shown that a redundant fault *must* be within a petal [10], hence one may argue, as in the case of example of Figure 3, that the gain in resynthesis of a corolla is due to removal of redundant logic. This need not be the case. The corolla in Figure 4 was extracted from benchmark t481, a prime and irredundant multi-level circuit synthesized from a PLA description. The improvements upon resynthesis are dramatic: from 42 to 25 gate inputs, from 8 to 6 fanout stems and from 24 to 12 fanout branches. When processing the corolla by a technology mapper, we reduce the transistor count from 38 pairs to 21 pairs. While algebraic synthesis of the circuit t481 produced an irredundant but highly inefficient circuit, applying the same synthesis tool to the smaller corolla partitions produced much better results overall: an area reduction of 54.6% and a delay reduction of 34.2% (Table 1). In this case, the search space of choosing kernels is smaller and the same kernel extraction heuristic works better.

It is useful to understand the structure of the circuit to explain additional experimental results in Table 1. A case in point is the benchmark c1355. This circuit is logically equivalent to c499, with all 104 XOR gates in c499 expanded into a 3-level circuit with four 2-input NAND gates. The challenge for the partitioning algorithm is to extract the 4-NAND gate clusters, so that the resynthesis process can identify them as XOR functions and remap them to CMOS circuits with least number of transistors. The second challenge is whether larger corollas with 8 known redundancies can be identified and resynthesized, thereby eliminating redundancies from c1355. As shown in Table 1, corolla-based partitioning worked well for c1355: all redundancies were removed and the reduction in layout area amounted to 47.3%.

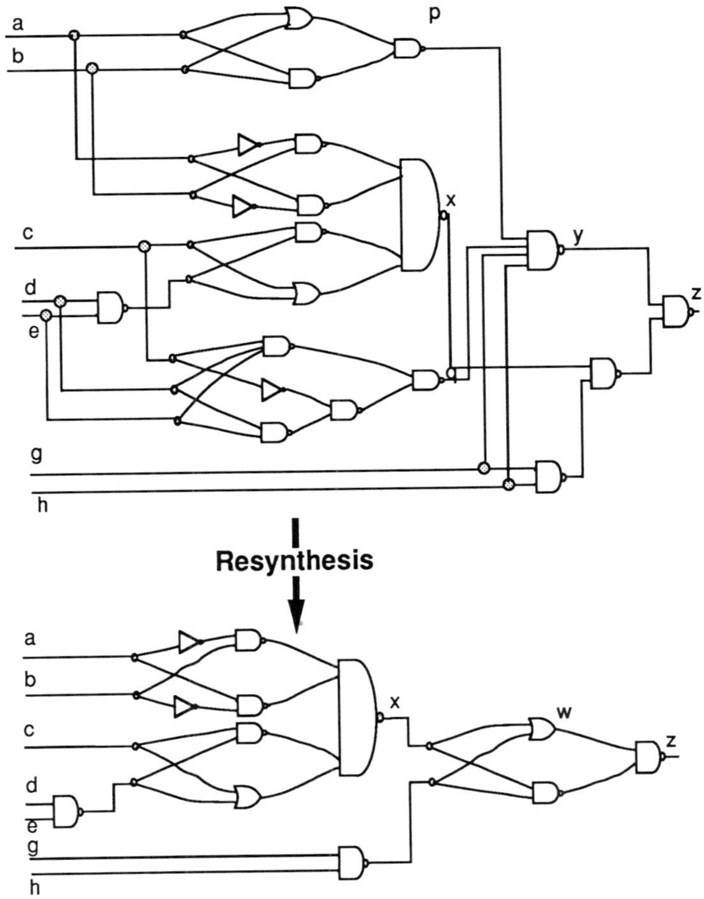

Figure 4. Irredundant corolla before and after resynthesis

5 Don't Cares in Corolla Resynthesis

It has been shown that don't cares play an important role in two level as well as multi-level logic minimization [8,9]. However, the don't care sets of the nodes of a large multi-level network can be huge and time consuming to compute. Of all the possible don't cares that can be extracted for a node, not all of them are useful in minimizing the boolean function at the node. We attempt to reduce the complexity of extracting don't cares by using the corolla partitions. We will extract don't care sets for nodes of the circuit where we expect to gain, and select a subset of the possible don't cares which we think will be useful to minimize the node.

5.1 Intermediate Fanin Don't Cares

The Intermediate Fanin Don't Care Set for a node x in the circuit graph, $IDC(x)$, is the boolean expression $x\bar{f_x} + \bar{x}f_x$. The Intermediate Fanin Don't Care Set for all the nodes in the graph is defined as, $IDC = \sum_{\forall x} IDC(x)$.

It has been shown previously [9] that a boolean function f_y at node y can be minimized by using the intermediate fanin don't care set, IDC. We are in effect expressing f_y in terms of boolean functions of other nodes (boolean resubstitution). However, for many circuits, the IDC set can be large. Computing the IDC set over all nodes and minimizing f_y using the whole IDC set may become impractical. Attempts have been made to find useful subset of IDC to minimize f_y by using filters to weed out unuseful cubes of the IDC [9]. However, this technique still computes the whole IDC set, and every time a node has to be minimized, the heuristic has to be used to find out the useful part of the IDC set. Corolla partitions provide an opportunity to target only a subset of nodes to minimize using don't cares, and to compute the useful subsets of IDCs only.

For a node x inside a corolla, we define f_x, and hence $IDC(x)$, in terms of C-Support(x). Note that $|\text{C-Support}(x)| \leq |\text{Support}(x)|$.

Property 3 *For IDC(x) of any node x to be useful in minimizing another node y, it is necesssary that C-Support(x)$\cap$ C-Support(y) $\neq \phi$.*
A node x satisfying property 3 is considered a good candidate for boolean substitution in f_y.

5.2 Minimization of Corollas Using IDC of Internal Nodes

Since a corolla has fewer nodes than the whole circuit, it is possible to find the IDC for all the internal nodes of the corolla, and use them to minimize other internal nodes of the corolla. Referring to the corolla of Figure 4, we notice that C-Support(x) = $\{a, b, c, d, e\}$, and C-Support(y) = $\{a, b, c, d, e, g, h\}$. Node x satisfies the property of being a good candidate for boolean resubstitution in y. Using IDC(x) to substitute x in y, we minimize f_y to get $f_y = x + \overline{gh}$ (node w in Figure 4). This reduces the corolla to the same minimized corolla that we had obtained earlier by flattening and resynthesis.

If a corolla C or I(C) is too large for flattening and resynthesis, we can still use the alternative approach of using intermediate fanin don't care sets to minimize internal nodes of the corolla, by finding internal nodes which satisfy property 3.

5.3 Minimization of Corollas Using IDC of other Corollas

Given a circuit partitioning into corollas, we have demonstrated gains by resynthesizing each corolla independently. We now attempt to take a more global view to further minimize the output functions of corollas. Partitioning a stem region into two or more petals generates two or more corollas with overlapping C-Support sets. To minimize an ouput y of a corolla, we can use the intermediate don't care sets of the corolla outputs which are good candidates according to property 3; that is, we can use IDC(x_i) of corolla outputs x_i which have overlapping C-Support sets with y. Note that there is no additional overhead in computation of IDC(x_i) since we have already computed and stored the onset/offset of each corolla.

Refering to Figure 5, output nodes x and y of Corollas C_1 and C_2 have overlapping C-Support sets, and hence we can further minimize f_y using IDC(x). In general, in synthesizing each corolla using IDCs of other corollas, we can improve over resynthesizing each corolla independently.

As an example of minimizing one corolla using the IDC set of another, we refer to Figure 6, which shows two corollas extracted from a larger circuit. Both the corollas are irredundant, and resynthesis of them taken together, using misII, produces no gain (has 23 transistor pairs). Since the C-Support sets of the two corollas overlap, we try to minimize f_y with the further assistance of IDC(x). In this case, f_y reduces to $\bar{c}\bar{x} + cx$, and the final reduced circuit has fewer stems, branches and transistor pairs (reduced to 15 transistor pairs).

In particular, the output nodes of a multi-output corolla C have overlapping C-Support sets, and hence are good candidates for boolean resubstitution in the output nodes of C. Refering to Figure 5, output nodes w and z have overlapping C-Support sets, and hence we can further minimize f_z using IDC(w). Also, since the C-Support sets of corolla outputs x and y are also

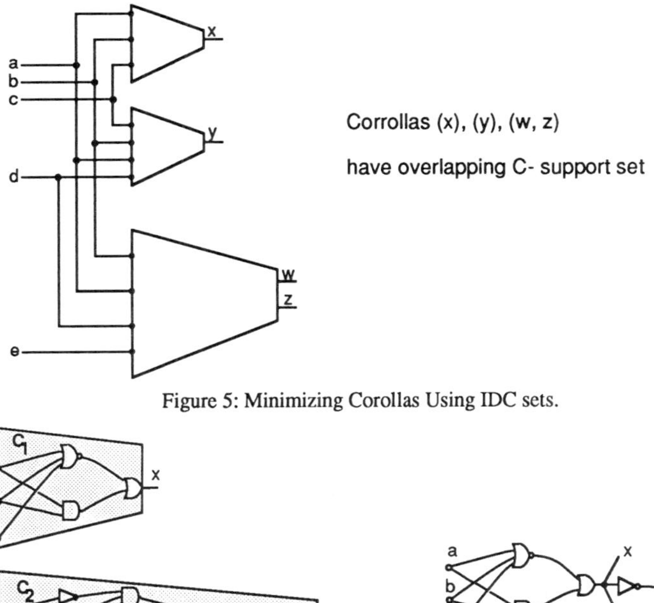

Figure 5: Minimizing Corollas Using IDC sets.

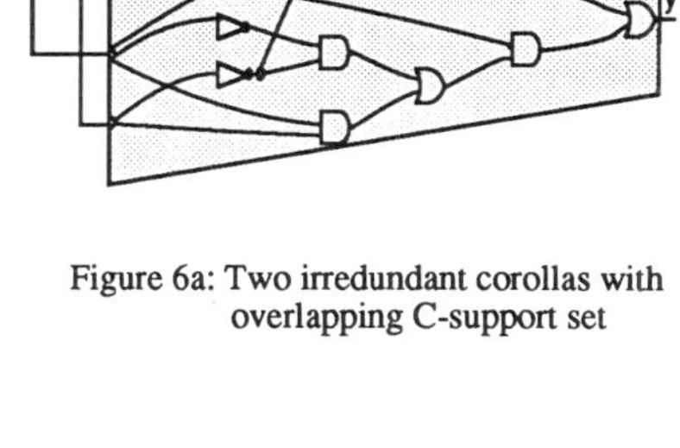

Figure 6a: Two irredundant corollas with overlapping C-support set

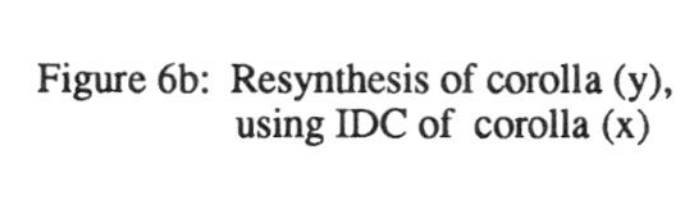

Figure 6b: Resynthesis of corolla (y), using IDC of corolla (x)

overlapping with the C-Support set of z, we can use IDC(x) and IDC(y) to minimize f_z. For example, we extracted a 2-output corolla *r44* from the ISCAS85 benchmark circuit c3540. The original multi-level circuit had a 36 transistor pairs, and misII resynthesized the 2-level r44 to produce a much larger circuit having 87 transistor pairs. Minimizing one output f_y using IDC(x) produces a much smaller circuit with 38 transistor pairs.

5.4 Observability Don't Cares

The Observability Don't Care set for a node y in the circuit, ODC(y), is the complement of OBS(y), the set of primary input cubes which makes y observable at one or more primary outputs of the circuit. However, for any node y belonging to a corolla C, $OBS(y)$, and $ODC(y)$, can now be measured in terms of $I(C)$ and $O(C)$ only, and hence the computation is limited to the corolla only.

Property 4 *For ODC(y) of any node y to be useful in minimizing f_y, it is necessary that* $Support(ODC(y)) \cap Support(y) \neq \phi$.

Property 5 *A node y has to be an interior node of a primary stem region (petal) for it to have an ODC set useful in minimizing f_y.*

Proof: Let y be a node which has an useful ODC set. Then there exists a boolean variable a which belongs to both the support sets of y and ODC(y). Since a drives y, as well as another node z in the path of y to a primary output (so as to make y unobservable), a has to be a reconvergent stem reconverging at z. By definition of stem region, y is an internal node of the stem region of reconvergent stem a.

Property 5 excludes nodes in fan-out free regions from further consideration for extracting and using ODC sets. If the circuit is partitioned into maximal corollas, nodes of corollas are the only nodes to be explored for ODC sets. Corollas which are too large to be flattened and resynthesized may be further partitioned into smaller corollas that are better matched to the capabilities of the synthesis tools. This results in loss of global information, and resynthesis of a smaller corolla C might not be as effective as resynthesis of the larger corolla could have been. However, use of ODC set of C, with the existing onset/offset of C, helps retrieve some of the global information for the smaller corolla.

As an example, assume that the corolla in Figure 3 is too large to be flattened and globally resynthesized. In this case, we would extract two interior corollas EQ and GT, and their individual resynthesis produces a corolla with a total of 27 transistor pairs (10 for GT, 14 for EQ). This is far worse than minimization of the larger corolla which had produced a circuit with 11 transistor pairs only. Since corollas EQ and GT are embedded in a larger corolla, we can minimize each of the individual corollas using ODC(y) $= f_z$ and ODC(z) $= f_y$. Resynthesizing using ODC sets, GT and EQ are reduced to 8 and 6 transistor pairs respectively, so the overall corolla now totals 17 transistor pairs. This is significant improvement over the resynthesis of the individual corollas without using ODC sets.

6 Conclusions

We have proposed a new technique to partition circuit graphs, and have demonstrated its usefulness by applying it to reduce layout area of large multi-level circuits through logic resynthesis. We have discussed the use of don't cares in further minimizing the corollas. We are still researching heuristics to select corollas which will resynthesize with even larger gain using don't cares.

References

[1] R. Brayton, R. Rudell, A. Sangiovanni-Vincentelli, and A. Wang. MIS: A Multiple-Level Logic Optimization System. *IEEE Transactions on Computer-Aided Design*, cad-6(6):1062 - 1081, Nov. 1987.

[2] R. Lisanke, G. Kedem, and F. Brglez. *DECAF: Decomposition and Factoring for Multi-Level Logic Synthesis, Version 1.0.* Technical Report TR87-15, Microelectronics Center of North Carolina, Research Triangle Park, NC, August 1987.

[3] J.A. Darringer, Jr. W.H. Joyner, C.L. Berman, and L. Trevillyan. Logic Synthesis Through Local Transformation. *IBM Journal of Research and Dev.*, 25(4):272 - 280, July 1981.

[4] D. Brand. Redundancy and Don't Cares in Logic Synthesis. *IEEE Trans. on Computers*, C-32(10), October 1983.

[5] L. Berman and L. Trevillyan. Improved Logic Optimization Using Global Flow Analysis. In *IEEE Intl. Conf. on Computer-Aided Design*, pages 102 - 105, Nov. 1988.

[6] H. Cho, G. Hachtel, M. Nash, and L. Setiono. BEAT_NP: A Tool for Partitioning Boolean Networks. In *IEEE Intl. Conf. on Computer-Aided Design*, pages 10 - 13, Nov. 1988.

[7] R. Camposano and R. Brayton. Partitioning before Logic Synthesis. In *IEEE Intl. Conf. on Computer-Aided Design*, pages 324 - 326, Nov. 1987.

[8] K. Bartlett, R. Brayton, G. Hachtel, R. Jacoby, R. Rudell, A. Sangiovanni-Vincentelli, and A. Wang. Mulit-Level Logic Minimization Using Implicit Don't Cares. In *IEEE Intl. Conf. on Computer Design*, pages 552 - 557, October 1986.

[9] A. Saldanha, A. Wang, R. Brayton, and A. Sangiovanni-Vincentelli. Multi-Level Logic Simplification using Don't Cares and Filters. In *26th ACM/IEEE Design Automation Conference*, June 1989.

[10] S. Dey, F. Brglez, and G. Kedem. *Corolla-Based Circuit Partitioning and Applications.* Technical Report , Microelectronics Center of North Carolina, Research Triangle Park, NC, May 1990.

[11] S. Dey, F. Brglez, and G. Kedem. Corolla Based Circuit Partitioning and Resynthesis. In *27th ACM/IEEE Design Automation Conference*, June 1990.

[12] F. Maamari and J. Rajski. A Reconvergent Fanout Analysis for Efficient Exact Fault Simulation of Combinational Circuits. In *18th Intl. Sym. on Fault Tol. Comp.*, June 1988.

[13] A. V. Aho, J. E. Hopcroft, and J. D. Ullman. *The design and Analysis of Computer Algorithms.* Addison-Wesley Publishing Company, 1974.

[14] R. Rudell. *Multiple-Valued Logic Minimization for PLA Synthesis.* Technical Report, University of California, Electronics Research Laboratory, Berkeley, CA, June 1986.

[15] R. Lisanke, F. Brglez, and G. Kedem. McMAP: A Fast Technology Mapping Procedure for Multi-Level Logic Synthesis. In *IEEE Int. Conf. Computer Design*, October 1988.

[16] *VPNR User Guide.* Technical Report, MCNC, RTP, NC, 1988.

[17] F. Brglez, D. Bryan, J. Calhoun, G. Kedem, and R. Lisanke. Automated Synthesis for Testability. *IEEE Trans. on Ind. Electronics*, 36(2):263–277, May 1989.

Low Level Synthesis II

Logic and Architecture Synthesis
P. Michel and G. Saucier (Editors)
Elsevier Science Publishers B.V. (North-Holland)

COMBINING SERIAL DECOMPOSITION WITH TOPOLOGICAL PARTITIONING FOR EFFECTIVE MULTI-LEVEL PLA IMPLEMENTATIONS

Tadeusz ŁUBA, Jerzy KALINOWSKI, Krzysztof JASIŃSKI, and Andrzej KRAŚNIEWSKI

Institute of Telecommunications
Warsaw University of Technology
Warszawa, Poland

In this paper an effective logic synthesis method aimed at multi-level PLA implementations is presented. It combines logic decomposition with topological optimization of the resulting components to minimize the silicon area of the implementation. Experimental results are provided and assessed in terms of the estimated circuit area.

1. INTRODUCTION

A lot of research effort has recently been directed to multi-level logic structures. Most of the proposed multi-level logic synthesis methods support the gate array design style [1, 2]. These methods are, however, not suitable for multi-level PLA structures that, due to their compact layout style, consume significantly less silicon area than gate array based circuits [3].

In this paper we propose an effective logic synthesis method based on decomposition. Its distinctive feature is that the decomposition process is carried out as the very first step in logic synthesis. This general design strategy illustrated in Fig. 1 can be applied for a variety of implementation styles, such as PLAs, gate arrays, or even programmable logic devices. In this paper we will focus on the serial decomposition aimed primarily at multi-level PLA structures and its potential capabilities to reduce the silicon area of the actual implementations.

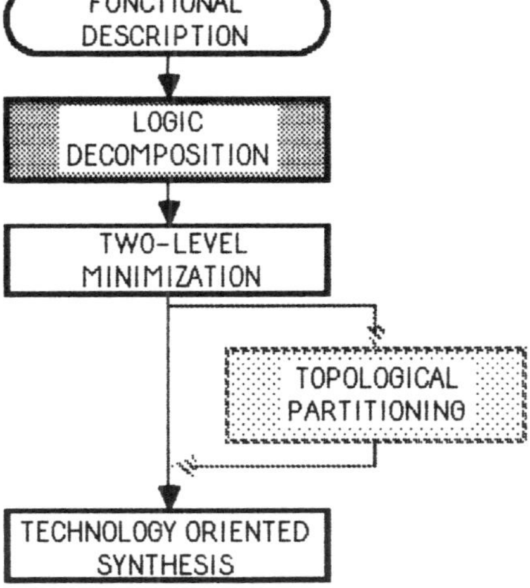

Fig. 1: Logic synthesis based on decomposition.

For a long time, the idea of Boolean decomposition, formulated originally by Curtis in 1962 [4], was applied rather infrequently in practical chip designs because of the lack of effective algorithms. In recent years, however, it is gaining more and more attention. Spectral methods of logic decomposition have been presented by Varma [5]. The complexity of the proposed algorithms precludes, however, their practical application. Devadas et al. have shown in [6] that re-encoding of selected inputs reduces the number of product terms in the logic function. The re-encoding network is serially connected with the PLA that implements the reduced function. The general disjunctive decomposition is the preferred technique of the GATEMAP logic synthesis system [7]. The GATEMAP synthesizer is used to partition complex logic equations. A similar approach, but based on an algebraic division method, has been presented in [8].

In this paper we propose a new algorithm that attempts to resolve the general problem of disjunctive functional decomposition of incompletely specified multiple-output Boolean functions. Our experience has shown that the serial decomposer is very powerful and can significantly reduce the silicon area of the actual implementation. As shown in Fig. 1, the results of the decomposition procedure can be directly employed to find an effective physical arrangment of the obtained components using topological partitioning. An automated synthesis system, called PLATO, developed at the Warsaw University of Technology, Institute of Telecommunications, implements the proposed design strategy based on logic decomposition. PLATO includes several logic optimization procedures that allow designers to examine various design strategies and select the one which is most suitable for a given design task.

This paper is organized as follows. First, the symbolic decomposition idea is formulated. An effective algorithm that finds serial decomposition of a logic function is then presented. This is followed by discussing the concept of topological partitioning of the resulting components. Finally, experimental results and their assessment are provided.

2. SYMBOLIC DECOMPOSITION

Consider a multiple-output Boolean function f: $\{0,1\}^n \rightarrow \{0,1,-\}^m$, where "-" is a don't care value. Let $X = \{x_1,..., x_n\}$ be the input support set for function f.

Serial decomposition uses a Symbolic Partition Description (SPD) that represents the relation among cubes in the cover of f. In the SPD representation, function f is described as a set of n two-block input partitions, $\{P(x_i)\}$, $_{i = 1, ..., n}$, and a multiple-block output partition, P_0.

<u>Example 1</u>: Consider function f: $\{0,1\}^5 \rightarrow \{0,1,-\}^3$ given by the set of cubes in Table 1. Each input x_i generates a two-block input partition $P(x_i)$ in which one block includes all cubes for which x_i = 0 and the other block all cubes for which x_i = 1. Note that the generated input partitions presented in Table 2 are not necessarily output-disjoint. A block in the multiple-output partition P_0 includes cubes which produce the same 3-bit output. The resulting 5-block output partition is also shown in Table 2. ◘

A general method of symbolic decomposition using the SPD representation was developed in 1986 [9]. For a given multiple-output Boolean function f, it finds two components, g and h, connected as shown in Fig. 2. The key concept in this method is to divide the input support set, X, into two disjoint sets, A and C, which are subsets of the input support sets for components h and g, respectively. The input support set for component h includes also the outputs of g. For component g, there may exist an additional set of inputs, B, that includes the inputs which are shared between g and h, i.e., $B \subseteq A$.

Table 1: Example Boolean function f.

cube	x_1	x_2	x_3	x_4	x_5	y_1	y_2	y_3
1	0	0	0	0	0	0	0	0
2	0	0	-	1	1	0	1	0
3	0	1	0	1	0	1	0	0
4	0	1	1	1	1	1	0	1
5	0	1	1	0	-	0	0	1
6	0	1	0	0	0	0	0	1
7	1	1	0	1	0	0	0	0
8	1	0	0	1	1	1	0	0
9	1	0	0	1	0	0	0	1
10	1	0	1	1	1	0	0	0

Table 2: Symbolic Partition Description of function f of Table 1.

$P(x_1) = (1,2,3,4,5,6;\ 7,8,9,10)$
$P(x_2) = (1,2,8,9,10;\ 3,4,5,6,7)$
$P(x_3) = (1,2,3,6,7,8,9;\ 2,4,5,10)$
$P(x_4) = (1,5,6;\ 2,3,4,7,8,9,10)$
$P(x_5) = (1,3,5,6,7,9;\ 2,4,5,8,10)$
$P_0 = (1,7,10;\ 2;\ 3,8;\ 4;\ 5,6,9)$

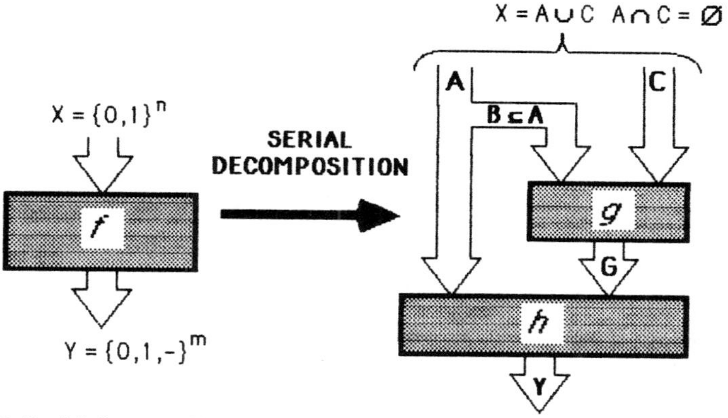

Fig. 2: Serial decomposition of a multiple-output Boolean function f into two components, g and h

<u>Theorem 1</u> [9]: Components g and h represent a two-level serial decomposition of function f iff there exists a multiple-block partition $\Pi_g \geq P_g = P(B \cup C)$, such that

$$P(A) \cdot \Pi_g \leq P_0 \qquad (1)$$

where all the partitions are defined over the set of cubes and $P(U) = \bigcap_{x_i \in U} P(x_i)$.

This theorem states the necessary condition for the serial decomposition in which $f = h(A, g(B \cup C))$. The interpretation is as follows. Partition Π_g represents component g and the product of partitions $P(A)$ and Π_g corresponds to h. The truth tables of the resulting components can be easily obtained from these partitions.

In this paper we propose an algorithm which finds decomposition in which the number of inputs for component h is equal to a given $r < n$. As the very first step, our algorithm determines a set of direct inputs to block h, $A \subseteq X$ (see Fig. 2), using an r-admissibility criterion.

<u>Definition 1</u>: A subset of k input variables, $A = \{x^A_1, x^A_2, \ldots, x^A_k\} \subseteq X$, which generates partition $P(A) = P(x^A_1) \cdot P(x^A_2) \cdot \ldots \cdot P(x^A_k)$ is r-admissible iff there exists a set of r-k partitions P_{k+1}, $P_{k+2}, \ldots, P_r$ such that

$$P(A) \cdot P_{k+1} \cdot P_{k+2} \cdot \ldots \cdot P_r \leq P_0 \qquad (2)$$

The r-admissibility relation can be interpreted in the following way. In a serial decomposition shown in Fig. 2, component h has r inputs: k primary inputs included in set A, and the remaining r-k inputs being the outputs of component g. The degree of admissibility for a k-element input support subset A can be practically found using the following formula:

$$r = k + \lceil \log_2 \mu(P(A)/P_0) \rceil \qquad (3)$$

where $\mu(P(A)/P_0)$ is the number of elements in the largest block of a quotient partition $P(A)/P_0$ [10].

Example 1(cont'd): For the function f of Table 1 consider a 3-element input support subset $A = \{x_1, x_2, x_3\}$. This subset generates the following quotient partition

$$P(A)/P_0 = ((1)(2);\ (3)(6);\ (4)(5);\ (8)(9);\ (7);\ (10))$$

with the largest partition block composed of 2 elements. Hence, $r = 3 + \lceil \log_2 2 \rceil = 4$, which means that function f can be represented as $f = h(x_1, x_2, x_3, g(x_4, x_5, C))$, where component h has 4 inputs (see Fig. 3). ◻

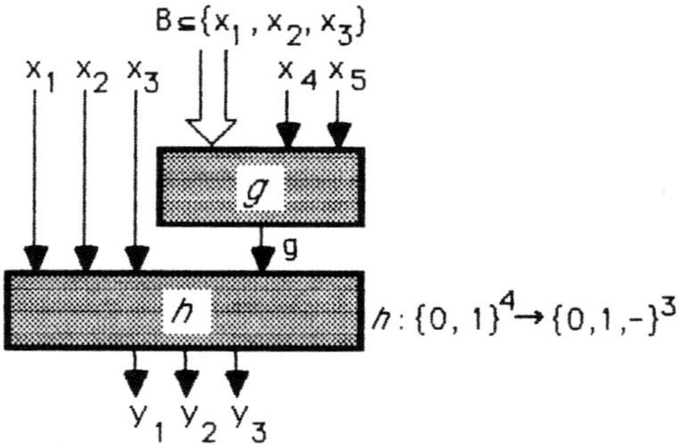

Fig. 3: Serial decomposition of function f of Table 1 generated by the 4-admissible input support set $A = \{x_1, x_2, x_3\}$

For a given r, our algorithm determines an r-admissible input support subset $A \subseteq X$. The next task is to find a subset of inputs for component g that serially connected with component h will implement function f (Fig. 2), i.e. to find P_g, such that there exists $\Pi_g \geq P_g$ that satisfies condition (1) in Theorem 1. To solve this problem, consider a subset of primary inputs, $D = B \cup C$, and an m-block partition $P(D) = (B_1; B_2; \ldots; B_m)$ generated by this subset. A relation of compatibility of partition blocks will be used to verify whether or not partition P(D) is suitable for serial decomposition.

Definition 2: Two blocks $B_i, B_j \in P(D)$ are compatible iff partition Π'_g obtained from partition P(D) by merging blocks B_i and B_j into a single block B'_{ij} satisfies condition (1) in Theorem 1, i.e., iff

$$P(A) \cdot \Pi'_g \leq P_0$$

Definition 3: A subset of n partition blocks, $\mathcal{B} = \{B^c_1, B^c_2, \ldots, B^c_n\}$, where $B^c_i \in P(D)$, is a class of compatible blocks for partition P(D) iff all blocks in $\mathcal{B}$ are pairwise compatible.

When evaluating an input support subset D with the corresponding partition P(D), we generate a set of all maximal classes of compatible blocks, $\mathcal{B}_{max}$. Then, from elements of $\mathcal{B}_{max}$ we form a minimal cover, $\mathcal{B}_{cov}$, of m blocks in partition P(D). Clearly, if card($\mathcal{B}_{cov}$) $\leq 2^{r-k}$, then partition $P(D)= P_g$ guarantees that function h can be implemented with only r arguments.

For a given r and earlier determined input support subset A, our serial decomposer first checks the condition for disjoint implementation of both components by assuming that $D = X - A$. When the result of this test is negative, we add to subset D a single variable $x \in A$. The selection of the variable x is based on heuristic criteria. This procedure is repeated until condition (1) in Theorem 1 is satisfied.

Example 1 (cont'd): Consider again the function f of Table 1 and the 4-admissible input support subset $A = \{x_1, x_2, x_3\}$. First, we will check for the existence of disjoint decomposition. The input support subset $D = X - A = \{x_4, x_5\}$ generates the 4-block partition $P(D) = (B_1; B_2; B_3; B_4)$ where $B_1 = \{1,5,6\}$, $B_2 = \{5\}$, $B_3 = \{3,7,9\}$, and $B_4 = \{2,4,8,10\}$. There are only two pairs of compatible blocks in partition P(D): (B_1, B_2) and (B_2, B_3). Thus, $\mathcal{B}_{cov} = \{(B_1,B_2), (B_2,B_3), B_4\}$, and since $card(\mathcal{B}_{cov}) = 3 > 2^{r-k}$, no disjoint decomposition exists.

Now, we will extend subset D by adding input variable $x_1 \in A$. For $D = \{x_1, x_4, x_5\}$ we obtain

$$P(D) = (B_1 = \{1,5,6\}; B_2 = \{5\}; B_3 = \{3\}; B_4 = \{7,9\}; B_5 = \{2,4\}; B_6 = \{8,10\}).$$

The resulting maximal classes of compatible blocks are

$$\mathcal{B}_{max} = \{(B_1,B_2,B_4), (B_1,B_2,B_6), (B_2,B_3,B_4), (B_2,B_3,B_6), (B_3,B_4,B_5), (B_3,B_5,B_6)\}$$

One of feasible minimal covers is $\mathcal{B}_{cov} = \{(B_1,B_2,B_4), (B_3,B_5,B_6)\}$; it produces partition $\Pi'_g = (1,5,6,7,9;\ 2,3,4,8,10)$. It is easy to check that $P(A) \cdot \Pi'_g \leq P_0$, and therefore,

$$f = h(x_1, x_2, x_3, g(x_1, x_4, x_5)) .$$

The truth tables of components g and h can be obtained from partitions P(A), Π'_g, and P_0. Encoding the first block of Π'_g as "0", and the second block as "1", we immediately obtain the truth table of function $g: \{0,1\}^3 \rightarrow \{0,1,-\}$; it is presented in Table 3. The truth table of function h can be derived by reencoding all cubes in Table 1 using an intermediate variable $g = g(x_1, x_4, x_5)$; for example, cube 1 in Table 1 can be represented as 0000 because $g(0,0,0) = 0$. The truth table obtained in this way is shown in Table 4. ◻

Table 3: Tabular representation of component g

x_1	x_4	x_5	g
0	0	-	0
0	1	0	1
1	1	0	0
0	1	1	1
1	1	1	1

Table 4: Tabular representation of component h

x_1	x_2	x_3	g	y_1	y_2	y_3
0	0	0	0	0	0	0
0	0	-	1	0	1	0
0	1	0	1	1	0	0
0	1	1	1	1	0	1
0	1	1	0	0	0	1
0	1	0	0	0	0	1
1	1	0	0	0	0	0
1	0	0	1	1	0	0
1	0	0	0	0	0	1
1	0	1	1	0	0	0

3. TOPOLOGICAL PARTITIONING

In the next step of logic synthesis, the components obtained from serial decomposition are minimized using one of the standard two-level minimizers. Then, the circuit can be further optimized using either folding, in the case when function f is to be implemented as a multi-level PLA structure, or another form of partitioning. PLA folding is aimed at reducing the silicon area consumed by circuit implementation. It examines properties of a Boolean function to find a physical rearrangement of PLA rows and columns in which unused PLA space is reduced. The optimal folding has been shown to be an

NP-complete problem [11], and most automatic methods use heuristic techniques to obtain near-optimal solutions [12, 13].

In our synthesis system we apply a technique that enables physical circuit optimization, without any computational effort. The results of the serial decomposition can be directly applied for the PLA partitioning. The components g and h obtained from serial decomposition can be implemented in a block-folded PLA. To illustrate this idea, consider a 3-bit adder [14]; the result of its serial decomposition is shown in Fig. 4. Two obtained PLAs: the 11-term PLA1 implementing component g and the 17-term PLA2 realizing component h can be arranged in two ways; using either column folding or row folding, as illustrated in Fig. 5. The appropriate structure can be selected depending on the circuit floorplan and other design constraints.

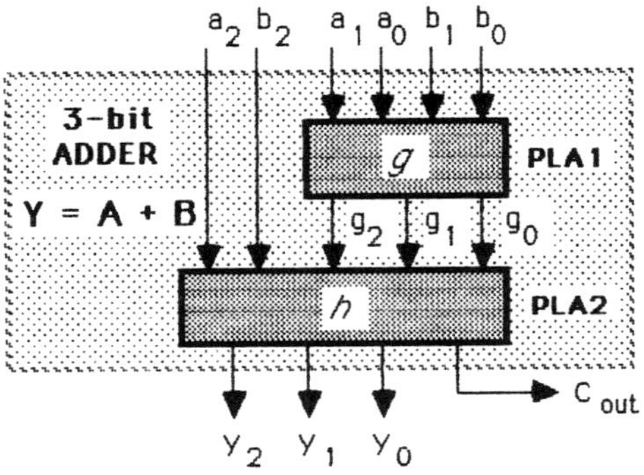

Fig. 4: Serial decomposition of the 3-bit adder.

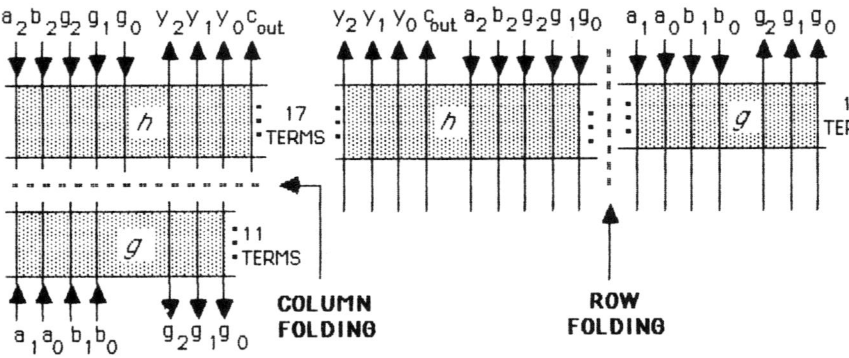

Fig. 5: Column folding and row folding for the multi-level PLA implementation of the 3-bit adder circuit of Fig. 4.

4. EXPERIMENTAL RESULTS

To assess the efficiency of our synthesis method, the procedures implementing serial decomposition have been excercised for several benchmark examples for logic synthesis and optimization [15]. The benchmarks are given in minimized tabular representations, which in fact restricts a class of optimal solutions in serial decomposition. To demonstrate real power of the serial decomposer, it would be necessary to have an opportunity to first decompose the original circuits, and then to minimize the obtained components.

Experimental results for selected circuits are summarized in Table 5. In the column headed *f*, the basic characteristics of the original function and the results of the conventional two-level minimization using ESPRESSO II [16] are given. The silicon area of the resulting PLA has been estimated as S = (2n + m)P, where n and m denote the number of inputs and outputs, respectively, and P is the number of product terms in a minimized form. Characteristics of the components resulting from the serial decomposition are provided in the columns headed *g* and *h* in a similar fashion.

Table 5: Results of logic synthesis based on serial decomposition combined with topological partitioning

	f				*g*				*h*				
	n	m	P	S	n	m	P	S	n	m	P	S	E
adr4	8	5	75	1575	6	4	42	672	6	5	25	425	30%
rd84	8	4	255	5100	5	3	22	286	6	4	35	560	83%
add4	9	5	135	3105	5	3	20	260	7	5	44	836	65%
add3	7	4	59	1062	5	3	23	299	5	4	11	154	58%
mod18add	10	5	224	5600	5	3	19	247	8	5	111	2331	54%
mod20add	8	5	139	3475	5	4	19	266	9	5	86	1978	35%
alu	8	4	58	1160	4	3	5	55	7	4	54	972	11%

Table 5 shows the estimated area saving factor, $E = 1 - S_c/S_f$, where S_c is the estimated area of PLAs resulting from the folding procedure. The area saving factor is as high as 83% for the rd84 circuit [15], while the standard folding applied to non-decomposed function cannot reduce the circuit area by more than a half. When the rd84 function is decomposed into 3 components connected as shown in Fig. 6, then the saving factor is 90% which is a much better accomplishment than the best result reported so far in the available literature [6].

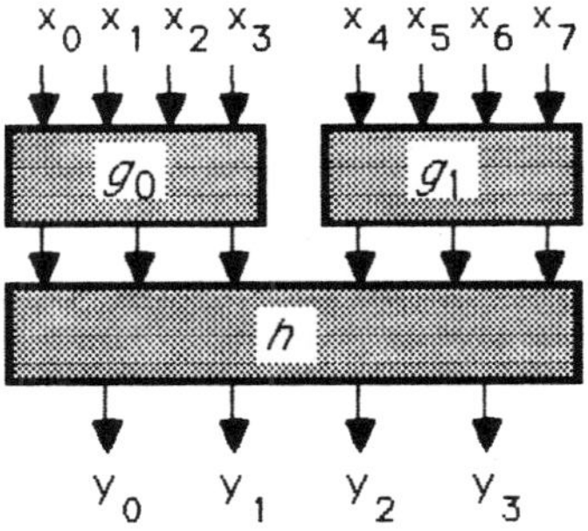

Fig. 6: 3-component decomposition of the rd84 function.

From the obtained results, one can conclude that the logic synthesis based on the serial decomposition significantly reduces silicon area of the actual circuit implementation. Although our approximations underestimate the silicon area taken by input/output buffers and border PLA cells, the obtained results are still very promising.

5. PLATO SYSTEM

The characteristic feature of the presented design methodology is that the decomposition process is carried out before logic minimization (Fig. 1). An automated design system, called PLATO, based on this design strategy, has been developed at the Warsaw University of Technology, Institute of Telecommunications. The organization of the PLATO system is shown in Fig. 7.

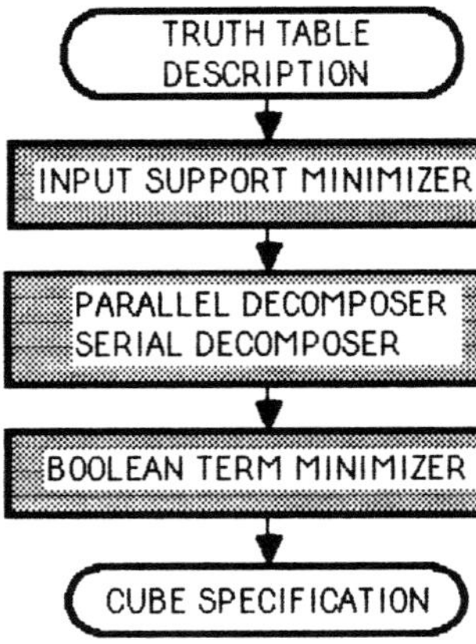

Fig. 7: Organization of the PLATO system.

PLATO uses a tabular representation (or any equivalent form) of a multiple-output Boolean function f. The tabular representation is then converted into a more compact form to reduce the computational complexity of the PLATO procedures.

As the first step of logic synthesis, PLATO reduces the number of input variables upon which function f depends. Our input support minimizer finds, for a given function f, its equivalent form with a minimal number of arguments. Clearly, this problem is NP-complete. The computational effort can, however, be substantially reduced using an argument representation (Petrick's equation) of Boolean functions. This is in fact a compacted form of the tabular representation. The complexity of the argument reduction can be further decreased by eliminating from the search all essential input variables, that is input variables that are arguments of any equivalent form of f. Our experience has shown that this step significantly reduces computational effort associated with the subsequent design steps without any remarkable loss of its optimality.

In the next step, the original function is decomposed into separately synthesized components. The minimization of the silicon area of the implementation is the primary objective of the PLATO decomposition procedures. We propose two basic logic decomposition strategies:

- parallel decomposition, and
- serial decomposition.

In the parallel decomposition illustrated in Fig. 8, the set of outputs is partitioned into two or more disjoint subsets. The goal is to find an output partition that minimizes either the global number of inputs for the resulting components or the number of inputs for a component with the largest input support set [17]. Usually, a two-block partition is the most effective with regard to the circuit area reduction.

When the parallel decomposer does not provide satisfactory results, we employ serial decomposition. The components obtained from logic decomposition can be further optimized using standard logic synthesis tools like the ESPRESSO minimization procedure. The results of benchmark experiments indicate that the design strategy implemented in the PLATO system is very effective.

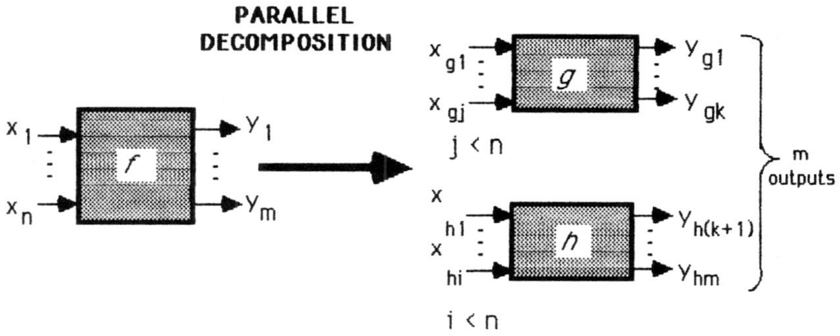

Fig. 8: Parallel decomposition of a multiple-output Boolean function *f* into two components, *g* and *h*.

6. SUMMARY

In this paper we have proposed a general method for decomposition of incompletely specified multiple-output Boolean functions. The symbolic decomposition algorithm is simple and, even for complex circuits, does not involve significant computational effort.

Our experiments have shown that, when a Boolean function is decomposed before the two-level minimization is applied, then the minimization process is more effective. By combining serial decomposition with topological partitioning we have obtained more than 50% of the silicon area savings for several benchmark circuits. Furthermore, the experimental results indicate that our method gives better results than most of the commonly used synthesis tools.

In the paper we have focused on the application of serial decomposition to synthesis of multi-level PLA structures. However, this concept can also be effectively used for logic synthesis based on programmable logic devices with limited number of inputs and outputs. Again, we have obtained much better results than those reported in the literature. Advantages of our approach can be demonstrated by comparing various PLD-based implementations of the rd84 benchmark function. Using the design method presented in [8], this function was reported to require 5 PALs from the MMI PAL library. Our design procedure yields a spectacular result: just one PAL device is sufficient. The implementation of the rd84 function decomposed as shown in Fig. 6 using an Am PAL22P10 circuit is illustrated in Fig. 9; a similar solution can be obtained using a pal20s10 device from the earlier mentioned MMI PAL library.

The serial decomposer has been implemented as part of the PLATO logic synthesis system. The PLATO system that exploits logic decomposition is in fact a very effective preprocessor for standard logic design tools. Currently, we are augmenting the PLATO system with procedures that support selection of optimal design strategy and improve circuit testability [18].

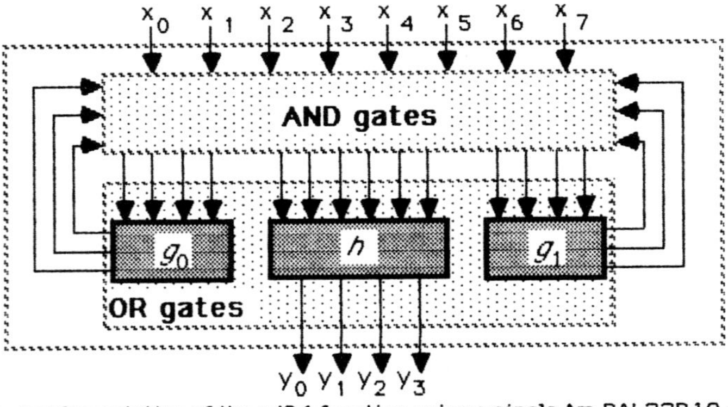

Fig. 9: Implementation of the rd84 function using a single Am PAL22P10 device

REFERENCES

1. Brayton R.K., Rudell R., Sangiovanni-Vincentelli A., Wang A.R., "MIS: Multiple-Level Logic Optimization System", IEEE Trans. on Computer-Aided Design, pp. 1062-1081, Nov. 1987.
2. Mathony H.J., Baitinger U.G., "CARLOS: An Automated Multilevel Logic Design System for CMOS Semi-Custom Integrated Circuits", IEEE Trans. on Computer-Aided Design, pp. 346-355, March 1988.
3. Malik S., Katz R.H., "Combining Multi-Level Decomposition and Topological Partitioning for PLAs", Proc. International Conf. on Computer Design, pp. 112-115, Sept. 1988.
4. Curtis H.A., "A New Approach to the Design of Switching Circuits", D. Van Nostrand Co., Princeton, 1962.
5. Varma D., Trachtenberg E.A., "Design Automation Tools for Efficient Implementation of Logic Functions by Decomposition", IEEE Trans. on Computer-Aided Design, pp. 901-916, Aug. 1989.
6. Devadas S.,. Wang A.R, Newton A.R., Sangiovanni-Vincentelli A., "Boolean Decomposition in Multi-Level Logic Optimization", Proc. International Conf. on Computer-Aided Design, pp. 290-293, Nov. 1988.
7. Pitty E.B., "A Critique of the GATEMAP Logic Synthesis System", Proc. International Workshop on Logic and Architectural Synthesis for Silicon Compilers, pp. 65-84, May 1988.
8. Saucier G. , Sicard P., Bouchet L., "Multi-Level Synthesis on PALs", Proc. European Design Automation Conference, pp. 542-546, March 1990.
9. Łuba T., "A Uniform Method of Boolean Function Decomposition", Rozprawy Elektrotechniczne (Journal of Polish Academy of Science), No. 4, pp. 1041-1054, 1986.
10. Torng H.C., "Switching Circuits - Theory and Logic Design", Addison-Wesley Publ. Co., 1972.
11. Luby M., Vazirani U., Vazirani V., and Sangiovanni-Vincentelli A., "Some Theoretical Results on the Optimal PLA Folding", Proc. IEEE Conf. on Circuits and Computers, pp. 165-170, 1982.
12. Egan J.R., Liu C.L., "Bipartite Folding and Partitioning of PLA", IEEE Trans. on Computer-Aided Design, pp. 191-199, July 1984.
13. Lecky J.E., Murphy O.J., Absher R.G., "Graph Theoretic Algorithms for the PLA Folding Problem", IEEE Trans. on Computer-Aided Design, pp. 1014-1021, September 1989.
14. Sasao T., "Input Variable Assignment and Output Phase Optimization of PLA's", IEEE Trans. on Computers, pp. 879-894, Nov. 1984.
15. Logic Synthesis and Optimization Benchmarks, 1989 MCNC International Workshop on Logic Synthesis.
16. Brayton R.K., Hachtel G.D., McMullen C.T., Sangiovanni-Vincentelli A., "Logic Minimization Algorithms for VLSI Synthesis", Kluwer Academic Publ., 1984.
17. Jasiński K., T. Łuba, Kalinowski J., "Parallel Decomposition in Logic Synthesis", Proc. 15-th European Solid-State Circuits Conf., pp. 113-116, Sept. 1989.
18. Kraśniewski A., "Design for Verification Testability", Proc. European Design Automation Conf., pp. 644-648, March 1990.

Logic and Architecture Synthesis
P. Michel and G. Saucier (Editors)
Elsevier Science Publishers B.V. (North-Holland)

Redesign and automatic error correction of combinational circuits

Masahiro Fujita, Taeko Kakuda, and Yusuke Matsunaga

Artificial Intelligence Lab.

FUJITSU LABORATORIES LTD.

1015 Kamikodanaka Nakahara_ku Kawasaki 211, JAPAN

E-mail: fujita%flab.fujitsu.co.jp

From the viewpoint of practical situation, there are many cases where new designs are slightly different from existing ones, and reuse of existing designs as much as possible is a key point to realize efficient and high-quality design synthesis. In this paper, we present two methods for efficient reuse of existing designs: The first method is effective, when the difference between the new design and the existing one is relatively small. It is a direct application of the "error compensation procedure" of the transduction method [1], and it first calculates errors in the circuit, if we directly use the existing design, and then it tries to compensate those errors by transforming parts of the circuit. The second method is effective, if the difference is relatively large. It first generates initial circuit for the new design by some fast method, and then that initial circuit coupled with the existing design is minimized by the boolean resubstitution procedure so that intermediate logics of the existing design be utilized as much as possible. If some intermediate logic of the existing design is also a good intermediate logic for the new design, it is reused by the boolean resubstitution procedure. We present preliminary results for the two methods.

1. Introduction

Recently not a little progress has been accomplished in logic synthesis field. Especially, logic synthesis with multi-level logic minimization mechanisms can generate circuits which are competitive with human designs, even in the design of functional units, such as ALUs [1~4]. However, there are several important problems that must be resolved relating to logic synthesis. Among them are efficient redesign for slightly modified specifications and automatic error correction. These two happen very frequently, because in practical designs, specification changes very often, and manual designs have errors although they are high quality ones. These problems are closely related in the sense that we should use existing designs as much as possible for efficient processing and high quality designs. Moreover, same procedures can be applied to both redesign and automatic error correction problems, because these two problems are essentially the same.

As for the automatic error correction problem, previous works can only detect errors, and can not correct them. From the view point of practical design environment, however, automatic error correction is highly desired to minimize designers' load. As for the redesign problem, previous works use some rule-based approaches and can only utilize some special hardware modules. They do not make use of internal sub-networks of existing designs.

In this paper, we present two methods which can be applied to both redesign and automatic error correction. The first method uses error compensation procedure in the transduction methods [1] so that it can correct errors automatically. If we consider designs with some errors as existing designs, we can apply this method to the redesign problem. The error compensation procedure uses permissible functions with errors which show how far the current incorrect designs are from the correct designs. Logic design errors can be efficiently corrected with the guidance of permissible functions with errors by transforming parts of the circuits.

The second method makes the most use of internal logics of existing designs so that it can generate new designs with high quality in much shorter time. Again, this method can be applied to the automatic error correction problem, although it is effective, when the difference between the current design and the correct design is large. This method use boolean resubstitution procedure [3,4,5] to make the most use of existing designs. Here we use the boolean resubstitution procedure based on permissible functions implemented in Binary Decision Diagrams [6] so that we can handle large circuits [5].

In the following, we present the two methods and some experimental results.

2. Redesign by automatic error correction using permissible functions with errors (PFE)

The first method uses error compensation procedure in the transduction methods [1] so that it can correct errors automatically with the guidance of permissible functions with errors. Here we first introduce permissible functions with errors and then explain how they are used to correct errors with an example (and also to redesign) and finally present some results.

Permissible functions with errors (PFE)

Here we explain permissible functions with errors and its calculation method. In this paper, logic functions are represented in a vector which corresponds to a truth table. (Note that the real implementation uses Binary Decision Diagrams [6] to represent permissible functions as well as logic functions to be able to handle large circuits)

If input variables are X1, X2, ..., Xn, the first element in a vector corresponds to the value when all input variables are 0 and the second element corresponds to the value when all input variables except Xn are 0 and the last element corresponds to the value when all input variables are 1. The number of element in a vector is 2^n, where n is the number of input variables. So the exclusive-or function of two variables X1 and X2 is represented in the vector [0,1,1,0].

The logics of outputs in a circuit may not change, even if we change some parts of the circuit because of the don't cares which come from the circuit topology. Permissible functions are defined for each net and gate and are sets of logic functions which are allowed for that net or gate on the condition that the output logics do not change. For example, in the exclusive-or gate, shown in Figure 1 (a), we can change the logic of G3 to [1,0,0,1] without changing the logic of output Y, although the current logic of G3 is [1,1,0,1]. In this case, permissible functions for G3 is (1,*,0,1), where * means a don't care value and permissible functions are represented in vectors surrounded by (and). Permissible functions can be calculated from outputs to inputs by traversing the circuit [1].

Permissible functions with errors (PFE) are also defined for each net and gate and show the difference between the correct logic and the current logic (which may have errors). PFE can have one of the five values; 0, 1, *, 0' (which represets that the correct value is 1 but the current value is 0 because of some design errors), and 1' (which represents that the correct value is 0 but the current value is 1 because of some design errors) [1]. 0' and 1' are called error elements.

The circuit in Figure 1 (b) is obtained from the circuit in (a) by deleting a net. The output logic of Figure 1 (b) is [0,1,1,1], and so the PFE for the output is (0,1,1,1').

PFE can be calculated from logic functions for nets and gates by traversing from outputs to inputs. The details can be found in [1]. For the example of Figure 1 (b), PFEs are shown in Figure 1 (c).

Redesign or automatic error corrections with the guidance of PFE

In order to correct design errors, we must remove error elements from all PFEs. Our strategy here is to apply circuit transformations which maximally minimize the number of error elements in the PFEs having maximum number of error elements or eliminate all error elements in some PFEs. The following four design errors are considered to cover most common design errors:

(1) Insufficient nets (conditions are too weak)
(2) Excessive nets (conditions are too strong)
(3) Polarity error
(4) Gate type error
(5) Connection error (Some nets are incorrectly connected)

Therefore, we should apply the transformations which compensate the above four types of errors and check how many error elements are eliminated by that transformation. Although some nets can only be added or removed on some conditions, those conditions can be efficiently checked by using PFEs [1,2].

As for the example in Figure 1 (c), here we try to correct design errors by adding a net. PFEs for gate G3, G4 have one error element. The only connectable net to gate G3 is the one from gate G1 (this can be checked by analyzing permissible functions [1,2]), and this connection eliminates the error element in PFE for G3. The result is a correct design.

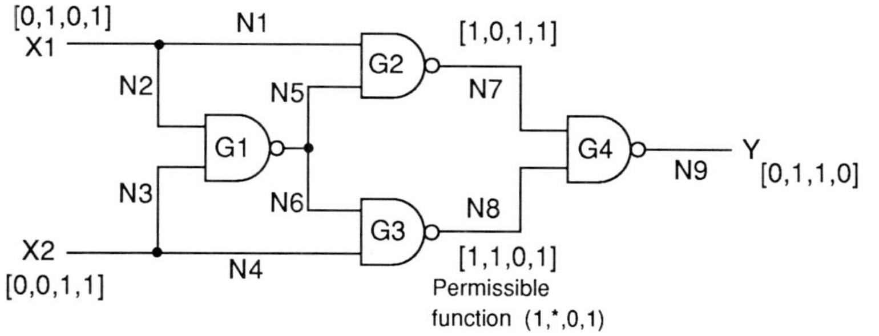

(a) An exclusive-or circuit

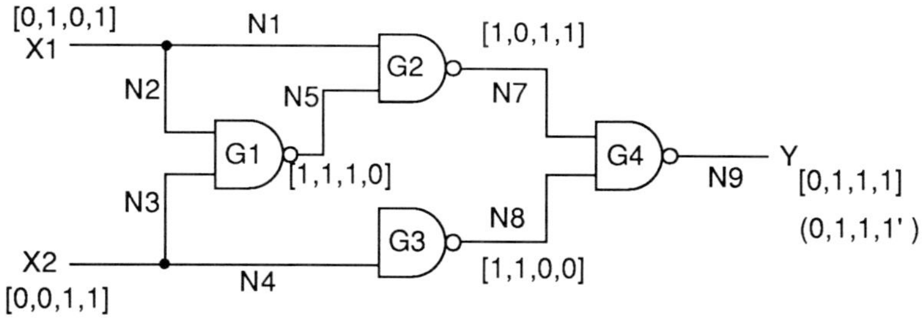

(b) A net is deleted from (a)

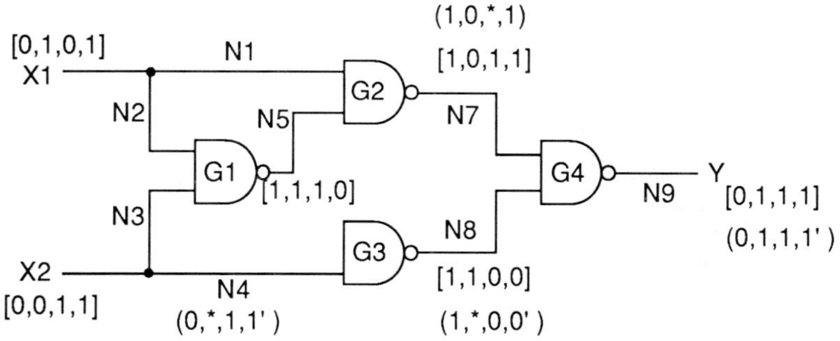

(c) Permissible functions with errors (PFE) for (b)

Figure 1. An exclusive-or circuit and permissible functions with errors

Of course we can try to correct design errors by other types of transformations above, such as, deleting nets, changing gate types, etc. So, we apply the above transformation one by one to compensate design errors, and when we get a correct design (completely compensated), we stop. Since this is a very naive method, we can consider more complex transformations for efficient processing. These are, however, among future research points.

Some experimental results

We implemented the above procedure and applied it to synthesis benchmark circuits. The results are shown in Table 1. The design are automatically generated using the transduction methods [1,2] from sum-of-product forms given in the benchmark circuit format. So, they have almost no redundancy and have comparable sizes compared with manual designs. The first column shows how many trials succeed when a extra net is randomly added. The second column shows how many trials succeed when a net is randomly deleted. And the third column shows how many trials succeed when a extra net is randomly added and another net is randomly deleted. We tried 10 times for each case, so the figures divided by 10 in the table means the ratio of correction success.

As can be seen from the table, 40 ~ 70% of randomly generated errors of types (1) and (2) were corrected by the procedure. The required CPU time on SUN3/60 is around several seconds.

We can say this method can be applied to redesign or automatic error correction of combinational circuits with not so much errors.

3. Redesign by utilizing parts of existing designs

One of the most difficult points in multi-level logic optimization is to find good intermediate logics. If we can find good intermediate logics, we can get not only high quality designs but also with much less computational time. Here we claim that intermediate logics for some very good designs, which are automatically generated or manually designed, are also very good intermediate logics for designs having slightly different specifications. So, we should use such intermediate logics in new designs which have similar specifications. This can be easily accomplished by using boolean resubstitution procedures [3,4,5].

Circuits	Added	Deleted	Added and deleted
5xp1	9	7	4
9sym	4	4	3
bw	5	8	4
dk27	5	8	4
f51m	6	4	2
rd53	9	8	4
rd73	8	2	3
rd84	7	7	4
sao1	7	4	3
sao2	7	3	2
sex	9	9	7
sqn	9	9	6
z4	5	8	5
average	6.9	6.2	3.9

10 trials were made for each case.

Table 1. Results of automatic error correction

Spec. No.	Literals	CPU time (SUN4/260 seconds)
1	156	868.82
2	124	966.00
3	128	768.63
4	140	930.80
5	152	877.52

Note 1: Without redesign, the synthesis result has around 180 literals, and the execution time is around 1400 seconds.

Note 2: Manual design has 135 literals.

Table 2. Results of redesign using Boolean resubstitution

Figure 2 shows our redesign method. We first generate initial designs (which can be very poor) for the new specification by using some fast procedure such as weak division. Then we minimize both the initial design and a very good design which has slightly different specification by using boolean resubstitution procedures. Boolean resubstitution procedures automatically make it possible to utilize intermediate logics of the existing design. The flow is shown in Figure 3.

Some experimental results

We implemented boolean resubstitution procedure [5] based on the minimizer presented in [2] and applied it to the minimization of a 4 bit ALU which is considered to be one of the most difficult benchmark circuits to be synthesized. We first generated an initial circuit by weak division procedure. It has much redundancy and is more than 4 times larger than a manual design. We used a manual design whose specification is slightly different. Two experiments were done.

In the first experiment, new specifications are created by modifying the intermediate logics of the existing design. The synthesis process started with the new specification in sum-of-products forms. The results are shown in Table 2. We tried 5 times. Each trial has different specification (some different intermediate logics were changed). Without our redesign method, it took around 1400 seconds on SUN4/260 to synthesize, and the results have around 183 literals. However, with the redesign method, it took only 770 - 970 seconds, and the results have only 124 - 156 literals, which are comparable with manual design (135 literals). These show that our redesign method's significance for both redesign and automatic error correction problems.

In the second experiment, specification was changed so that the outputs of ALU have different logic function from the original ALU on only a certain condition. In this case, without our redesign method, it took 1024 seconds to synthesize, and the result has 169 literals, while with our redesign method, it took only 565 seconds to synthesize, and the result has only 126 literals.

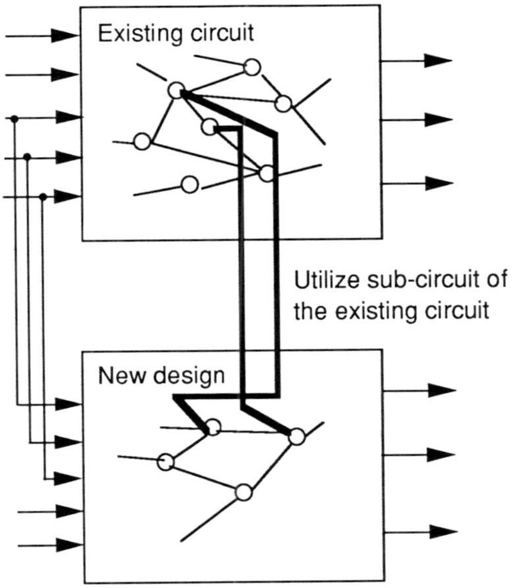

Figure 2. Redesign by utilizing parts of existing designs

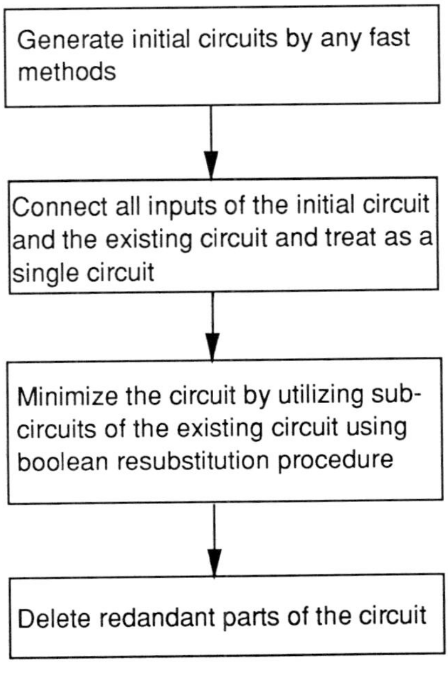

Figure 3. Flow of redesign by utilizing parts of existing designs

4. Future work

In this paper, two redesign methods and their preliminary results, which are very promising, were presented. Future work includes more rigid evaluation of the methods in real designs. In the design of macro modules in a data path, usually rule-based synthesis methods are used. We plan to use our redesign methods in such synthesis systems.

References

[1] S. Muroga, Y. Kambayashi, H.C. Lai and J.N. Culliney, ''The Transduction Method - Design of Logic Networks based on Permissible Functions'', *IEEE Trans. Comput., Vol.C-38, No.10, pp.1404-1424*, October 1989.

[2] Y. Matsunaga and M. Fujita, ''Multi-level Logic Optimization Using Binary Decision Diagrams'', *Proc. ICCAD '89*, November 1989.

[3] A. Saldanha, A.R. Wang, R. Brayton, and A. Sangiovanni-Vincentelli, ''Multi-Level Logic Simplification using Don't Cares and Filters'', *Proc. 25th DAC*, June 1989.

[4] D. Bostick, G.D. Hachtel, R.M. Jacoby, M.R. Lightner, P. Moceyunas, C.R. Morrison, and D. Ravenscrofit, ''The boulder optimal logic design system'', *Proc. ICCAD '87*, November 1987.

[5] H. Sato, Y. Yasue, Y. Matsunaga, and M. Fujita, ''Boolean resubstitution with permissible functions and Binary Decision Diagrams'', to appear in DAC, June 1990.

[6] R.E. Bryant, ''Graph-based algorithms for boolean function manipulation'', *IEEE Trans. Computer, C-35(8):667-691*, August, 1986.

Technology Mapping

Logic and Architecture Synthesis
P. Michel and G. Saucier (Editors)
Elsevier Science Publishers B.V. (North-Holland)

Automatic Layout of CMOS Cells

Robert JAMIER, Jean FREHEL
SGS-Thomson Microelectronics
Avenue des Martyrs
BP217 - 38019 Grenoble Cedex - FRANCE

Abstract

This contribution presents a software system for the automatic generation of the layout of CMOS cells of small or medium complexity from transistor networks. Contrary to already published approaches, which seem to ignore the access problems and allow inputs to be accessible only on polysilicon, we allow inputs and outputs to be accessible on metal2, as it is often required when designing a library of standard cells. Cells are automatically designed using a fixed grid approach, which takes into account the insertion of vias to have access to inputs and outputs and allows us to be easily adaptable to a new technology. This grid approach is particularly interesting to deal with technologies for which stacking rules exist, as it makes it very easy to avoid stacking polysilicon, metal1 and metal2. All the different steps of the layout generation from a transistor netlist are successively presented. Two steps are particularly original. First an algorithm, which allows to place effectively the vias to have access to transistor gates, is proposed. Secondly, an optimal transistor pairing mechanism is described for dual CMOS gates described at the transistor level. The system has been implemented in LE-LISP [8] and in an in-house procedural layout language named STYX [13]. Designs up to hundred transistors can be generated in a few minutes. Automatically generated layouts are comparable to manual layouts designed using the same layout style.

1 Introduction

The interest of a "technology adaptable" random logic cell generator is twofold :

1. It can be used first to generate cells of small complexity or medium complexity (up to 100 transistors), which are not available in the desired technology, when time is a critical factor but not density.

2. Secondly the cell generator can be used as part of a global automatic block layout generation system [4].

Besides the obvious interest to automate layout design, a random logic cell generator offers many advantages in comparison with a standard cell library :

1. Since the cell generator is technology adaptable, there is no need to wait for the design of a new cell library, when the technology is refined.

2. Any cell can be generated therefore mapping onto a library of standard cells can be avoided as well as the resulting loss of space due to functionality underutilization.

3. Since the number of logic levels of a complex logic gate is not limited by the generator, a mapping onto complex logic gates can save transistors in comparison with a mapping onto standard logic gates.

4. A cell can be generated on the fly with transistor sizes chosen for the specific application, as opposed to standard cells that have fixed transistor sizes and therefore potentially larger areas.

In the near future, if the automatic cell layout generation technique is improved sufficiently to produce cells as dense as hand carved cells, it may eliminate manual design of standard cells.

2 Overview of the CMOS gate layout generator

The random cell compiler presented in this paper, generates the layout of CMOS cells from transistor networks. At the present time, three input modes are available :

1. the schema of the flattened circuit with transistor sizes can be entered graphically at the transistor level,

2. the flat sized transistor netlist can be described in EDIF,

3. or finally the netlist can be described in a denser lisp-like textual format. A logic gate must be specified as its decomposition into sized transistors in parallel or in series. Internal nodes within a series are omitted. This last format is generated by an in-house synthesis tool [4].

The user can specify transistor widths, power line widths and cell heights. This generator is written in a technology independent way.

The cell compiler is applicable to any CMOS technology with two metal layers. Minor modifications may be necessary to adapt the generator to a new technology. In the present version, it can generate circuit layout for five in-house CMOS technologies. For two of them, stacking rules (stacking of metal2, metal1 and polysilicon) must be taken into account.

Input and output ports are vertically accessible on metal2. Power and ground are brought horizontally on metal1. The order of the I/O ports is computed by the generator so as to optimize total area.

The CMOS cell layout compiler can generate dense layout for cells containing several interconnected complex logic gates or transmission gates. Presently, a logic gate is accepted if and only if it has the same number of P transistors and of N transistors. The N and P networks do not need to be dual. The complexity in terms of number of levels of a complex logic gate is not limited. The complexity in terms of number of transistors is also not limited.

3 Target model

The layout style we adopted is the one used commonly in the layout of the functional cells in a standard cell library. It is often referred to as the "line of diffusion" style. Transistors (see figures 1 and 2) can be arranged in a 2 by n array, where 2 represents the number of diffusion rows (one N-type diffusion row and one P-type diffusion row) and n represents the common number of transistors in the N network and in the P network. In 1981, Uehara and Van Cleemput [14] proposed to systematize this layout strategy. Many papers have adopted this layout model : [1], [2], [6], [7], [9], [10], [11], [15].

Transistors belonging to the same column share when possible the same gate signal so as to reduce the number of crossovers of transistor gate connections. Transistors are placed in such a way that a maximum number of successive transistors have common diffusion ports. Thus, connections can be made by abutment and diffusion utilization as well as the number of metal1 to diffusion contacts are therefore minimized.

Besides another important aim is to avoid crossings of source and drain connections which are not made by abutment so as to use only metal1 to realize the connections thus minimizing further the area of the cell.

Although the aforementioned papers have adopted the same layout model, they differ in many aspects.

1. Some papers only consider dual CMOS gates [9], [10], [14].
2. The cells may be assembled automatically into rows [2].
3. Some algorithms are able to handle global physical constraints, such as pin placement [5], [9].
4. Some algorithms allow to reorder single transistors or transistor groups in series to decrease the number of diffusion chains [9], [10].
5. The layout targets also differ with regard to many important details :
   (a) Complementary transistor gates must be aligned [10], [14], [15] or may be connected by river routing [1], [2], [5], [7], [9], [11].
   (b) Transistor gates may be bent [7].
   (c) Routing over diffusion area may be permitted [10], [11], [14]. Multiple metal1-to-diffusion contacts may be placed in the source or drain region of a transistor thus minimizing the resistance but eliminating the posssibility to route over transistors [1], [10].
   (d) Different layout styles may be proposed [10].
   (e) Too large transistors may be split into parallel fingers [2], [5], [10], [11].
   (f) In all approaches inputs are vertically accessible on polysilicon [1], [2], [5], [7], [9], [10], [11], [14], [15], and outputs on metal1 vertically [1], [2], [5] [9], [15] and horizontally [14] or on polysilicon [7], [10], [11].
   (g) The position of power and ground busses may also differ. [10] allows power and ground busses to be placed at the center of the cells.
6. Most approaches seem to favour symbolic layout generation [1], [5], [7], [9], [11], [15]. Some directly generate the layout [2], [10].

3.1 Detailed layout target

In all the papers referenced above, inputs could only be accessed in polysilicon and outputs in metal1 or polysilicon. It is of course much simpler to have polysilicon inputs since inputs are connected to transistor gates which are in polysilicon and to have metal1 outputs since outputs are located both on the N-diffusion side and on the P-diffusion

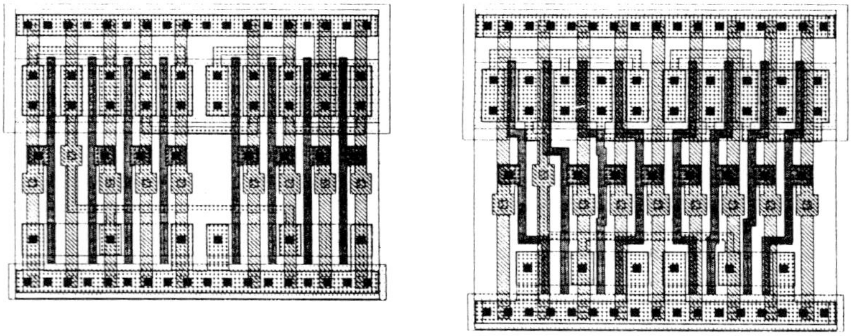

Figure 1: Nor4and2 - left : gates aligned - right : gates not aligned

side and they are usually connected by metal1 wires. If inputs are accessed on metal2, one via and one metal1 to polysilicon contact must be added for each input. Two layout targets, which allow to insert easily vias are proposed below.

Inputs and outputs are accessible vertically in metal2 both from the top and from the bottom. Metal2 wires cross over the N-type row or the P-type diffusion row to reach the input and output vias which are located in between. The VDD line is placed above the P diffusion row whereas the VSS line is placed below the N diffusion row. Body ties are placed along the supply lines. The space between the different strips is used to route intracell connections. The routing is done mainly between the P diffusion row and the N diffusion row. The space between the power lines and the transistor rows is reserved for source-drain connections, which could not be routed by abutment.

Transistors, vias and contacts are laid out using a fix grid, unless the cell includes only one dual logic gate.

3.1.1 Layout target for dual CMOS gates

Figure 1 shows the layout of a dual CMOS complex gate :
(nor (and a b) (and c d) (and e f) (and g h).

The contacts and vias which are necessary to have access on metal2 to inputs can be aligned horizontally to form two connected parallel rows whose width does not exceed the width of the diffusion rows. Moreover no router is needed for a logic gate with independent inputs. This solution seems to be optimal in density.

To allow further diffusion minimization, common gates may be non-aligned. Thus the spacing between gates of transistors in series can be minimal. Complementary gates are routed in polysilicon whereas N-to-P connections are routed in metal1. The metal1 to polysilicon contact row is located just above the via row. P-type gates as well as N-type gates are connected to these contacts by river routing.

3.1.2 More general layout target for logic gates

Figure 2 shows the schematic diagram of a D flip flop and its automatically generated layout.

Gates of paired transistors are aligned. I/O connections are routed within a rectangular channel between the N-row and the P-row. Complementary gates are routed in polysilicon whereas N-to-P connections are routed in metal2. Horizontal connections are realized in metal1. A multiple row routing strategy is used. A via to have access to

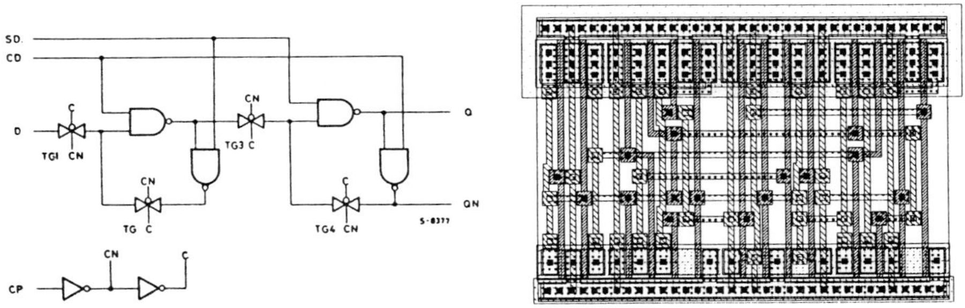

Figure 2: D flip flop - left : schematic - right : automatic layout

an output is placed in the same column as the output contacts of the transistor drain. Vias to have access to inputs are placed in the remaining columns between the gates. Since a grid has been adopted, the stacking of polysilicon, metal1 and metal2 can be avoided very easily. Technologies requiring that some stacking rules must be respected, do not raise special problems for the cell generator.

4 The cell synthesis algorithm

The detailed steps of the transformation of a sized transistor netlist into a geometric layout are successively presented. CMOS gates are recognized if necessary. Then they are ordered. Paired transistors are chained. Chains are possibly ordered and mirrored. Vias to have access to inputs are placed. Connections are routed differently according to their types. Finally the geometric layout is generated.

4.1 Recognition of CMOS gates

Cell elements : logic gates, transmission gates and pass transistors, are extracted from the transistor netlist [5]. Dual logic gates must also be recognized. Transistors in parallel are placed in the same order as the transistors in series in the dual network. The order of the complementary transistors becomes thus identical in the series-parallel decompositions of the two networks.

4.2 Order of CMOS gates

There is no point in ordering transistors independently of their belonging to CMOS gates, therefore the CMOS gates pertaining to the same cell are ordered first. The aim of the linear placement is to minimize the routing area as well as the number of transistor chains. A transistor chain is a series of transistors such that adjacent transistors have common diffusion ports. As a transistor chain can spread out from one CMOS gate to its neighbor, CMOS gate placement has an impact on the formation of transistor chains.

The order of the cell elements is computed using a linear mincut algorithm [3]. The partitioning procedure is applied recursively (n-1 times if they are n cell elements) to obtain a complete order of the cell elements.

A transistor chain can continue from a logic gate to its neighbor, through their common connections to the power and ground nodes if the neighbor gate is a logic gate, or through their common N-to-P connection nodes if the adjacent gate is a transmission gate. To favour the continuation of chains between logic gates and transmission gates, we create one additional connection between a transmission gate and the logic gate delivering the signal it must transmit.

4.3 Transistor pairing

A transistor pair consists of one p-type and one n-type transistor such that they either both have the same gate, or they have a common diffusion port. The aim of pairing is to avoid unnecessary crossings when connecting N-type and P-type transistor gates (dual logic) or when connecting sources and drains of the transistors of a transmission gate. Transistor pairing has a great impact on transistor chaining since transistors are chained by pair and not independently in the N and P networks. Chaining paired transistors allows to decrease the number of cases to be explored.

Contrary to other approaches, transistors belonging to a dual CMOS gate, are not paired in priority if they have the same gate, but if they are dual transistors. Several transistors in the same network can be driven by the same signal and a pairing mechanism not based on duality may lead to disastrous results. If a CMOS gate is input at the logic level, there is no problem. If it is specified at the transistor level, complementary transistors can be easily recognized if they are ordered in the same way in the series-parallel decompositions of the two dual networks (see subsection 4.1). Within a non dual logic gate, transistors which have the same gate are paired. Transistor gates of a transmission gate are paired.

4.4 Transistor chaining

The aim of the chaining process is twofold : it is not only to minimize the number of chains but also to allow minimal planar metal1 routing of source drain connections within these chains. Paired transistors are chained simultaneously in the N network and the P network. The transistors of a logic gate may have to be split up into several chains of transistors. A chain is allowed to extend to the next cell element (according to the mincut linear placement) only if the current cell element has been completely covered by the current chain.

As the size of the gate is usually small due to electrical constraints, our chaining algorithm is based on heuristics. Chains of paired transistors are formed starting from necessary chain ends, if such ends exist. In a dual network, there is only one way to continue a chain once it is started. The longest chains are thus formed. New chains are formed until the whole CMOS gate is covered. If several solutions are obtained, the one which results in minimal routing area is chosen. Our algorithm gives the optimal solution for all normal size CMOS gates. An optimal chaining algorithm can be found in [15].

4.5 Placement of transistor chains covering one CMOS gate

Several transistor chains may be necessary to cover one CMOS gate. These chains are first ordered to minimize the number of tracks necessary to route source-drain connections. As an average logic gate can be covered by a very small number of chains, a combinational approach is possible. Then, chains are possibly mirrored so as to minimize the sum of the lengths of the source-drain connections. Once the order is known, the effect of mirroring a given chain on the routing is independent of the mirroring of the other chains. Moreover the global length of a net is only affected by the mirroring of extremity chains. For each chain, we compute for each net beginning or ending with this chain, the difference in interconnection length and we compute the sum of these differences to decide whether a chain has to be mirrored or not.

4.6 Mirroring of sets of source-drain interconnected chains

A set of source-drain interconnected chains represents one CMOS gate or possibly several CMOS gates connected by one diffusion chain. The chain set can possibly comprise only one chain. Each set of connected chains is possibly mirrored, so as to minimize the sum of internal and external I/O connection lengths. The orientation of a diffusion chain connecting several CMOS gates must correspond to the order computed during the first phase of the cell synthesis algorithm. The mirroring algorithm is quite similar to the one used in subsection 4.5.

4.7 Allocation of vias to free columns

As a grid has been adopted to lay out transistors, there is a limited number of free positions for access vias. Vias to have access to external inputs have to be allocated to free columns between transistor gates. Some columns between transistor gates are already occupied by metal2 N-to-P connections. If necessary, columns are added between chains or on the sides of the cell, wherever they are most needed. Since a via cannot always be placed next to the input it is connected to, an allocation mechanism must be devised to minimize the distance between an input gate and its access via.

Since there are usually several inputs and several possible positions for the vias, the global distance minimization can be costly if it is computed in a pure combinational way. If the number of free columns is equal to the number of external inputs and if there are only single terminal inputs, then the following algorithm is used : the via to have access to the ith input will be allocated to the ith free column.

When the number of free columns exceeds the number of inputs or/and when there are multiple terminal inputs, this algorithm will not give good results. We have devised a simple algorithm to compute the order of attribution of vias to inputs. For each input, for each free column the distance between them is computed. Then for each input, the free columns are sorted according to increasing distance and then for each pair of successive columns in the ordered list, the interconnection length increase is computed as the difference between the distance to the second column and the distance to the first one. The input for which the increase in interconnection length if we take the second column instead of the first, is the greatest, is treated first. If there is a tie, we consider the increase in interconnection length if the third column instead of the second column is chosen. The ties are broken recursively if necessary. The access via is allocated to the first free column in the ordered list. The same procedure is applied recursively until all inputs have been treated.

4.8 Routing of transistor sources and drains

It is possible to route planarly on metal1 the source-drain connections which could not be routed by abutment, provided that the decomposition of the dual graph into chains has been done carefully. A simple single row router is used to decide whether connections have to be placed above or below each diffusion row.

4.9 I/O signal routing

A specialized fixed-grid router based on the well known greedy channel router [12], was developped in order to route all the remaining connections. The greedy router gives very good results for the small or medium size channels encountered in cell synthesis and it is very easy to implement. There are 2 main reasons why the router has to be tuned to our specific problem :

1. It is a pseudo three layer routing problem. Two non overlapping layers are used vertically. Polysilicon is used to connect N transistor and P transistor gates whereas metal2 is used to connect N and P outputs as well as to connect inputs to the external world.

2. To minimize diffusion utilization, the grid pitch is smaller than the spacing between two contacts. Hence specific constraints must be taken into account during the routing phase and during the layout realization phase.

rule 1 : two consecutive horizontal segments belonging to different nets must be separated by at least one grid column.

rule 2 : 2 horizontal segments of length 1 must be separated by at least 2 columns if the second column of the first segment and the first column of the second segment are polysilicon columns.

rule 3 : if an horizontal segment of length 1 has one of its columns in polysilicon, then the metal1 to polysilicon contact must be shifted away from the via.

rule 4 : if the following conditions are satisfied :

- when two horizontal segments are separated by one grid column,
- when one of the segments is of length 1,
- when the last column of the first segment and the first column of the second segment are on polysilicon,

then the metal1 to polysilicon contact of the longer segment must be shifted in the same direction as the contact of the smaller segment (see rule 3 above).

4.10 Layout generation

Layout generation is based on the STYX symbolic layout language developed in SGS-THOMSON [13]. STYX provides the designer with a layout programming environment, which allows the design of parameterized layout generators. Contrary to approaches using a compactor, the design rules must be taken into account while writing the generator. This approach is very flexible and it allows without any problem to use a fixed grid approach to lay out the transistors and the contacts of the cell.

5 Conclusion

It does not seem that there is a general solution that is adapted to all layout design problems. To obtain high density, specific algorithms adapted to specific layout targets must be developed. The solution greatly depends, of course, on the specific access constraints. We allow inputs to be accessed on metal2. Existing products usually ignore the problem of metal2 access and allow inputs to be accessed only on polysilicon or sometimes on metal1. We have devised an original algorithm to allocate vias to free columns, in order to have an efficient layout. Cells with about 150 transistors (see figure

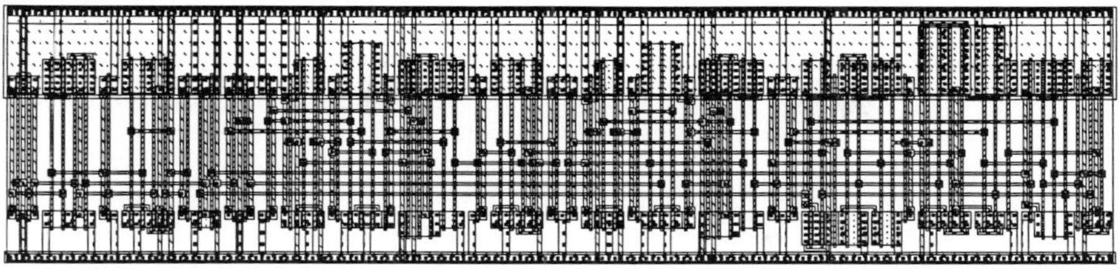

Figure 3: Bit slice of a simplified data path : 148 transistors

3) have been generated in 5 minutes on a SUN3/60 workstation. It takes a few seconds to generate a cell containing a couple of inverters in series.

The random logic cell compiler gives nearly optimal results compared to manual layout designed using the same layout style. The layout target needs still to be improved to be able to reach the density of very compact standard cells. Single contacts can be used instead of multiple contacts. Thus it will be possible to route over the transistors. The number of strap contacts can be reduced. Thus the power and ground lines could also be placed over the transistors or some gate to gate connections could be realized in polysilicon under the power and ground rails. Moreover transistors can be bent to reduce both the height and the width of the diffusion strips.

Anyway the cell generator can generate very quickly reasonably dense layouts in five CMOS technologies. The density is especially close to manual layout when transistor widths as well as power and ground line widths are small and when all transistors of the same type have the same width.

6 Acknowledgements

The greedy router was implemented by Thomas Lauxmann. The linear mincut algorithm was implemented by Jacques Chambon. They are gratefully acknowledged.

References

[1] C. C. Chen and S.-L. Chow, "The Layout Synthesizer : An Automatic Netlist-to-Layout System", 26th DAC, pp. 232-238, 1989.

[2] A. Domic et al., "CLEO : a CMOS Layout Generator", ICCAD-89, pp. 340-343.

[3] C. M. Fiduccia and R. M. Matteyses, "A linear-time heuristic for improving network partitions", 19th DAC, pp. 175-181, 1982.

[4] J. Frehel, "Logic synthesis and optimization for random logic and automatic layout in the ESPRIT 88 AIDA project", ESPRIT '88, Putting the technology to use, Part 1, pp.186-201, Proc. of the 5th ESPRIT Conf., November 14-17, 1988.

[5] C. W. Kapral, J.J-Y. Chen, J. Chen, J. Do and C. R. Fawcett, "Layout Synthesis : An Automatic Cell Generator", Silicon Compiler Systems.

[6] M. Lefebvre and C. Chan, "Flexible boundaries for layout synthesis", VLSI-89, pp. 465-474.

[7] M. Lefebvre, C. Chan and G. Martin, "Transistor Placement and Interconnect Algorithms for Leaf Cell Synthesis", 1st EDAC, pp. 119-123, 1990.

[8] LE-LISP, INRIA, Version 15.22, Reference manual, 1989.

[9] J. Madsen, "A New Approach to Optimal Cell Synthesis", ICCAD-89, pp. 336-339.

[10] F. Mailhot and G. De Micheli, "Automatic layout and optimization of static CMOS cells", ICCD-88, pp. 180-185.

[11] C.-L. Ong, J.-T. Li and C.-Y. Lo, "GENAC : An Automatic Cell Synthesis Tool", 26th DAC, pp. 239-243, 1989.

[12] R.L. Rivest and C.M. Fiduccia, "A greedy channel router", 19th DAC, pp. 418-424, 1982.

[13] STYX, SGS-THOMSON, Version 2.0b, Reference manual, 1989.

[14] T.Uehara and W.M. vanCleemput, "Optimal layout of CMOS functional arrays", IEEE Trans. Computers, vol. C-30, pp. 305-312, May 1981.

[15] S. Wimer, R. Y. Pinter and J. A. Feldman, "Optimal chaining of CMOS transistors in a functional cell", IEEE Transactions on CAD, Vol. CAD-6, No. 5, September 1987.

Logic and Architecture Synthesis
P. Michel and G. Saucier (Editors)
Elsevier Science Publishers B.V. (North-Holland)
IFIP, 1991

BAGDAD: An Oriented Layout Cell Synthesizer

J. Cloutier and J. Zahnd

Ecole polytechnique Fédérale de Lausanne
Laboratoire de Systèmes Logiques
EL Ecublens, CH-1015 Lausanne
Switzerland

Abstract

This article describes a cell synthesizer for CMOS VLSI circuits, using the technique of orthogonal layers known as "oriented layout". Circuits generated by this tool are decomposed into slices and cells. Each cell is a transistor matrix with a very adaptable structure. The synthesizer compiles the symbolic layout of each cell in a module, from a net list, minimizing the width and height of the module without using routing channels. The shape of each cell is thus adapted to its environment in order to optimize the global area of the circuit. Our approach permits us to generate relatively large circuits without using large computing resources.

1.Introduction

Cell synthesizers are CAD tools that automatically generate transistor-level layouts for complex logic functions, starting from a net list or a schematic. They can compile the layout of leaf cells in a hierarchical structure so that adjacent cells are best adapted to each other. It has been shown [Meyer88, Aladdin88] that this approach may considerably improve upon the results obtained by traditional standard cell methodologies. Cell synthesizers are usually technology-independent and permit symbolic editing.

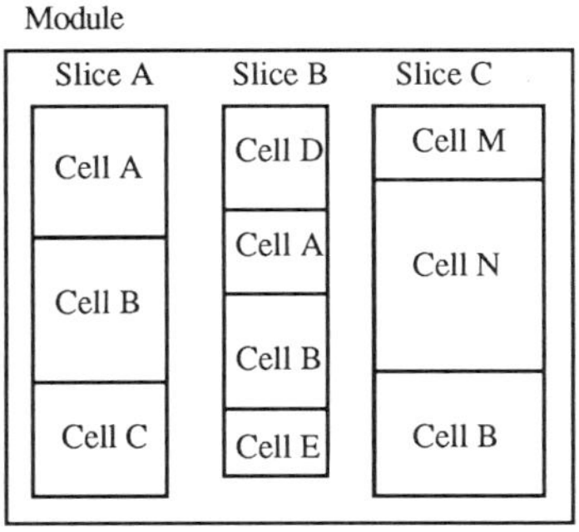

Figure 1. Module's composition

BAGDAD is a cell synthesizer for CMOS circuits, producing a special kind of matrix layout [Piguet84] that will be described in the next section. The starting point of the synthesis is the floor plan of a module, hierarchically decomposed into slices and cells, as shown in fig. 1, together with a net list of transistors for each cell. Slices are linear arrays (columns) of cells. It is supposed that the placement of the cells has already been performed (later versions of BAGDAD will include tools for this operation).

The algorithms presented in this paper concentrate on the problem of optimizing the global circuit by adapting the layout of individual leaf cells to their environment. The symbolic layout of the whole circuit is finally passed to a compactor which performs the translation from symbolic to geometric layout.

2. Oriented layout symbolic description

In order to build automatic layout tools capable of producing circuits of random logic with high density, a particular layout structure has been proposed for CMOS logic circuits [Piguet84]. It is called *oriented*, because two conducting layers (e.g. polysilicon and metal or diffusion and metal) are concentrated along two sets of orthogonal lines, building rows and columns. Moreover, a restriction is put on the topological structure of a cell at the transistor level. Not every series-parallel connection of transistors is allowed, as could correspond to arbitrary boolean expressions, but only those which correspond to sum of products (i.e. parallel composition of series of transistors), called *branches*. An example of two different symbolic oriented layouts for the same cell (nand gate with output-enable) is showed in fig. 2. This circuit realizes the following functions (where nL is not L):

$C = \overline{AB}$ $F = 0$ if $\overline{C} \cdot L = 1$, $\quad = 1$ if $C \cdot L = 1$, $\quad = Z$ (high impedance) if $L = 0$.	The corresponding branch description is: P branches list: (+ (A) C); (+ (B) C); (C (nL) F) N branches list: (- (A B) C); (C (L) F).

A branch description of a circuit is equivalent to a net-list. A branch is composed of two terminators (contacts) and a list of transistors in series. A branch in the symbolic layout is a horizontal line with contacts and transistors represented respectively as crosses and circles. For example, the branch (F (nL) C) is placed on the first line of fig. 2-a, while branches (F (nL) C) and (C (B) +) are placed in the first row of fig. 2-b. Variables are assigned to each column so every branches of a cell have their transistors variables between their two contacts variables. Such assignment is called *realizable* for a cell when all its branches can be placed.

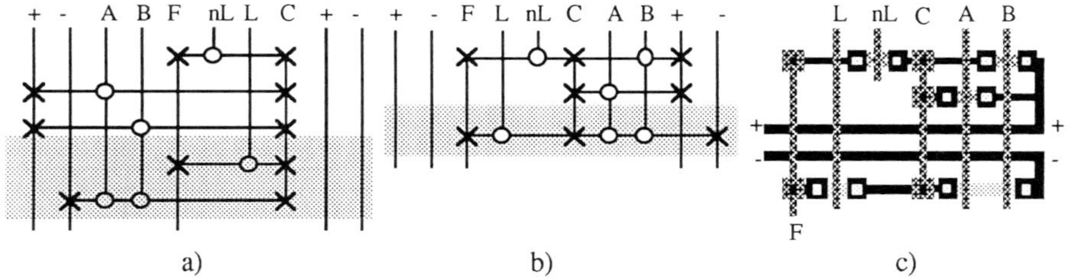

Figure 2. Two symbolic and one geometric oriented layouts of a nand gate with output-enable

The synthesizer imposes four restrictions on the oriented layout of a cell:

1- only one variable is assigned per column;
2- a variable is assigned to only one column;
3- '+' (5 volt) and '-' (0 volt) power lines are available on both sides of a cell;
4- a row is composed of branches of the same type (n or p).

It is quite clear that a geometrical layout corresponding to this kind of layout (fig. 2-c), for an individual cell, is not the most compact possible. It presents however decisive advantages for the *global* optimization of a circuit made of many cells, due to its *adaptability*. It is easily seen that any two rows in the same doping area may be interchanged. Variables assignment may often be permuted in many ways, because the ordering of A, B, say, in a branch like (+ (A B) C) is unimportant, and because any branch may be *reversed*, like (C (A B) +). Moreover, the structure presents the greatest horizontal and vertical *permeability*: an external wire can be routed anywhere through the cell with the least possible increase of area (one row or one column). Vertical composition of cells, called *slices*, are built by taking advantage of this adaptability to minimize the amount of routing between cells, by making adjacent cells assemble by abutment.

3. Cell's graph representation (for slice synthesis)

The crucial part of slice synthesis is to find a realizable variable assignment for each cells allowing slice composition by cell abutment. We have seen previously that this assignment depends only on the set of branches, which defines a partial-ordering on variables. This partial-ordering can be conveniently represented by a precedence graph.

Since ground (-) and power (+) lines are available on both sides of a cell, branch constraints such as (+ (A) B) or (- (A) B) are always satisfied an may be discarded. Similarly, branches with two ore more transistors, but terminated with - or +, like (- (A, B) C), are satisfied by a column ordering if the A-column and the B-column are on the same side of the C-column. The graphs representing partial-ordering of variables for branches (- (A B) C) or (C (A B) -) are showed in fig. 3-a and 3-b. The labeled graph of fig. 3-c represents both graphs: edges having the same label may be reversed simultaneously to yield a new partial-ordering.

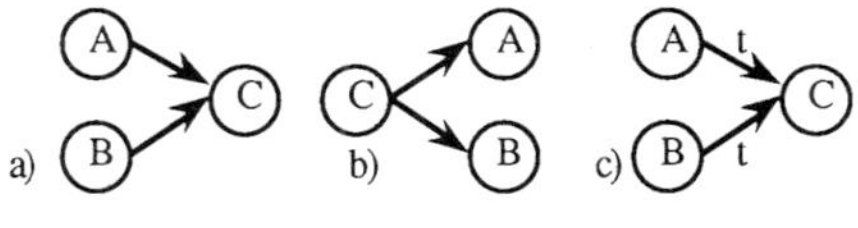

Figure 3. Precedence graph representations of a branch constraints

A set of n branches may yield as much as 2^n graphs, since every branch may be inverted independently. However, since only acyclic graphs represent realizable variable assignments, the number of acceptable graphs is greatly reduced. We may build a partition P of branches that must be reversed simultaneously to avoid any cycles. This partition can easily be computed so we shall not give here a detailed algorithm (empirical results obtained so far indicate that the number of graphs is rarely greater than 8). The partition P determines the *minimal subsets of simultaneously reversible arcs* in G. The graph G and the partition P determine a set of graphs that we call *minimal constraint graphs*.

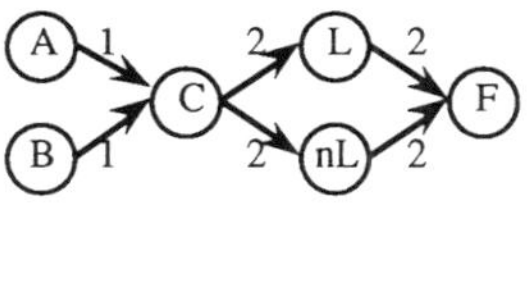

Figure 4. Minimal constraint graphs for the circuit of fig. 2

A labeled graph representing all variable partial-orderings of the nand gate with ouput-enable is showed in fig. 4. Such a labeled graph is used to represent both the graph and the partition of edges. We can derive from this labeled graph four different acyclic graphs from which realizable variable assignments can be obtained.

As stated earlier, a slice is a composition of abutted cells. Cells may be abutted only if their column ordering (variable assignment) permits their correct interconnections. The goal of the slice synthesizer is to combine one precedence graph from each cell to form a global acyclic graph for the entire slice. This global graph represents realizable variable assignments (it must be acyclic) for all variables. Restrictions imposed on the cell's layout permit to achieve this goal without exponential explosion of the solution domain.

In order to determine the "best" graph from the set of minimal constraint graphs of a cell, we define a simple heuristic cost function that evaluates if a graph is *well balanced.* A graph is well balanced when for every node we have approximately the same number of arcs going in and out. Empirical studies have shown us that well balanced graphs yield solutions with more branches on the same row, thus increasing cell's density. We call this heuristic function the *balanced-degree* of a precedence graph. It is defined as

$$\text{Balanced-Degree} = \sum_{\text{all vertex } v} \left| \begin{array}{c} \text{(number of edges having v as their start-vertex) -} \\ \text{(number of edges having v as their end-vertex)} \end{array} \right| .$$

A graph having a small balanced-degree is well balanced.

4. Slice Synthesizer

The slice synthesizer builds a global precedence graph for the entire slice by selecting a graph for each cell and merging them together. The global graph must satisfy two criteria:

1- being acyclic;
2- having a global variable assignment which induces a slice's width smaller than a fixed limit (number of columns).

We use a *greedy algorithm* to achieve the above goals. We build a global graph by processing the cells from top to bottom. We select a graph for the current cell from its set of minimal constraint graphs (with the lowest possible balanced-degree) and merge it with the global graph. If the above criteria are satisfied we may process the next cell; otherwise, we split the slice in two different slices which are realized separately. This implies that the two slices cannot be abutted and that a special routing cell must be inserted between them to achieve correct connection of the complete slice. A variable assignment is then extracted from the global graph, and enables one to perform a branch placement for all cells.

4.1. Example of slice synthesis

An example of a slice synthesis is described in this section. The floor-plan of a full-adder is described in fig. 5-a. It is build from the interconnection of 3 cells: a nand gate and two complex functions (exclusive-or and a nand gate). Their branch descriptions are given in figure 10. Minimal constraint graphs for each cells are showed in fig. 5-b. Edges having the same label must be inverted simultaneously to insure that the graph remains acyclic. Two nodes having the same name represent the same circuit's variable.

Graphs 1, 2 and 3 are respectively the minimal constraint graphs of the cell 'nxorn1', 'nxorn2' and 'nand'. Graph 0 is added to represent an alignment constraint. This constraint is given with the floor-plan to insure that the carry-in and carry-out connections are aligned. This permits to create a n-bit adder by repeating n 1-bit adders. Such an alignment constraint induces an edge in the graph that may not be reversed contrary to other edges.

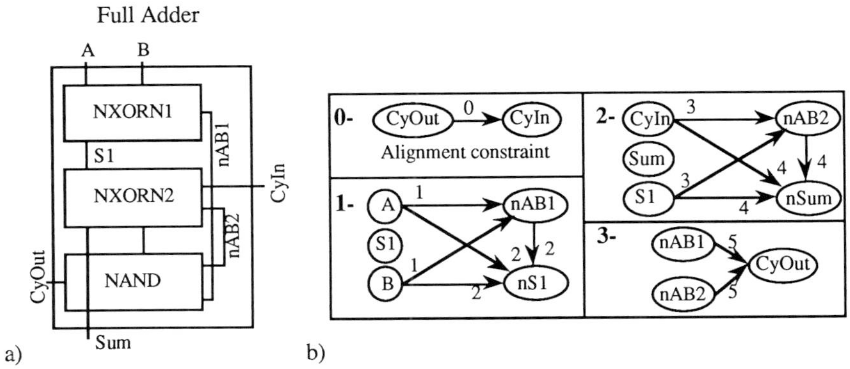

Figure 5. A slice floor-plan and its cell's minimal constraint graphs

The synthesizer combines those minimal constraint graphs in order to create a global graph. Nodes having the same name are merged. As stated before, global graph generation must insure that no cycles are created and that the slice's width is smaller than a fixed number of columns (for example, the maximum number of variable in a cell).

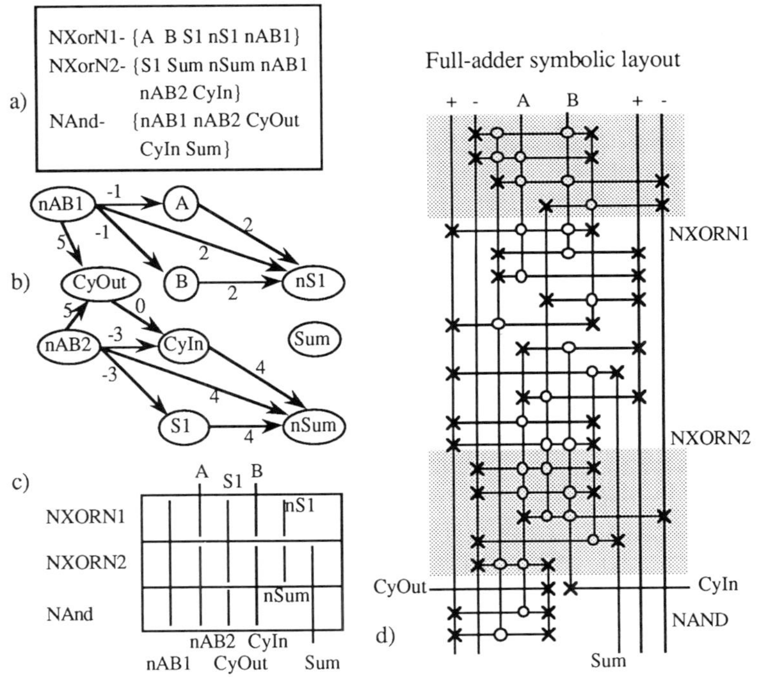

Figure 6. A global graph, its variable assignment and corresponding symbolic layout

A *utilization vector* of the slice is extracted from the floor-plan interconnections. It is a vector of sets which specify the variables passing through or used by a cell. The utilization vector of our example is given in figure 6-a. This vector is used concurrently with the global graph of figure 6-b to compute the variable assignment of figure 6-c. This assignment insures that a branch placement for each cell may be done successfully. The corresponding symbolic layout is showed in fig. 6-d.

4.2. Minimum slice width

We have at this stage a global graph representing the partial-ordering of variables for a slice (fig. 6-b) and its utilization vector (fig. 6-a). Our goal is to determine the minimum slice width which permits the placement of all variables. We show in this section that this problem is NP-complete. An efficient heuristic method for its solution is then proposed.

Problem's formulation

We have as input a global graph M=<V,A> where V is the set of variables, a utilization vector C[1..c] of sets of variables where *c* is the total number of cell, and an integer *s* specifying the maximum number of columns of the slice. An example of a graph and utilization vector are showed respectively in fig. 7-a and 7-b.

We may state the minimal slice width problem as follows:
Does a placement of variables exists such that it satisfies graph M and vector C, and has a number of columns less or equal to *s* ?

With each variable i we associate an integer interval [b(i), t(i)], where b(i) and t(i) are the cell numbers of the bottom most and the top most cells using variable i. We then construct a constraint graph G representing an intersection graph of this system of intervals. Furthermore, an edge is added (dotted lines in fig. 7-c) for each pair of nodes occuring on the same path in M. We show in fig. 7-c the constraint graph corresponding to the vector of fig. 7-b and the graph of fig. 7-a. The constraint graph G has an edge between any two nodes that cannot share the same column in the final variable's assignment.

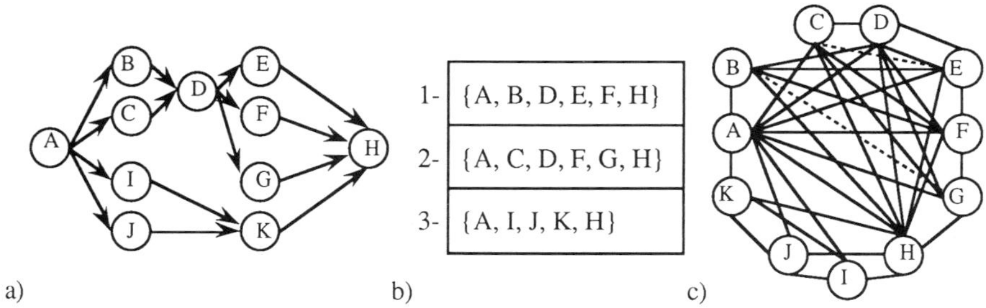

Figure 7. A global graph (a), a utilization vector (b) and their corresponding constraint graph (c)

We are now ready to reformulate the minimum slice width problem as follows:
Given a constraint graph G and the maximum number of column *s*, is $\gamma(G) \leq s$ (where $\gamma(G)$ is the chromatic number of G)?

This is a classical problem of graph theory which is known to be NP-complete [Even79, Gondran79].

Heuristic algorithm for variable's assignment

Since the minimal slice width problem is NP-complete, it is most likely that the complexity of any algorithm that insures an optimal solution of our problem is exponential. Having a large number of variables (often more than 50), an exhaustive algorithm is not practical, consequently an heuristic algorithm must be used.

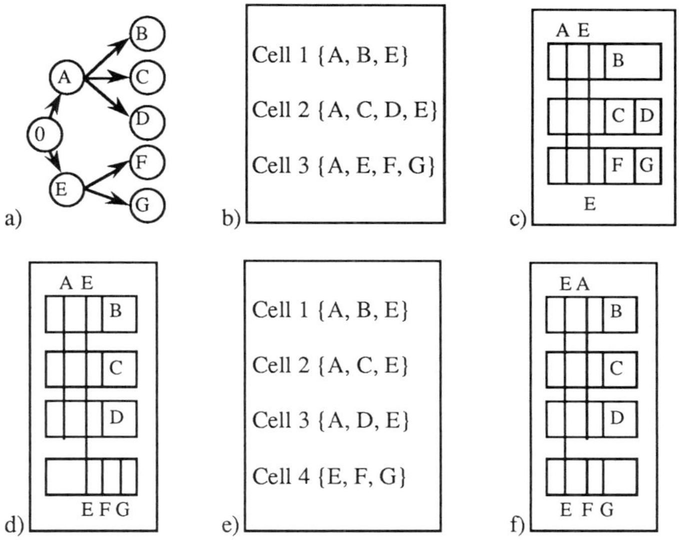

Figure 8. Optimal and sub-optimal variable's assignment

The placement of each variable is done in order to compute the minimal slice width. This placement is performed by a special topological sort of the global graph where each variable is given a priority (heuristic function). Many variables may be placed on the same slice's column. At each step the algorithm determines the utilization of a column by trying to put on it as many variables as possible, starting with the variable having the highest priority. A variable having many successors in the graph is preferably placed before another one having less successors. This must be done without violating the partial-ordering of the global graph.

We show in figure 8 two examples for which the same global graph is used (fig.8-a). The first example uses the utilization vector of fig. 8-b and the algorithm finds the optimal solution of fig. 8-c. Variable *A* is placed before variable *E* since it has more successors (3 against 2). The second example uses the utilization vector of fig. 8-e for which the algorithm finds the sub-optimal solution of fig. 8-d. The optimal solution for this problem is showed in fig. 8-f. Fortunately, experience has showed that the algorithm usually finds an optimal solution for "real life" problems. The algorithm has a worst case time complexity in O(max(e, n log n)) where n is the number of variables and *e* is the number of edges of graph M.

4.3. Cycle's detection

Cycle's detection is done by a depth first search through the global graph. A cycle is detected when the same variable is found twice on the same search path. This algorithm is straightforward and is not described in this paper. Its time complexity is in O(max (e, n)) [Brassard87] where e and n are respectively the number of edges and nodes of M.

5. Module Synthesizer

The module synthesizer calls the slice synthesizer and afterwards performs the alignment of the regions (n or p) and the connections of the slices. The slice synthesizer is invoked only once for slices that are repeated many times. Oriented layout flexibility and adaptability are used to modify slices' layout to insure that their abutment constitutes a correct module.

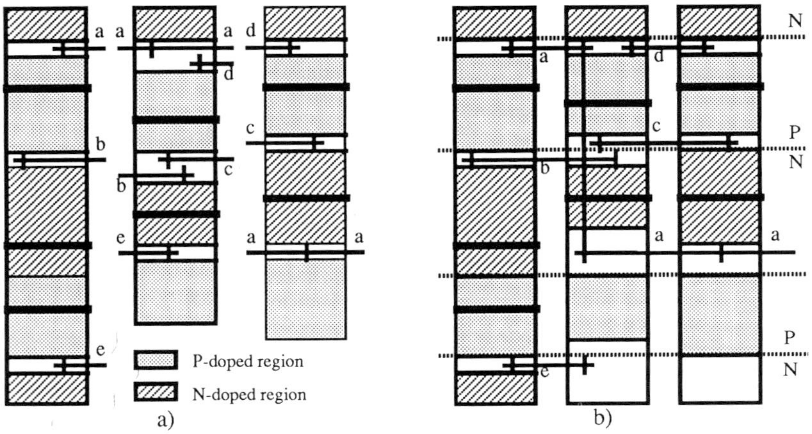

Figure 9. Regions and connections alignment of a module's slices

Once the slices have been synthesized, we have a symbolic layout for each slice composing the module. Different slices do not necessarily have the same number of rows or regions, but have the same number of ports (left and right) as their neighbor slices as showed in fig. 9-a. The goal of the alignment is to make sure that only one type (n or p) of transistors is found in the same row of the module and that connections between slices are aligned. Those two requirements must be fulfilled concurrently. Figure 9 shows an example of slices before and after region and connection alignment. White regions without wires represent empty rows added to obtain the desired alignment.

6. Results

BAGDAD is implemented in Common Lisp (35K lines) on a MacIntosh II with 5 MB of memory. The results obtained so far are very encouraging. Relatively large circuits were realized in comparison with other cell synthesizers. Usually cell synthesizers may handle up to about 500 transistors [Meyer88]. We were able to generate (as showed in table 1) circuits with more than 2400 transistors in acceptable computing time.

Two circuits were used to compare BAGDAD's synthesized oriented layouts with a standard cell and a manually oriented layout implementation (table 1). The circuit used for the standard cell comparison is a 8-bit version of the 4-bit circuit showed in figs. 10, 11 and 12. It is an adder-comparator coupled with a register. The geometric oriented layout is more than twice denser than the standard cell layout. The second comparison is done with the layout of a large 8-bit processing unit taken from the WORP implementation [Iseli90]. BAGDAD's circuit density outperformed the manual design (2 man-months) by 11.5%. Results from the implementation of a small 8-bit processing unit is also showed in table 1. This implementation was done by an unexperienced designer [Grigorie90] who was able to learn the BAGDAD environment and to implement the circuit in about 40 hours.

Circuit	Program	#transistors	CPU time (sec.)	MOS/mm^2
8 bit adder	VTI standard cells	720	n/a	1450
" "	BAGDAD	672	9.1	3107
small 8 bit processing unit	BAGDAD	1317	125.0	1765
large 8 bit processing unit	manually implemented oriented layout	2486	n/a	1582
" "	BAGDAD	2486	204.8	1823

Table 1. Results of various layout implementations in 2μm technology

Acknowledgements

This work was supported by the Swiss Commission pour l'encouragement de la recherche scientifique (CERS grant #1398) and by the Département d'informatique et de recherche opérationnelle de l'Université de Montréal (Canada). The authors also give special thanks to Eric Fournier for his work on the cell compactor (fig. 12).

References

[ALADDIN88] C. Piguet, C., G. Berweiler, C. Voirol, E. Dijkstra, J. RTijmenants, R. Zinszner, M. Stauffer, M. Joss, "ALADDIN: A CMOS Gate Matrix Layout System", IEEE ISCAS, June 1988, Espoo, Finland.

[Brassard87] G. Brassard et P. Bratley, "Algorithmique: Conception et Analyse", Masson et les Presses de l'Université de Montréal, 1987.

[Grigorie90] Mihaela Grigorie, "Processeur 8 bits intégré", student project, Ecole polytechnique fédérale de Lausanne, winter 1990.

[Even79] Shimon Even, "Graph Algorithms", Computer Science Press, 1979.

[Gondran79] Michel Gondran et Michel Minoux, "Graphes et Algorithmes", Editions Eyrolles, Collection de la direction des études et recherches d'électricité de France, 1979.

[Iseli90] C. Iseli, "A Microcode compiler for the watch-oriented RISC processor", Software practice and experience, to be published.

[Meyer88] E. Meyer, "Cell synthesis in action", VLSI Systems Design, May 1988.

[Pedron88] C. Pedron, A. Stauffer, "Analysis and Synthesis of Combinational Pass Transistor Circuits", IEEE Transactions on Computer-Aided Design, vol. 7, no 7, July 1988.

[Piguet84] C. Piguet, J. Zahnd, A. Stauffer and M. Bertarionne, "A Metal-Oriented Layout Structure for CMOS Logic", IEEE Journal of Solid-State Circuits, vol sc-19, no. 3, June 1984.

```
(DEFUNSLICE FA_Comp_Reg_1_Bit
        (In Cp CarryIn Reset CompIn)
        (CompOut CarryOut Accu) ()
  (INSTANCE Comp Comp)
  (INSTANCE NXORN HA1)
  (INSTANCE NXORN HA2)
  (INSTANCE NAND  NAND)
  (INSTANCE MASDR MA)
  (INSTANCE SLADR SA)

  (CONNECT
        (In Comp.A HA1.B)
        (Cp MA.CP SA.CP)
        (Reset MA.CDN SA.CDN)
        (CarryIn HA2.A)
        (CarryOut NAND.C)
        (Accu SA.Q HA1.A Comp.B)
        (CompIn Comp.Ei)
        (CompOut Comp.Eo)
        (HA1.nAB NAND.A)
        (HA2.nAB NAND.B)
        (HA1.S   HA2.B)
        (HA2.S   MA.D)
        (MA.A SA.A)
        (MA.E SA.E))
  (CONSTRAINT
        (TOP Accu In)
        (LEFT  CompOut CarryOut)
        (RIGHT CompIn  CarryIn)
        (LEFTRIGHT CP Reset)
        (ALIGNED     (CompOut CompIn)
                     (CarryOut CarryIn))
  )
)
```

```
(DEFUNMODULE FA_Comp_Reg_4_Bit
      (in[0.3] Accu[0.3] CP Reset CyIn CompIn)
      (CyOut CompOut) ()

  (INSTANCE FA_Comp_Reg_1Bit FA[7.0])

  (CONNECT

        (FA[0].CompIn CompIn)
        (FA[3].CompOut CompOut)
        (FA[1.3].CompIn FA[0.2].CompOut)

        (FA[0].CarryIn CyIn)
        (FA[3].CarryOut CyOut)
        (FA[1.3].CarryIn FA[0.2].CarryOut)

        (FA[0.3].Reset Reset)
        (FA[0.3].CP CP)

        (FA[0.3].In in[0.3])
        (FA[0.3].Accu Accu[0.3]))

  (CONSTRAINT
        (TOP Accu[0.3] In[0.3])
        (LEFT CyOut CompOut)
        (RIGHT CyIn CompIn)
        (LEFTRIGHT  CP Reset)
        (TECHNOLOGY vti2mu)
        (Use-lower-constraint)
  )
)
```

```
(DEFUNCELL Inverter  (A) (nA) ()

 (PBRANCH   (VDD (A) nA))
 (NBRANCH   (GND (A) nA))
)

(DEFUNCELL COMP
               (Ei A B)  (Eo) (nEi nAB)
               {input   output  internal}
 (MACRO Inverter (Ei) (nEi) ())
 (MACRO NAND (A B) (nAB) ())
 (PBRANCH   (VDD (nAB nEi) Eo)
            (VDD (A B nEi) Eo))
 (NBRANCH   (GND (nEi) Eo)
            (GND (nAB A) Eo)
            (GND (nAB B) Eo))
)
(DEFUNCELL SLADR
        (CDN CP A E) (Q) (QN)
                 .
;slave part of a flip-flop with reset 'CDN'
)
```

```
(DEFUNCELL NAND  (A B) (C) ()

 (PBRANCH   (VDD (A) C)
            (VDD (B) C))
 (N       GND (A B) C))
)
(DEFUNCELL NXORN
              (A B) (S nAB) (nS)

 (MACRO Nand (A B) (nAB) ())
 (MACRO Inverter (nS) (S) ())
 (PBRANCH   (VDD (nAB) nS)
            (VDD (A B) nS))
 (NBRANCH   (GND (nAB A) nS)
            (GND (nAB B) nS))
)

(DEFUNCELL MASDR
        (CDN CP D) (A E) (B)
                 .
;master part of a flip-flop with reset 'CDN'
)
```

Figure 10. Description of a circuit (its synthesis is showed in fig. 11)

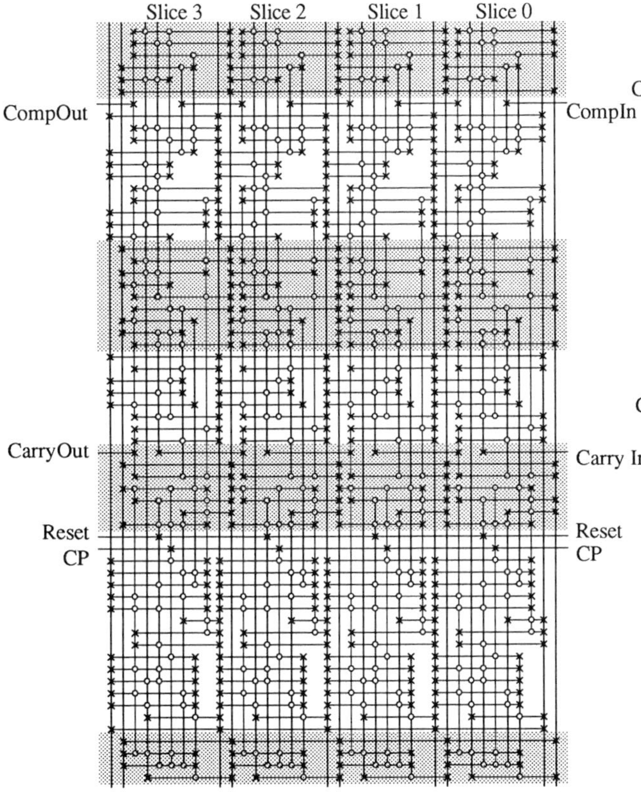

Figure 11 Result of the synthesis of fig. 11 circuit's description

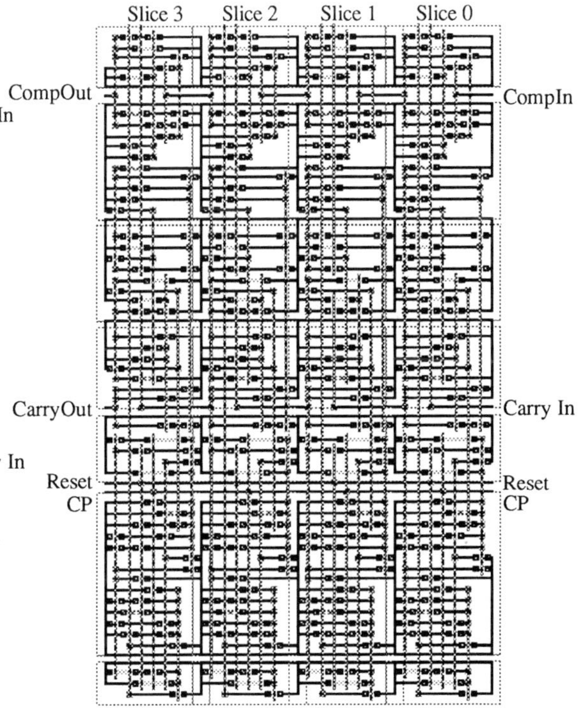

Figure 12 Geometrical layout of fig. 12 symbolic layout

Logic and Architecture Synthesis
P. Michel and G. Saucier (Editors)
Elsevier Science Publishers B.V. (North-Holland)
IFIP, 1991

Flexible and Optimizing ALU Synthesis

Frank Buijs, Peter Vogelgesang, Thomas Lengauer

Cadlab
Bahnhofstr. 32
4790 Paderborn
West Germany

Abstract

We present a new top-down approach for flexible and optimizing ALU synthesis which exploits the regular structure of ALUs. The ALU is first decomposed into a vector of bit-slices, and subsequently the individual bit-slices are optimized. Therefore the optimization is efficient and produces regular circuitry. Another novel feature is that we handle the encoding of the control signals of the ALU flexibly. Our approach is implemented in CLASSY (CadLab ALU Synthesis SYstem). The input of CLASSY is a behavioral ALU description, in which every combination of operations from a large set is allowed. The output of CLASSY is a regular layout that is made up of bit-slices.

1 Introduction

Usually, an Application Specific Integrated Circuit (ASIC) is realized using modules such as RAMs, ROMs, PLAs, ALUs, registers, multipliers, etc. Therefore most silicon compilation systems are based on a set of module generators for such modules [1, 2, 3]. These module generators generate the layouts of the various modules from a structural or behavioral description.

Traditional module generators exploit the regularity of the modules. They generate layout by abutting predesigned cells according to a parameterizable floorplan. For example, an ALU generator produces layout by abutting predesigned bit-slices in a row, where the number of bit-slices is a parameter.

For ALU generation, this bottom-up approach suffers from the following disadvantages:

Inflexible Even though some ALU generators can incorporate design options and variants (such as the inclusion or exclusion of overflow handling) into the generation process, it is not possible to select the operations the ALU must perform from a large collection of arithmetic and logical operations. This happens because the available bit-slices are fixed and no logic generation is performed.

Non optimal The bit-slices are optimized for the required operations. If only a subset of these operations is used, e.g. only the operations ADD and AND, the ALU is too powerful for its purpose. Vandeweerd et al. [4] present an approach that selects portions of the bit-slices to taylor the ALU to the application, but this approach cannot compete with general logic optimization of the bit-slices.

Flexible and optimizing ALU synthesis requires a top-down approach. A well-known approach is the use of logic optimization techniques. The Yorktown Silicon Compiler [5] uses logic optimization techniques for the synthesis of all combinational modules. However, general logic optimization neglects the regular structure of ALUs. One of the consequences is that logic optimization has to be performed on the ALU as a whole. Hence the resulting circuit is usually more irregular than necessary, and the optimization process takes inordinately long [6].

This paper presents a new top-down approach that *exploits* the regular structure of ALUs. First the ALU is decomposed into a vector of bit-slices. For this decomposition, we use knowledge-based techniques. Subsequently logic optimization is performed on the individual bit-slices. Because logic optimization is performed on the bit-slices instead of the entire ALU, the optimization is efficient and produces regular circuitry.

Another novel feature is that we handle the encoding of the control signals of the ALU flexibly. An appropriate encoding of the control signals is a major factor in the optimization of the ALU, but existing ALU generators assume a fixed encoding that is dictated either by the predesigned bit-slices or by the high-level synthesis. In order to handle the encoding flexibly, our approach first decomposes the ALU into a vector of bit-slices and performs logic optimization on the bit-slices. During these phases the values of the control signals are represented by symbols. We developed powerful logic optimization algorithms that can handle these symbols and exploit the fact that the encoding of the control signals is not fixed. After the logic optimization, an appropriate encoding of the control signals is performed.

Our approach is implemented in CLASSY (CadLab ALU Synthesis SYstem), a new tool that performs flexible and optimizing ALU synthesis *and* exploits the regular structure of ALUs. The input of CLASSY is a behavioral ALU description in which every combination of operations from a large set is allowed. The output of CLASSY is a regular layout that is made up of bit-slices.

Section 2 gives an overview of the bit-slice architecture of CLASSY and introduces the concept of a conditional bit-slice which enables us to handle the encoding of the control signals flexibly. Section 3 describes the ALU synthesis. Finally, section 4 gives our conclusions.

2 Concepts

2.1 Bit-slice Architecture

CLASSY exploits the regular structure of ALUs. This allows an efficient optimization. We define the regular structure of ALUs as follows:

An ALU consists of a vector of bit-slices. Each bit-slice operates on one bit of the operands. The bit-slices are arranged from the least-significant bit to the most-significant bit. A bit-slice has two kinds of inputs and outputs, namely data I/O and carry I/O. Carry wires communicate between a bit-slice and its neighbor. Another input for the bit-slices are the control wires. Via these wires, the control signals are distributed across the ALU.

This bit-slice architecture is a "ripple" architecture. In the future we will extend our techniques such that they can synthesize more general architectures such as carry look-ahead logic.

2.2 Conditional Bit-slice

The bit-slices can perform all operations described in the input of CLASSY. The values of the control signals determine which operation is performed. The assignment of appropriate control signal values to the various operations is a major factor in the optimization of the bit-slices.

In order to take this aspect into account, CLASSY first decomposes the ALU into a vector of bit-slices and performs logic optimization on the bit-slices. During these phases the values of the control signals are represented by symbols called *operation labels*, one for each operation. In a succeeding phase, appropriate control signal values are assigned to the operation labels.

The introduction of operation labels leads to the concept of a conditional bit-slice. Figure 1 shows an example of a conditional bit-slice. In this example the inputs of the bit-slice are a, b, and c_{in}. The outputs are out and c_{out}. The variables p and g are intermediate variables. The operation labels are `add` and `subtr`.

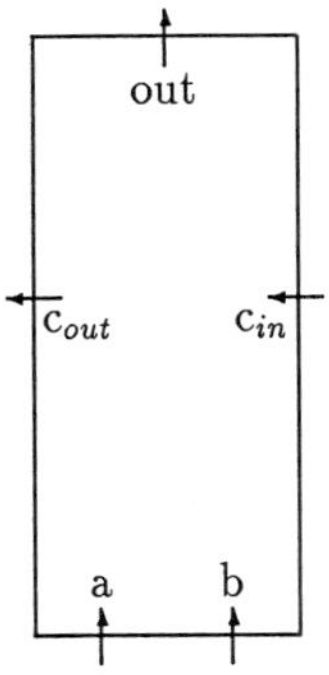

out	<-	{add, subtr}	$c_{in}\,\overline{p} + \overline{c_{in}}\,p$
c_{out}	<-	{add, subtr}	$g + p\,c_{in}$
p	<-	{add}	$a\,\overline{b} + \overline{a}\,b$
		{subtr}	$a\,b + \overline{a}\,\overline{b}$
g	<-	{add}	$a\,b$
		{subtr}	$a\,\overline{b}$

Figure 1: Example of a conditional bit-slice

In the general case, we have the following definitions:
A *conditional bit-slice* is specified by a set of conditional Boolean equations.
A *conditional Boolean equation* is the assignment of a number of conditional Boolean functions to a variable.
A *conditional Boolean function* consists of a condition followed by a Boolean function.
A *condition* is a set of operation labels. Within a conditional Boolean equation, the conditions of the conditional Boolean functions must be disjoint. The conditions act as case-labels. For operation label add the bit-slice of Figure 1 performs addition. For operation label subtr this bit-slice performs subtraction.

3 ALU Synthesis

The ALU synthesis is divided into three phases. The input of CLASSY is a behavioral ALU description. In the first phase the ALU is decomposed into a vector of conditional bit-slices. Subsequently, in the second phase, logic optimization is performed on the conditional bit-slices. Because the logic optimization is performed on the individual bit-slices, the optimization is efficient and produces regular circuitry. Finally, appropriate control signal values are assigned to the operation labels, the logic is mapped onto a technology, and layout is generated.

The following subsections give a description of these phases and illustrate the operation by an example. This example is a small ALU that can perform an addition, a logical shift left, and in one operation a bit-wise OR and a comparison (less-or-equal). The operation labels for these operations are add, sll, and or_leq. Figure 2 shows a behavioral description of the ALU.

```
IN d1, d2:(7:0);
OUT res:(7:0);                      (* result *)
OUT ovr:(8);                        (* overflow *)

res <- CASE
       add:    d1 "+" d2;           (* addition *)
       sll:    "SHIFTLL" d1;        (* shift left logical *)
       or_leq: d1 "OR" d2;          (* bit-wise OR *)
       END;
ovr <- CASE
       add:    d1 "O@+" d2;         (* overflow at addition *)
       sll:    d1(7);               (* overflow at shift *)
       or_leq: d1 "<=" d2;          (* test on less or equal *)
       END;
```

Figure 2: Behavioral ALU description

3.1 ALU Decomposition

In the first phase, the ALU is decomposed into a vector of conditional bit-slices. For this decomposition, we use knowledge-based techniques. The knowledge we use is the knowledge of how operations can be realized. For most operations there are several alternative realizations. E.g. a shift left operation can be realized directly, or can be realized by an addition (`"SHIFTLL" a` $\equiv$ `a "+" a`).

The knowledge is stored in the *decomposition library*. This library contains for each operator one or several decomposition procedures. A decomposition procedure generates for an operation a vector of bit-slices which perform the operation. Such a vector is called a *realization* of the operation.

The ALU decomposition consists of two steps:

Step 1 For each operation the ALU has to perform, the appropriate decomposition procedure is selected, and a realization of the operation is generated by this procedure. Selecting the appropriate decomposition procedure means selecting an optimal realization. The optimality of a realization depends on the amount of logic which the realization shares with the realizations that are selected for the other operations. For example, the optimal realization for the shift left operation of Figure 2 is the realization by an addition, because the ALU has to perform an addition anyway.

So, the problem is to find an optimal selection of the realizations. This problem is $\mathcal{NP}$-complete, but efficient branch and bound criteria can be formulated, and the problem sizes in this domain are not that large. Therefore, we use a branch-and-bound algorithm to solve this problem.

Step 2 The realizations of the various operations are merged to one vector of conditional bit-slices. The realizations are merged in such a way that redundant logic is prevented (e.g. carry-chain logic is shared by the operations). First, the realizations are transformed to vectors of conditional bit-slices. The realization of an operation is transformed to a vector of conditional bit-slices by attaching a condition to each Boolean function. This condition contains only the operation label of the operation. After these transformations, the vectors of conditional bit-slices are merged to one vector. In order to prevent redundancy, conditional Boolean equations that assign to the same variable are merged to one conditional Boolean equation.

Figure 3 shows the result of ALU decomposition for the ALU of Figure 2, and also a specification of the bit-slices in the middle. The specifications of the leftmost bit-slice and the rightmost bit-slice are omitted for space reasons.

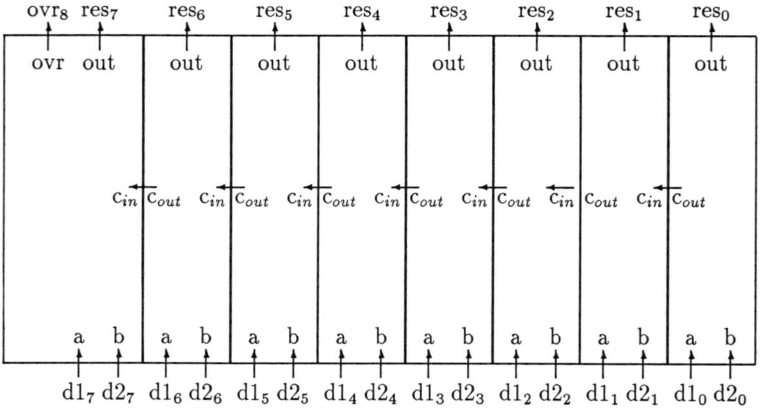

out	<-	{add, sll}	$c_{in}\,\overline{p} + \overline{c_{in}}\,p$
		{or_leq}	$a + b$
c_{out}	<-	{add, sll, or_leq}	$g + p\,c_{in}$
p	<-	{add}	$a\,\overline{b} + \overline{a}\,b$
		{sll}	0
		{or_leq}	$a\,b + \overline{a}\,\overline{b}$
g	<-	{add}	$a\,b$
		{sll}	a
		{or_leq}	$a\,\overline{b}$

Figure 3: Result of ALU decomposition for the ALU of Figure 2

3.2 Logic Optimization

Next, logic optimization is performed on the bit-slice specifications. This optimization differs from the usual logic optimization, because a set of *conditional Boolean equations* is optimized instead of a set of ordinary Boolean equations. We represent the specification of a conditional bit-slice as *multi-level* logic with a *symbolic input* (a definition of symbolic input can be found in [7]). The operation labels are the values of the symbolic input.

Prior work on multi-level logic with a symbolic input (e.g. [8, 9, 10]) has been directed to the encoding of the symbolic input, not to the optimization of the logic. Therefore we developed a number of powerful algorithms for the optimization of multi-level logic with a symbolic input:

- We developed a weak division algorithm for logic with a symbolic input (weak division is the process of extracting common subexpressions).

- We extended the logic minimization algorithm described in [11], and also some of the algoritms described in [12], in such a way that they can handle the symbolic input and our specific cost function, which incorporates estimates of the costs of the control wires.

Our algorithms operate on sum-of-terms representations of the conditional Boolean equations. A *term* is a product of literals and a condition. A sum-of-terms representation of the Boolean equation for g in Figure 3 is {add} a b + {sll} a + {or_leq} a $\overline{b}$. Our logic minimization algorithm minimizes this to {add, sll} a b + {sll, or_leq} a $\overline{b}$.

For the weak division, we developed a new algorithm for the "algebraic" division of such sums-of-terms (here "algebraic" indicates that the algorithm neglects the Boolean properties $x\ x = x$ and $x\ \overline{x} = 0$) and we modified and extended the kernels algorithm of Brayton [12] accordingly. Furthermore, we extend the rectangle covering method of Brayton [12] in such a way that it can be used for logic with a symbolic input. Independent of our development, Malik et al. [10] present a similar extension of the rectangle covering method.

In the bit-slice specification of Figure 3, CLASSY finds the following common subexpression:

{add, sll} b + {sll, or_leq} $\overline{b}$

This is a subexpression of p (divisor: {add, or_leq} $\overline{a}$), and a subexpression of g (divisor: {add, sll, or_leq} a). For this common subexpression the new intermediate variable t_1 is created. This leads to the bit-slice specification of Figure 4.

out	<-	{add, sll} c_{in} $\overline{p}$ + {add, sll} $\overline{c_{in}}$ p + {or_leq} a + {or_leq} b
c_{out}	<-	{add, sll, or_leq} g + {add, sll, or_leq} p c_{in}
p	<-	{add, or_leq} a $\overline{t_1}$ + {add, or_leq} $\overline{a}$ t_1
g	<-	{add, sll, or_leq} a t_1
t_1	<-	{add, sll} b + {sll, or_leq} $\overline{b}$

Figure 4: Extraction of a common subexpression

Another algorithm that we use eliminates intermediate variables of low "value". In the bit-slice specification of Figure 4 the variable g is eliminated.

The logic optimization is guided by a cost function which estimates the area of the logic, including the area for the control wires. As usual (e.g. [12]), area will be estimated in terms of the total number of literals in the factored form representations of the conditional Boolean equations. A *factored form* is either a sum-form or a product-form. A *sum-form* is a sum of product-forms. A *product-form* is a term or a product of (a term and) sum-forms. A factored form representation of {add, sll} a b + {sll, or_leq} a $\overline{b}$ is {add, sll, or_leq} a ({add, sll} b + {sll, or_leq} $\overline{b}$).

In order to incorporate estimates of the area for the control wires in the cost function, we assume that each incomplete condition (in the previous factored form {add, sll} and {sll, or_leq}) requires a control signal. We do not consider the possibility to save a control signal by using a Boolean function of other control signals, because, for an ALU in general, the area for a control wire is small compared to the area of additional logic in each bit-slice.

3.3 Technology Binding and Layout Generation

After the logic optimization, appropriate control signal values are assigned to the operation labels, the bit-slice specifications are mapped onto a technology, and layout is generated. Our current layout procedure is based on the use of standard cells.

First, appropriate control signal values are assigned to the operation labels. For this purpose we use the factored form representations described above. In the factored forms CLASSY replaces each incomplete condition by a control signal (cs_0, cs_1, ...). The result of this step is shown in Figure 5.

$$\begin{array}{ll} \text{out} & \texttt{<-}\ cs_0\ (c_{in}\ \overline{p} + \overline{c_{in}}\ p) + \overline{cs_0}\ (a + b) \\ c_{out} & \texttt{<-}\ a\ t_1 + p\ c_{in} \\ p & \texttt{<-}\ cs_1\ (a\ \overline{t_1} + \overline{a}\ t_1) \\ t_1 & \texttt{<-}\ cs_0\ b + cs_2\ \overline{b} \end{array}$$

Figure 5: Bit-slice specification

Subsequently, the bit-slice specifications are mapped onto standard cells. We use a tree matching method similar to the method described in [13].

An additional feature at this stage is that CLASSY can add latches, flip-flops, and tristate gates to the circuitry. CLASSY can also add a decoder to the ALU. This allows the specification of *control inputs* that are not the same as the control signals.

Finally, the ALU is laid out as a row of bit-slices. Each bit-slice is realized by one (or rarely several) columns of standard cells. Figure 6 shows the layout that is generated for the ALU described in Figure 2. In this example we used the SCMOS standard cell library [14].

Figure 6 shows that our layout algorithms, which exploit the regularity of the generated circuitry, produce compact layouts. In this example only 25% of the area is used for routing instead of the usual 50%. This corresponds to a reduction by 33% of the total area.

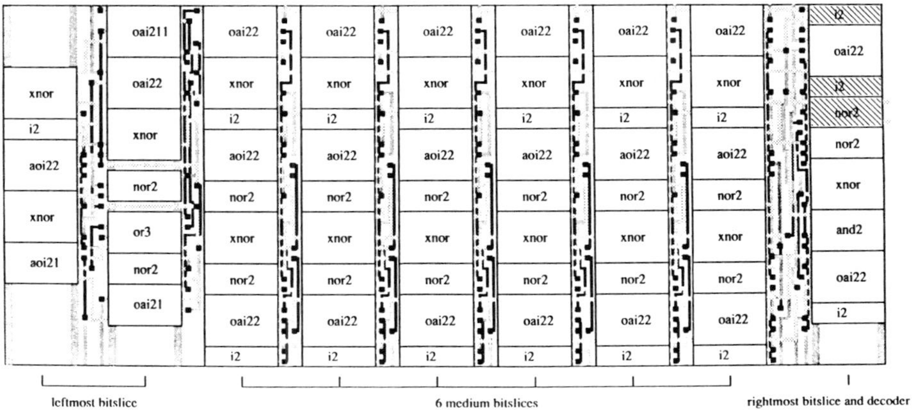

Figure 6: Layout generated by CLASSY for the ALU of Figure 2. In this example CLASSY added a small decoder that decodes two control inputs to the three control signals. The standard cells of the decoder are highlighted.

4 Conclusions

- CLASSY is flexible. In its input — a behavioral ALU description — any combination of ALU operations from a large set is allowed. Our current set contains 51 operators. This set can easily be extended by adding decomposition procedures to the decomposition library.
- CLASSY optimizes an ALU for the particular set of operations it has to perform. For this purpose, CLASSY uses knowledge-based techniques, employs powerful logic optimization algorithms, and handles the encoding of the control signals flexibly.
- CLASSY exploits the regular structure of ALUs. This allows an efficient optimization that produces regular circuitry. Our layout algorithms exploit this regularity to generate very compact layouts.

Acknowledgments

This research has been supported by the German Ministery for Research and Technology (BMFT) under grant number NT 2850 6. It was carried out within Cadlab, a joint venture of the University of Paderborn and Nixdorf Computer AG. We wish to thank Bryan Preas for useful remarks on earlier drafts of this paper. Jürgen Driller implemented the logic optimization algorithms. We also thank the other members of the Cadlab synthesis group for their support.

References

[1] P. Six, I. Vandeweerd, K. Croes, L. Rijnders, "Interactive Module Generator Design Based on Symbolic Layout", *ICCD '87*, IEEE, 1987, pp. 138-141.

[2] J. Schuck, N. Wehn, M. Glesner, G. Kamp, "The ALGIC Silicon Compiler System: Implementation, Design Experience and Results", *24th DAC*, ACM/IEEE, 1987, pp. 370-375.

[3] N. Dutt, "GENUS: A Generic Component Library for High Level Synthesis", *Tech. Rep. 88-22*, UC Irvine, Sept. 1988.

[4] I. Vandeweerd, Kris Croes, L. Rijnders, Paul Six, Hugo J. De Man, "REDUSA: Module Generation by Automatic Elimination of Superfluous Blocks in Regular Structures", *IEEE Trans. on CAD*, Vol. 8, No. 9, Sept. 1989.

[5] R.K. Brayton, R. Camposano, G. De Micheli, R.H.J.M. Otten, J. van Eijndhoven, "The Yorktown Silicon Compiler System", in: Daniel D. Gajski (ed.), *Silicon Compilation*, Addison-Wesley, Reading, MA, 1988, Chapter 7.

[6] Michael Lightner, Wayne Wolf, "Experiments in Logic Optimization", *ICCAD '88*, IEEE, 1988, pp. 286-289.

[7] Giovanni De Micheli, "Symbolic Design of Combinational and Sequential Logic Circuits Implemented by Two-Level Logic Macros", *IEEE Trans. on CAD*, Vol. 5, No. 4, Oct. 1986.

[8] Srinivas Devadas, Hi-Keung Ma, A.Richard Newton, A. Sangiovanni-Vincentelli, "MUSTANG: State Assignment of Finite State Machines Targeting Multilevel Logic Implementations", *IEEE Trans. on CAD*, Vol. 7, No. 12, Dec. 1988.

[9] Bill Lin, A.Richard Newton, "Synthesis of Multiple Level Logic from Symbolic High-Level Description Languages", *VLSI 89*, Munich, 1989, pp. 187-196.

[10] Sharad Malik, Robert K. Brayton, Alberto Sangiovanni-Vincentelli, "Encoding Symbolic Inputs for Multi-Level Logic Implementation", *VLSI 89*, Munich, 1989, pp. 221-230.

[11] Karen A. Bartlett, Robert K. Brayton, Gary D. Hachtel, Reily M. Jacoby, Christopher R. Morrison, Richard R. Rudell, Alberto Sangiovanni-Vincentelli, Albert R. Wang, "Multilevel Logic Minimization Using Implicit Don't Cares", *IEEE Trans. on CAD*, Vol. 7, No. 6, June 1988.

[12] R.K. Brayton, "Algorithms for Multi-Level Logic Synthesis and Optimization", in: G. De Micheli, A. Sangiovanni-Vincentelli, P. Antognetti (eds.), *Design Systems for VLSI Circuits, Logic Synthesis and Silicon Compilation*, Martinus Nijhoff Publishers, Dordrecht, 1987, pp. 197-248.

[13] Ewald Detjens, Gary Gannot, Richard Rudell, Alberto Sangiovanni-Vincentelli, Albert Wang, "Technology Mapping in MIS", *ICCAD '87*, IEEE, 1987, pp. 116-119.

[14] Microelectronics Center of North Carolina (MCNC), *VPNR System Package*, MCNC, 1988.

Logic and Architecture Synthesis
P. Michel and G. Saucier (Editors)
Elsevier Science Publishers B.V. (North-Holland)

Building Block Generation considering the inherent Hierarchy of Arithmetic Operations

Andreas Münzner

Institut für Theoretische Nachrichtentechnik und Informationsverarbeitung
University of Hannover
Appelstr. 9A, 3000 Hannover 1, FRG

Abstract

In order to shorten the design time parameterizable generators for arithmetic building blocks are provided by ASIC vendors creating the netlist with the elements of their cell library. Because these generators produce a fixed circuit structure which cannot be adapted to a given specification their application is limited. This paper presents a synthesis strategy called *Stepwise Refinement* which is based on the inherent hierarchy of arithmetic operations. It consists of a global optimization to choose the appropriate building block structure and a local optimization to exploit the special cells of a given library. It will be shown that the generated circuits are comparable in terms of delay and area to those being handoptimized although the presented optimization is not bound to a certain CMOS cell library.

1 Introduction

Design time of integrated circuits can be shortened by the use of building block generators. A building block is defined as a parameterized functional unit. In the area of digital signal processing (DSP) building blocks for arithmetic operations, like addition, multiplication etc., are frequently needed. Several kinds of building block generators are known in the literature and partly offered by ASIC vendors. They can be divided into two groups, one is producing the physical layout, the other creates a netlist using the cells of a semicustom library. Most of the generators lack under the following disadvantages:

- the generated structure is fixed, e.g. an adder is always generated as ripple-carry-adder,
- the implemented structure is optimized to a certain criterion, therefore the performance parameters are not variable,
- the optimization may not be optimal for all generator parameter values, and
- the adaption to different cell libraries is not supported.

These facts may prohibit the application of such a building block generator and cause troubles if a design has to be transfered to another cell library.

This problem is attacked by the CAD-Tool BADGE — a **B**uilding Block **Ad**viser and **Ge**nerator, which can be characterized by the following features:

- a hierarchy of generators for arithmetic building blocks is provided,
- building blocks for a given arithmetic function are available with alternative structures, like adders with conditional sum, carry lookahead, Brent-Kung structure etc.,

- estimations of the maximal delay and the needed area support the selection of building blocks,
- the building block synthesis consists of a global and a local optimization step,
- constraints can be specified to obtain a desired performance, and
- any CMOS libraries could be supported through the use of a built-in generic library.

This paper presents the implemented synthesis strategy called *Stepwise Refinement* in section 2. Section 3 describes the structure and the contents of the built-in library database and the representation and integration of generators. Section 4 covers the base of generator estimates and their use in building block structure selection. Sections 5 and 6 then describe employed algorithms for the global and local optimization and their impact on the synthesis result. Section 7 provides an overview about the BADGE system and explains a design session. Eventually section 8 finishes with the conclusions and an outlook on the future work.

2 Synthesis Strategy *Stepwise Refinement*

The goal of the synthesis strategy of *Stepwise Refinement* is to find the optimal building block structure in order to fulfill a given specification in terms of maximal delay and gate count. This new approach introduces a *generic* library which defines the interface and the function of arithmetic building blocks. These blocks as depicted on table 1 are selected by an inspection of various building block structures published in [3, 5, 7].

Arithmetic Building Block Types					
		InputOperand-		OutputOperand-	
BB-Type	Function	Wordwidth	Count	Wordwidth	Count
N-EXOR	Modulo-2 Adder	1	n	1	1
HA	Halfadder	1	2	2	1
FA	Fulladder	1	3	2	1
CSA	Carry-Save Adder	n	$2 \vee 3$	n	2
CSAT	Carry-Save Addertree	n	m	$n + Level$	2
				with $Level = f(m)$	
ADD	Two-Operand Adder	n	2	$n+1$	1
N-ADD	Multioperand Adder	n	m	$n + Level + 1$	2
				with $Level = f(m)$	
MUL	Multiplier	n,m	2	$n+m$	1
SUB	Subtractor	n	2	$n+1$	1
DIV	Divider	n,m	2	$n - m \wedge m$	2

Table 1: Arithmetic Building Block Types

Sorted by their complexity and dependency a hierarchy can be derived which is indicated in the first column of table 1. This column contains the abbreviation for each building block type and its horizontal position within the column reflects the degree of dependency, meaning a building block may consist of other building blocks appearing more left within the column. As an example the building block CSA may consist of FA, HA and N-EXOR building blocks. The remaining columns describe the function and the interface of each building block type leaving the internal structure unspecified.

Every building block generator generates its circuit with these *generic* library parts, i.e. creates the interface of its building block, the interfaces of the *generic* parts and the interconnections between them. It is now possible to assign *real* library parts to the *generic* ones which dispose of the same interface but represent different implementations combined with different performance parameters. Obviously the performance of the building block depends strongly on this assignment — further called *configuration plan* — which is performed by an optimization process called *configuration*. Based on estimates this process determines the appropriate assignment to achieve a desired performance. For this purpose all possible configurations are represented with a *configuration tree* consisting of AND- and OR-nodes which can be matched with the *generic* and *real* library parts respectively. For illustration the configuration tree of a Braun-Multiplier is partly depicted on figure 1. Its structure is determined by the generators defining the connections from the *real* to the *generic* parts, and the contents of the employed library affecting the possible connections from the *generic* to the *real* parts.

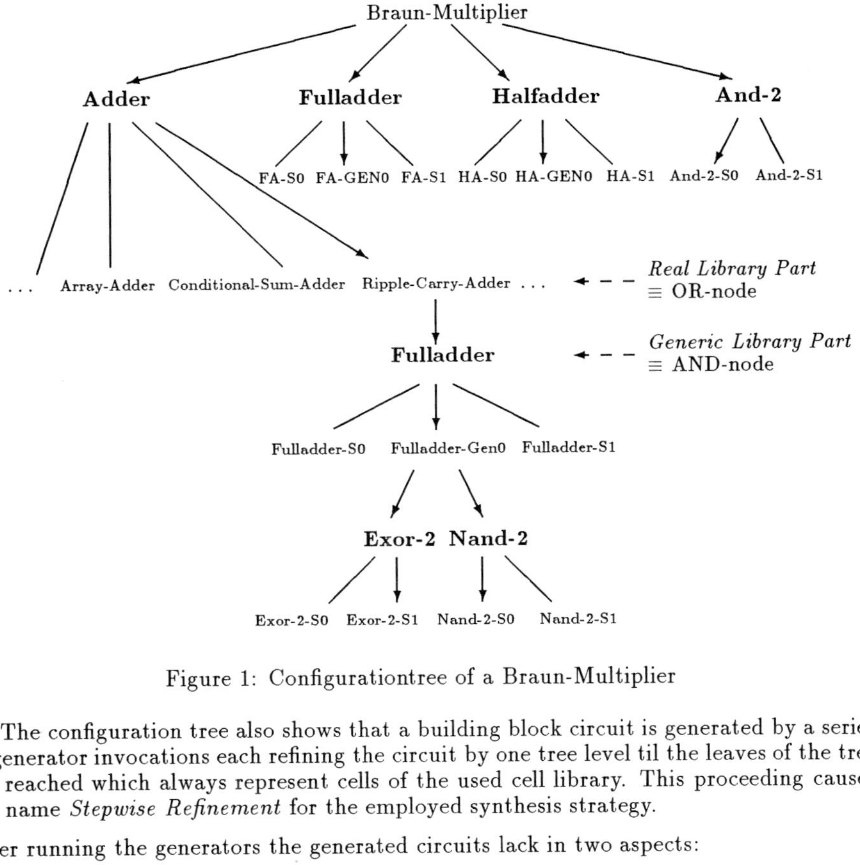

Figure 1: Configurationtree of a Braun-Multiplier

The configuration tree also shows that a building block circuit is generated by a series of generator invocations each refining the circuit by one tree level til the leaves of the tree are reached which always represent cells of the used cell library. This proceeding causes the name *Stepwise Refinement* for the employed synthesis strategy.

After running the generators the generated circuits lack in two aspects:

1. the circuits are built using only a restricted set of basic gates because the *generic* library is restricted to assure portability of the generators,
2. the driving capability of cells is in some cases not adapted to the FanOut.

Therefore a second optimization step follows optimizing the circuits locally. This is performed using the method of local transformations [1] which uses a set of rules describing subcircuits to be substituted by a complex gate or points within the circuit to be adapted to their load capacitance.

3 Structure of the Library Database

The library database of the BADGE system contains two groups of elements, the *generic* and the *real* library parts. The latter group consists of the elements of the used CMOS cell library, the generators, and the building blocks of former synthesis runs.

The generic library parts are connected among each other forming a *library tree* and all remaining library parts are attached to this tree as leaves. The tree has three main tasks:

1. It defines via inheritance the interfaces and the datastructures to represent real library parts.
2. It preserves a library interface which allows fast and uniform access to real library parts.
3. It enables the use of different cell libraries by exchanging the leaves of the tree.

All nodes of the library tree are represented by a *frame* datastructure or equivalently called *schema*. Each schema carries a unique name and consists of a set of slots which are uniquely named within the schema. Each slot in turn can carry values which may be of type number, string, symbol or function. Any value can now be accessed using the unique schema name together with the slot name as address. Regarding the schema describing a Conditional-Sum-Adder generator on figure 2 it is possible to get a list of supported number systems with the address: Conditional-Sum-Adder, Number-System. This schema conveys two further features to ensure the above mentioned characteristics of the library tree.

{{ **Conditional-Sum-Addierer**	
is-a:	Generator Adder
Inherited by *Adder* Prototype	
Wordwidth-1:	1...64
Additional-Inputs:	C-IN
Number System:	Two's Complement, Unsigned
etc.	
Inherited by *Generator* Prototype	
Pre-Action:	Adder-2Operand-Pre-Action
Main-Action:	CSA-GENERATOR-FN
Post-Action:	Adder-2Operand-Post-Action
Inputs:	··· to be generated ···
etc.	}}

Figure 2: Datastructure of a Conditional-Sum-Adder

1. The generator function is stored in the slot *Main-Action* and can be invoked with a message *(call-method 'conditional-sum-adder 'main-action)* send to the schema. This slot is also called a *method* of the schema. Because all generators dispose of this method a uniform interface is provided.
2. The slot *is-a* defines prototype schemata from which slots and default values are inherited. To integrate a new generator only generator specific values have to be stored in the appropriate slots.

Together with the specification of prototype schemata the possible assignments between real and generic library parts are defined. This means that all alternative structures are connected to their generic library part and can be retrieved easily. Therefore the library tree reflects parts of configuration trees.

Building Block Generators			
Function	Generator Name	Function	Generator Name
Adder	Ripple-Carry Fast-Carry-Chain Manchester-Carry-Chain Carry-Look-Ahead (CLA) Binary-Lookahead-Carry Decomposition Conditional-Sum	Adder Trees	Carry-Save Wallace-Tree
		Multiplier	Braun-Array Modified-Booth-Array Modular-Multiplication Pezaris-Array
		Subtractor	s.a. Adder

CMOS Cell Libraries		
Vendor	Channel-Length	Technology
Lsi-Logic	$1.5\,\mu m$	CMOS Gate Array
Matra	$2\,\mu m$	CMOS Gate Array
Valvo	$2\,\mu m$	CMOS Standard Cell
Valvo	$1.2\,\mu m$	CMOS Standard Cell

Table 2: Contents of the Library Database

In order to cover a broad range in the area/delay design space alternative structures for the arithmetic building blocks have been studied, implemented as generators and integrated into the library database. Table 2 provides a summary about the more complex building block generators and lists also the CMOS cell libraries adapted to the library database.

4 Estimation of Building Block Performance

The estimation of the building block performance concerns the maximal delay and the transistor or gate count as a measure for the needed chip area. The estimations are used for two purposes:

1. to give a system engineer a feeling about the achievable performance in order to support his architecture decisions,
2. to enable the building block configuration before generating the circuits.

Because the performance parameters of a building block depend heavily on its configuration and on the cell specification of the used CMOS library the estimations must consider these requirements. This can be achieved with estimation formulas using the same generic library parts as the generators do. As an example figure 3 lists the formulas for a Braun-Multiplier. In a mathematical view the generic parts serve as variables to which values or formulas can be assigned to depending on the configuration.

Therefore a formula interpreter is provided to evaluate these equations. It needs as input a building block specification and a configuration plan which may be a default plan. This plan defines a path through the configuration tree on which the formula interpreter works in postorder. This is done because formulas are always evaluated to a certain value based on the specifications of cell library elements before they are needed for the evaluation of other formulas.

- Transistor Count:

$$N_{BRM} = (n*m)*N_{and-2} + (n-1)*N_{halfadder} + (n-1)*(m-2)*N_{fulladder} + N_{adder-2op-nwidth(n\ (n-1))}$$

- Delay:

$$\begin{aligned} T_{BRM} &= T_{and-2}(FAN_{halfadder(inputA)}) + T_{halfadder}(FAN_{fulladder(inputA)}) \\ &+ (n-3)*T_{fulladder}((path(A\ C_{out}))\ FAN_{fulladder(inputA})) \\ &+ T_{fulladder}(FAN_{adder-2op-nwidth(wordwidth(n-1))}) \\ &+ T_{adder-2op-nwidth}((wordwidth\ (n-1))(path(A0\ C_{out}))\ FAN_{output}) \end{aligned}$$

- Fan-In Capacitance:

$$FAN_{BRM} = max(n,m) * FAN_{and-2}$$

Figure 3: Estimation formulas for a $n*m$ Braun Multiplier

Obviously the estimation formulas are generator specific and they depend on generator parameters like the wordwidth, the number of operands etc.. This suggests to implement their generation as a method attached to each generator schema which also has the advantage of a uniform interface to the formula interpreter.

5 Building Block Synthesis

As explained in section 2 the building block synthesis consists of three steps:

1. a global optimization to compute a configuration plan,
2. generator invokations according to this plan and
3. a local optimization.

The *global* optimization determines the configuration plan for a building block to achieve a desired performance in terms of area and delay. It works on the configuration tree, starting with a default plan, and updating the plan with the goal to minimize the costs which are computed employing the above mentioned estimations. To avoid the evaluation of the whole tree a *Branch-and-Bound* algorithm is implemented. This algorithm usually takes advantage of an estimated lower bound for the costs of a solution which allows to cut off subtrees if their lower bound is higher than the solution found so far. This proceeding guarantees to find the global minimum if the lower bound is a monotone increasing function of l with l being the level of the search tree. This requirement is not met by the estimation functions used for the global optimization because assignments at the leaves of the configuration tree can lower the estimated configuration costs. This is the reason why the implemented algorithm uses heuristic rules to determine the order of node selection or exclusion.

The base for all decisions are cost functions which exist for AND- and OR-nodes. Because all AND-nodes have to be included in the configuration plan, this cost function is used to find an optimal order of evaluation. The cost function consists of terms representing the gravity which the AND-node contributes to the whole configuration as shown in the following equation:

$$Q_{and} = \left(N_{OP} * \left(\frac{N_{and}}{N_{conf}}\right)^2 + T_{OP} * \left(\frac{T_{and}}{T_{conf}}\right)^2 \right)^{\frac{1}{2}}$$

with priority parameter $N_{OP}, T_{OP} \in \{0 \ldots 1\}$

The OR-node cost function, however, is used to judge alternatives which means it has influence on the order of node evaluation as well as on the exclusion of subtrees. It consists of a term which increases the costs, if constraints T_{max} — maximal delay — and N_{max} — maximal gate count — are not met, and a second part allowing to judge further improvements.

$$\begin{aligned} Q_{or} &= \left(N_{OP} * \left(\frac{N_{or}}{N_{def}}\right)^2 + T_{OP} * \left(\frac{T_{or}}{T_{def}}\right)^2\right)^{\frac{1}{2}} \\ &+ \left(N_{CP} * \left(\frac{N_{conf}}{N_{max}}\right)^2 * max\left(0, sign\left(N_{conf} - N_{max}\right)\right)\right. \\ &\left. + T_{CP} * \left(\frac{T_{conf}}{T_{max}}\right)^2 * max\left(0, sign\left(T_{conf} - T_{max}\right)\right)\right)^{\frac{1}{2}} \end{aligned}$$

The flowgraph 4 gives an overview about the proceeding of the implemented algorithm. During the way down the configuration tree alternatives are selected, estimated and depending on the result and the selected search strategy accepted for further refinement or excluded. In a second step the algorithm backs up the tree and selects from the inspected alternatives the best configuration for each OR-node.

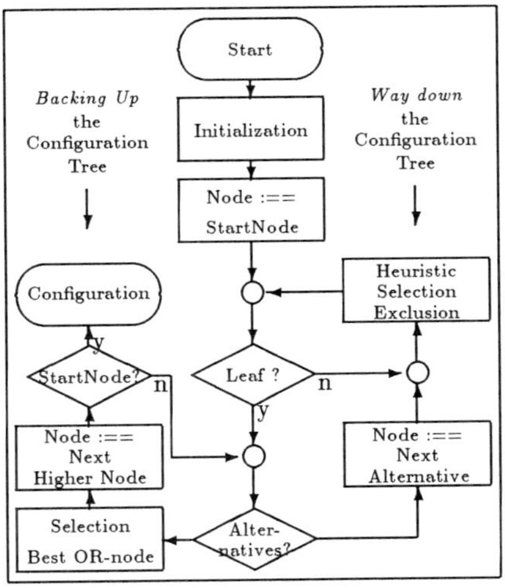

Figure 4: Flowgraph of the global optimization algorithm

Three search strategies are implemented which can be selected by setting parameters. These are

Hill-Climbing-Search which is fast because it considers only the best estimated alternative.

Beam-Search which is slower but considers a group of alternatives and usually results in better solutions.

Backtracking-Search which finds the optimal solution by an exhaustive search.

The generators produce circuits with a high degree of interconnections so that the local optimization has to work on a graph rather than on a tree representation of the circuits. Besides the circuits consist of multiple input gates which allow many different, functionally equivalent representations for complex gates. Local transformation systems reported so far [2] need for each possible representation a transformation rule which is contradictory to an easy integration of cell libraries.

Therefore a graph matching algorithm is implemented that works in a object oriented fashion. It sends a boolean description to a start gate which knows how to decompose and transform it in order to find a successful match. After processing the input this gate sends subterms of the input description to the gates connected to its in- or outputs which in turn know how to proceed with them. For this reason only one rule is necessary to integrate a complex gate for local optimization.

Rules are located in the library database as part of the schemata representing the complex gates.There are two slots to store a rule, the slot *Condition* carries the description of the circuit to be substituted and the slot *Replacement* the library part or circuit to replace it. The format to describe the circuits is similiar to SLIF[1] and allows the specification of multiple output gates as well as circuits with internal signals.

Similiar to [1] rules are classified depending on their effect into a group **DA** containing rules decreasing delay **and** area, a group **D** decreasing only the delay, and a group **A** decreasing only the area.

The local optimization proceeds in three steps:

1. Reduce the delay on the critical path using the rules of group **DA**. This may cause the location of the critical path to change. If the time constraint is met continue with step 3.
2. If the time constraint is not met repeat step 1 using the rules of group **DA** and **D** otherwise proceed with 3.
3. Reduce the area regarding only gates off the critical path using rules of group **A** and **DA**.

To avoid ending in a local optimum this proceeding is combined with a lookahead strategy which "simulates" a series of rule applications before accepting a transformation.

6 Results

The influence of the global optimization on the building block performance is demonstrated by the optimization of various 32 bit multipliers. For the examples listed on table 3 emphasis was put on reducing the maximal delay. The results obtained using a *beam-search* and a *hill-climing-search* are compared with a start configuration which consists of a ripple-carry-adder, a fulladder and halfadder generator, and the smallest version for each basic gate.

A comparision of the results shows that the beam-search produces better solutions than the hill-climbing search which is not surprising because it considers a group of OR-nodes as alternatives for an AND-node whereas the hill-climbing uses only the best estimated OR-node. However, as the example of the Pezaris multiplier reveals the beam-search does not necessarily end with the better solution. But the results are quite acceptible if the enhancement is regarded in absolute figures. The optimization of a Wallace-Tree-Multiplier starts with a delay of $124\,ns$ and ends with $32\,ns$. Thereby decreasing the area too because the used cell library dispose of area efficient implementations for full and half adders.

The improvement of the local optimization depends heavily on the structure of the building block. If it is very regular like that one of the Braun or Pezaris multiplier then little to no improvements are achieved. But if the building block structure is more irregular

[1]Stanford Logic Interchange Format

Results with LSI-LOGIC 1.5 μm Gate-Array Library:					
Search Strategy		Beam		Hill-Climbing	
Nr.	Typ	$\Delta_{T[\%]}$	$\Delta N_{[\%]}$	$\Delta_{T[\%]}$	$\Delta N_{[\%]}$
1	Braun-Multiplier	-39,9	+11,8	-16,5	+1,2
2	Carry-Save-Multiplier	-31,7	-6,7	-10,2	-8,4
3	Pezaris-Multiplier	-3,1	-7,8	-7,4	-8,7
4	Booth-Multiplier	-45,5	-24,3	-24,3	-41
5	Wallace-Tree-Multiplier	-74,4	-3,0	-31,7	-5,7

Table 3: Results of Global Optimization for 32 bit multiplier

so that the load capacitances of many gates depend on the building block parameters then the possible enhancement of the delay reaches up to 70 %. Most impressive is the example of a 40 bit Conditional-Sum-Adder generated with the LSI-LOGIC 1.5 μm Gate-Array Library. The generated circuit has a delay of 36 ns whereas the locally optimized circuit needs only 13.5 ns to perform an addition. This result is comparable in terms of delay with a handoptimized 32 bit adder generator offered by the vendor with a reported delay of 14.2 ns [6].

7 Overview of the BADGE

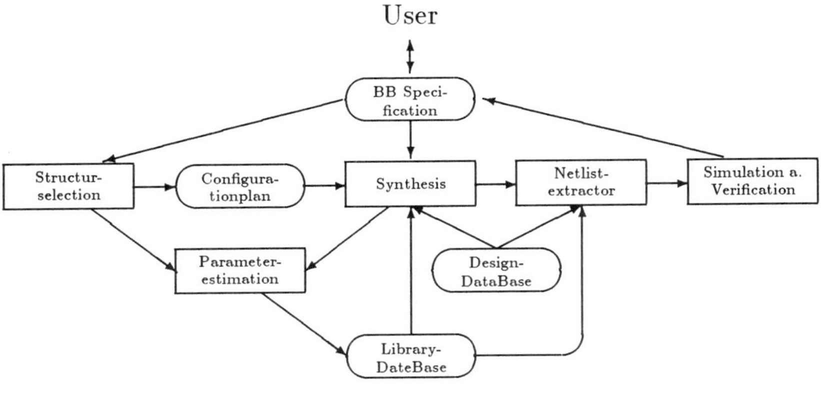

Figure 5: Overview of the BADGE system

The presented algorithm is part of the *B*uilding Block *Ad*viser and *Ge*nerator — BADGE — developed at Hannover University. The program like depicted on figure 5 takes as input a building block specification to which an appropriate building block structure can be interactively found in the library database using the *parameter estimation*. This *structure selection* supports system engineers while conceiving their system architecture and results in a configuration plan which fulfills a given specification. The configuration plan can also be obtained by the global optimization of the *synthesis* program. The synthesis itself proceeds in the way described in this paper and creates an internal representation of the building block in the *design database* which allows to generate a building block description containing the maximal delay for each output, the critical path, and the gate count. Also netlists can be extracted in EDIF 2 0 0 and HILO 3 format to interface to logic simulators. The evaluation of the simulator results proceeds automatically with a program called EXDIS[4]. BADGE is written in COMMON-LISP using CRL[2] as basic datastructure and consists of about 25 000 lines of code. It is part of a design assistent which supports the synthesis of systolic arrays from an algorithmic description.

8 Conclusions and Future Work

It has been shown that the quality of the generated arithmetic building blocks using the synthesis strategy of *Stepwise Refinement* can be compared with that of handoptimized circuits. This is achieved by establishing a *generic* library of building blocks which form a hierarchy of arithmetic operations. Functionally equivalent *real* library parts — elements of a cell library and generators — with different performance are bound to the respective *generic* library element. And because all implemented generators generate their circuits using only generic elements, totally different building block structures can be created by an appropriate assignment of real library parts. It has been shown that all possible assignments can be represented by a configuration tree on which a Branch-and-Bound algorithm works to find the problem adapted building block structure.

In order to preserve portability of the generators special cells like complex gates are only considered through the use of local transformations.

The object oriented structure of the built-in library database supports the integration of further generators each of which will improve the quality of the generated circuits because the global optimization automatically considers all structures located in the library database.

Further work is currently concentrated on the investigation of generators employing new number systems and on the implementation of divider generators. To improve data throughput the automatic insertion of registers into the building block circuits is being implemented.

References

[1] A.J de Geus and W.W Cohen, "A Rule Based System for Optimizing Combinational Logic". *IEEE Design & Test*, pp. 22–32, August 1985.

[2] E. Detjens, G. Gannot, R. Rudell, A. Sangiovanni-Vincentelli, and A. Wang, "Technology Mapping in MIS". *Proc. IEEE Intern'l Conf. on Computer Aided Design*, pp. 116–119, 1987.

[3] K. Hwang. *Computer Arithmetic.* John Wiley, 1979.

[4] A. Münzner and J. Thobe, "EXDIS – Rechnergestützte Simulationsauswertung". *4. E.I.S. Workshop, GMD-Studien Nr. 155*, pp. 281–290, Februar 1989.

[5] T.G. Noll and W. Ulbrich, "Architektur und Realisierung von Pipeline-Multiplizierern". *NTG-Fachberichte*, Vol. Bd. 87, 1985.

[6] V.G. Oklobdzija, "Simple and Efficient CMOS Circuit for fast VLSI Adder Realization". *IEEE Proc. of ISCAS*, pp. 235–238, 1988.

[7] N. Weste and K. Eshraghian. *Principles of CMOS VLSI Design.* Addison Wesley, 1985.

[2] Carnegie Representation Language

Logic and Architecture Synthesis
P. Michel and G. Saucier (Editors)
Elsevier Science Publishers B.V. (North-Holland)
IFIP, 1991

A fast and effective technology mapper on an autodual library of standard cells

K. Sakouti, G. Saucier

Institut National Polytechnique de Grenoble / CSI
46, Avenue Félix VIALLET
38031 Grenoble Cedex FRANCE

The technology mapper presented here works in 2 phases. The first phase performs pattern matching identifying OR/AND or AND/OR subtrees corresponding to the standard cell library. The second phase defines the technology mapping, transforming previous patterns in real standard cells. The optimization criterion is a severe limitation of the depth in terms of layers of the circuit and a minimal number of cells. The results show that in a very short computation time, a very dense circuit with an optimized critical path is obtained. If more severe critical path optimization is requested an additional phase working on the longest path brings it to a nearly optimal solution.

INTRODUCTION

The technology mapping phase in a multivel design process starts from a set of factorized boolean expressions and covers them by standard cells of a library. The result is a network of standard cells. The optimization criteria are usually first the performance then the area. A large number of technology mappers have been presented in the last years. One of the classical reference, the Dagon sytem [3] implemented in Mis [2] is based on a tree matching method. The weakness was a cumbersome decomposition procedure, applied on both the standard cells and the functions, into equivalent AND_OR trees of nodes having a 2_input fan_in. Some later work improved this approach by looking faster and more directly for an isomorphism between portions of the initial tree [4] [1] and standard cells of the library. The method presented here is closer to this latter approach. It looks for direct isomorphism between tree portions and standard cells and applies graph isomorphism methods [8]. A greedy approach starting from the root of the factorized tree is adopted. The cost of the standard cells is expressed by a factor taking into acount the computation power and the connection absorption of the cell. The mapping is guided by performance consideration as the introduction of layers of standard cells during the mapping is severely minimized.

This mapping is decomposed into 2 phases. A first phase looks for pattern matching in terms of OR(AND)/AND(OR) subtrees recognition. No inverter exists in this phase. This phase. This matching is called a "depth decreasing " matching as it avoids any unnecessary depth increase from an initial compacted AND/OR representation. The second phase is the real technology mapping phase. It is based on the autodual property of the library. A library is autodual if the existence of an AOIXYZ cell implies the existence of an OAIXYZ cell. This facilitates drastically the mapping in which no inverter has to be handled. The previous patterns are directly mapped into standard cells.

The results show first that this technology mapper is very fast and gives a low area circuit in a very short time. As far as critical paths have been investigated, it gives also an optimized critical path if the initial factorized expression is adequate. A last optimization step may improve this critical path.

1. PATTERN MATCHING PHASE

1.1. From cells to patterns and reduced patterns

The technology mapping phase starts from the set of factorized trees in which no inverter appears except at the leaves. The first phase will then consist of identifying patterns in the trees made up of elementary AND(OR) and OR(AND) trees corresponding to cells in the library. We shall see later on that the mapping to final gates introduces naturally inverting gates without additional inverter handling. A cell in a library identified by AOIXYZ is then reduced to AOXYZ . The library being autodual for this class of gates , the AOXYZ pattern implies the OAXYZ patterns. Therefore a pattern is identified as a 2-level OR/AND or AND/OR tree whose fan ins correspond to a cell of the library.

As an example, the reduced pattern shown in figure 1 corresponds either to the cell OAI221 or to the cell AOI221 and will be mapped to one of these 2 cells in the final phase

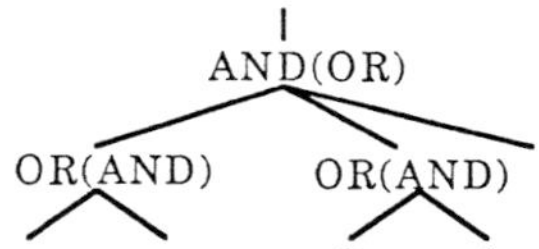

Figure 1: The reduced pattern corresponding to the AOI221 (OAI221) cells

1.2. Depth decreasing pattern matching

1.2.1. Pattern matching

Pattern matching consists of covering the factorized trees by reduced patterns. A pattern AOxyz (OAxyz) or Ax (Ox) can be assigned to a subtree in which the fanins of the nodes are larger than the parameters XYZ of the cell this allows a classical graph embedding.

For example, consider the tree represented in figure 2, the pattern AO22 is embedded in this tree as indicated in the same figure.

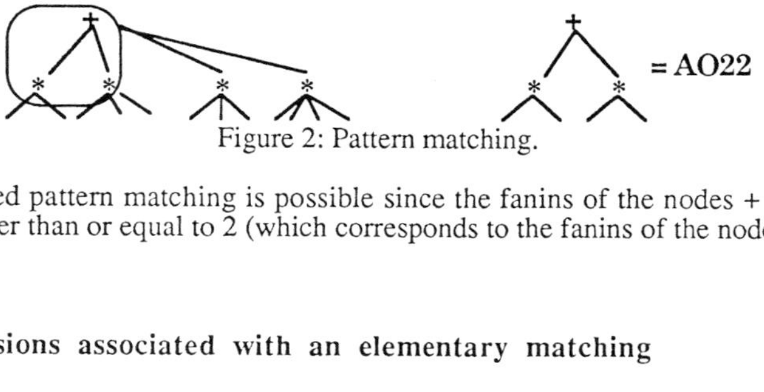

Figure 2: Pattern matching.

The indicated pattern matching is possible since the fanins of the nodes + and * of the graph are greater than or equal to 2 (which corresponds to the fanins of the nodes +, * of the AO22 pattern).

1.2.2. Expansions associated with an elementary matching

Each time a pattern embedded in a tree does not utilize all the input connections of the node of the initial tree, the node has to be split into several nodes thus introducing a new layer. This operation is called an expansion and introduces supplementary layers.

Consider as an example, a part of a factorized tree shown in figure 3:

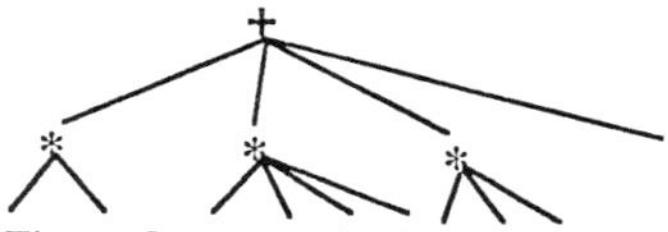

Figure 3: a part of a factorized tree

If we decide to match a cell AO22 as shown in figure 3a, we introduce 2 expansions one for the root and one for an AND node as shown in figure 3a:

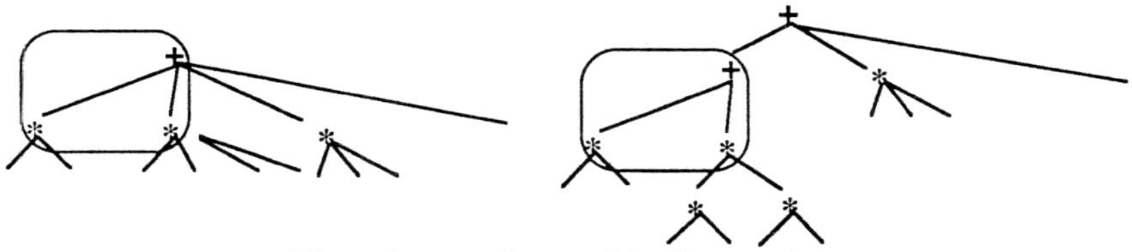

Figure 3a: mapping requiring 2 expansions

If we decide to match a cell as shown in figure 3b one expansion is generated.

Figure 3b: mapping requiring 1 expansion

If we now match a cell AO221 as shown in figure 3c, no expansion is generated.

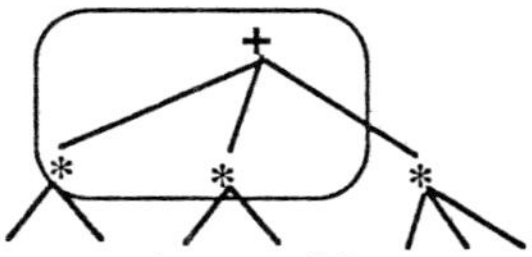

Figure 3c: mapping requiring no expansion

1.2.3. Nodes generating an obligatory expansion

A node in the factorized tree having a fanin larger than the largest fanin of an AND/OR gate available in the library will lead to an obligatory expansion introducing new layers.

The minimal number of layers introduced is equal to $\lceil \mathrm{Log}_{N2} N1 \rceil$ where N1 is the fanout of the node and N2 the largest fanout of the AND/OR gate available in the library, and the bracketsmean $\lceil \quad \rceil$ the smallest integer superior to .

Consider as an example N1 = 17 , N2 = 3, $\lceil \mathrm{Log}_3 17 \rceil = 3$

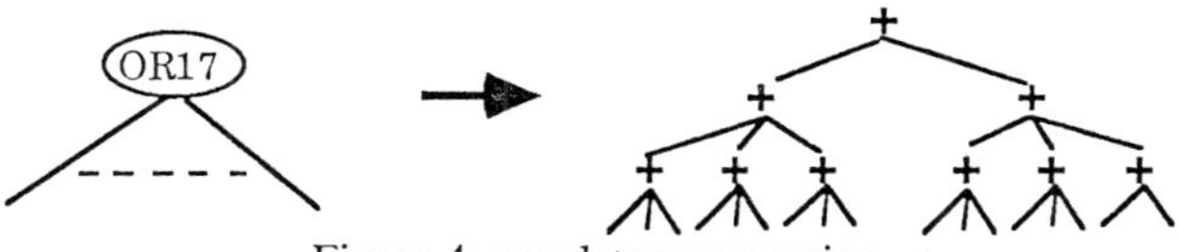

Figure 4: mandatory expansion

1.2.4. Reference depth of a leaf of a factorized tree

A leaf is connected to the root of the tree by a path crossing N operators. Suppose that N' of these operators generate obligatory expansions, and let Di be the number of layers for the operator i. We take as reference depth of the leaf:

$$\sum_{i=1}^{N'} Di + (N - N') .$$

This would be the number of cells crossed by a path going from the root to the leaves if a flat pattern matching is performed. A flat pattern matching associates, if possible, to each node in the initial tree, a single cell in the library. The reference depth of the leaves is shown for the example of figure 5. It is assumed that the largest fanin of the available AND/OR gate in the library is 4. Therefore the first OR node introduces 2 layers.

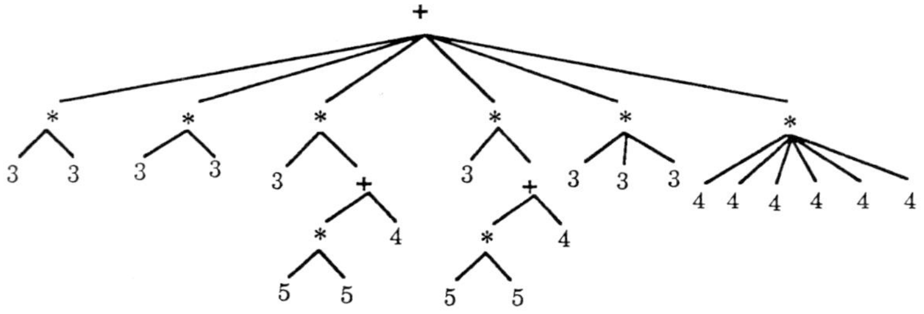

Figure 5: a factorized tree with indication of the reference depth on the leaf

1.2.5. Depth decreasing pattern matching

A pattern matching covers the factorized trees by patterns corresponding to the library and performing the required expansions.

A depth decreasing pattern matching is a matching for which the number of patterns crossed by a path going from a leaf to a root is minimized and in any case inferior to the reference depth. This new depth is called the matching depth. Consider the example of figure 5. The reference depth is indicated for each leaf. figure 6 shows the results of a depth decreasing mapping with the matching depths indicated on the leaves..

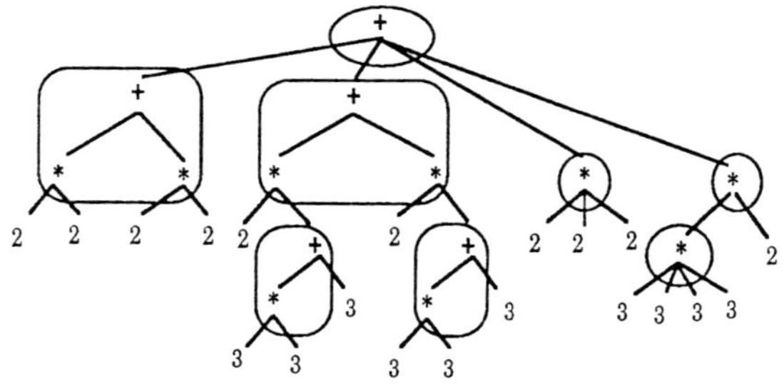
Figure 6: depth decreasing pattern matching

1.2.6. The algorithm applied to reach a depth decreasing pattern matching

To reach the depth decreasing pattern matching, local matchings are performed as long as they decrease or don't increase the number of layers. They will be called depth decreasing local matchings. The solutions decreasing the depth of the leaves are chosen first.

Consider as example the tree portion of figure 7. The depth decreasing local matching is shown.

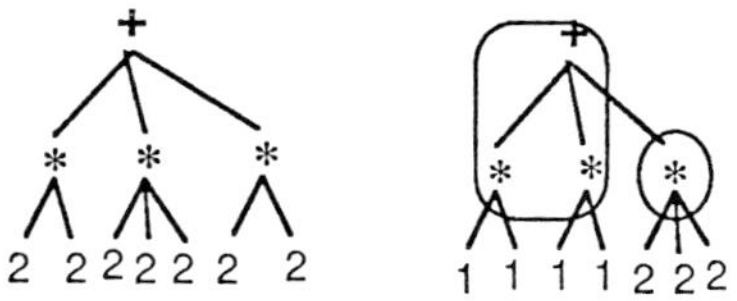

Figure 7: Depth decreasing matching.

The pattern matching is performed either from the root or from the leaves and the best solution is taken.

It can be noticed that each time an OR(AND) gate is the ancestor of 2 AND(OR) gates with fanouts corresponding exactly to an AO(OA) cell of the library, this mapping has to be chosen. This is automatically done in our greedy strategy.

The general algorithm of the first phase is the following:

Initialisation : current node = root of the tree then the leaves

Starting from the current node

Step1 :

Find a reduced pattern, that has the current node as root, without generating any splitting of the sons of this node

Step2 :

Expansion of the current node if necessary.
The mapped patterns are assimilated to blank nodes.

Step3 :

The sons of the current node become one after the other the current node.

Step4 :

At this step all nodes of the tree have been processed.
We restart the algorithm with the root as current node.

Figure 11 shows a factorized tree. Figure 12 shows the results of the pattern matching.

1.3. Expansion phase

The expansion phase is performed in 2 steps. First the expansion realizing the splitting due to pattern matching using partially the input connection of a node is done. This is shown in figure 8.

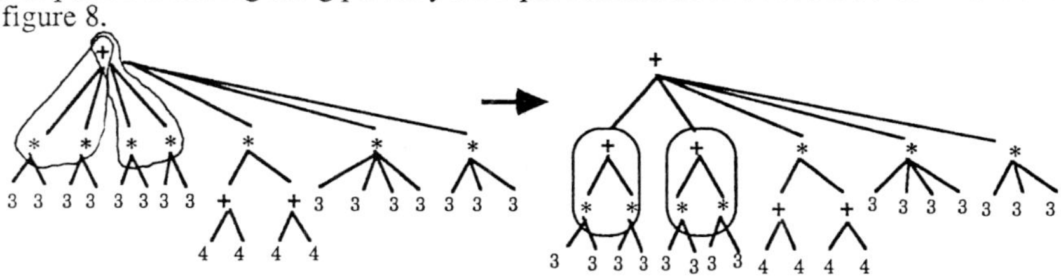

Figure 8: First phase expansion.

In a second step, expansion concerns now all the nodes having a too large fanin. This expansion will push in deeper layers the nodes having a small depth reference and preserve the depth of nodes having a large depth reference. Suppose in our previous example that the limitation of the fanout is 4. figure 9 shows that the leaves having as depth reference 3 still equal to 3 during the mapping.

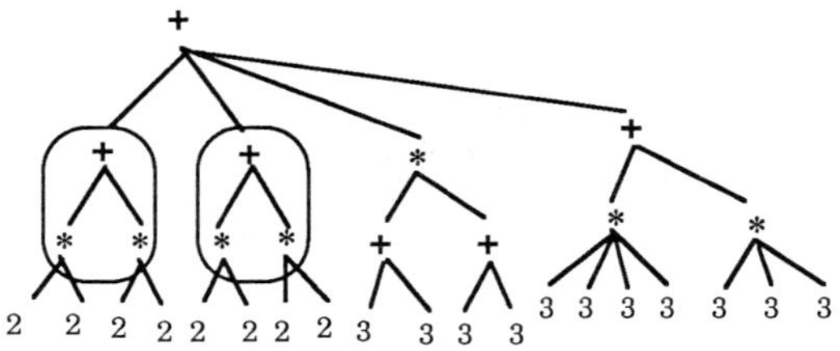

Figure 9: Final expansion.

2. TECHNOLOGY MAPPING PHASE

The technology mapping is very simple. It is a layer organized algorithm:
For layer i : the reduced pattern is transformed in the dual form with virtual inverter inclusion

For layer i+1: the inverters of the previous layers are absorbed and a direct mapping is possible. The figure 10 shows the result of the mapping.

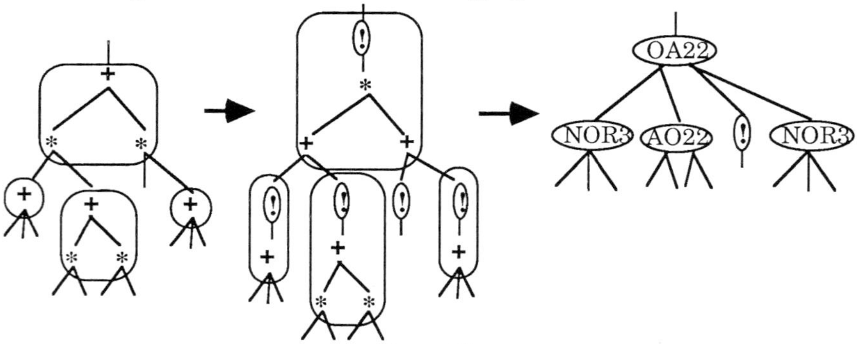

Figure 10

At this point the whole technology mapping has been performed without any inverter handling. Only at the leaves some input variables appear in a complementary form.

For the example of figure 11, the definitive mapping is given in figure 13.

3. RESULTS

To compare our mapper to MIS, we used the same library of standard cells (VSC100 1.5μ CMOS portable library from VLSI technology) and we start also from the identical factorized expression obtained by MIS Release #2.1. For the technology mapping we used the area oriented minimization command (map -m 0). On a set of standard benchmarks [6], we yield an area optimization at most equal to MIS results but 80 times faster (CPU time). We also indicate results obtained by DAGON [3], but for these results, the library and the initial factorized expression are different and unknown. Nevertheless, it is probable that the library is more or less similar to VLSI Technology and the initial expression a MIS expression with a minimal amount of literals Therefore the mapper can be considered as a good and fast mapper and cannot be responsible of any distorsion our results.

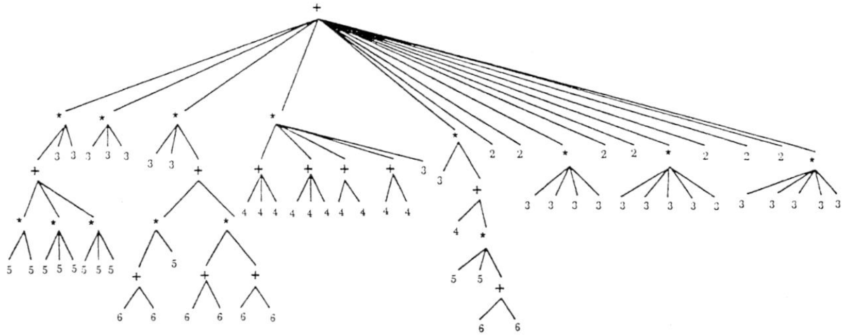

Figure 11: An initial tree with depth reference

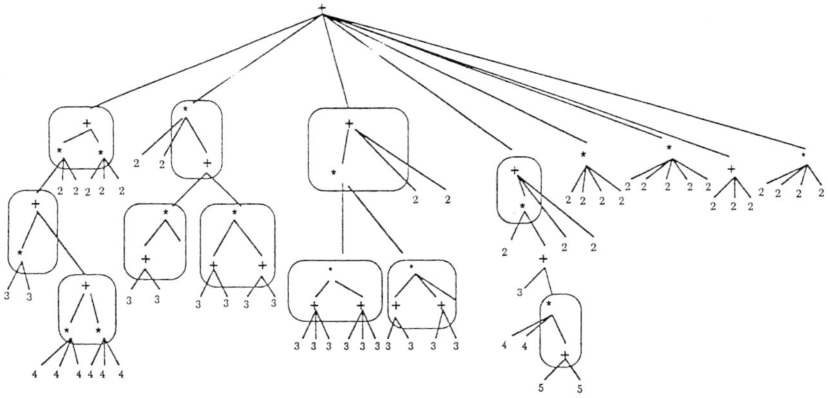

Figure 12: Pattern matching and matching depth

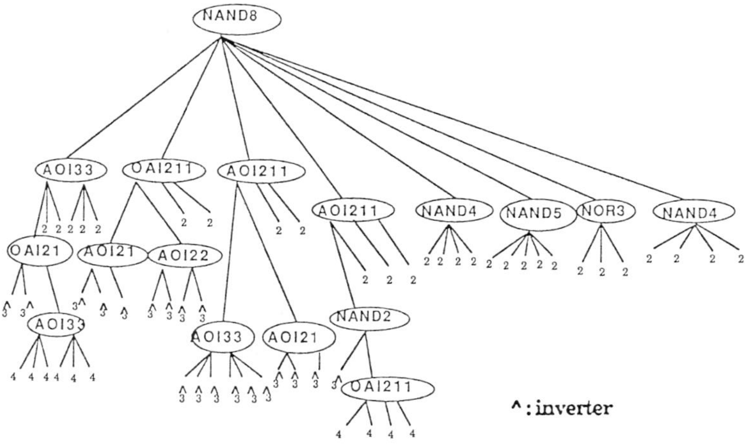

Figure 13: Final mapping

Benchmark	nb of gates including inverters			nb of inverters		gate area (K um**2)		CPU		
	ASYL	MIS	DAGON	ASYL	MIS	ASYL	MIS	ASYl (1)	MIS (1)	DAGON(2)
conl	10	12		4	6	14	16.59	0.1	5.6	
misex2	55	61		21	26	87.7	90.7	0.5	57.6	
duke2	249	245	251	35	39	394.8	397.1	11	271.2	1.2
5xpl	62	61		15	12	88.8	89.2	1	44.2	
rd73	41	44	47	10	8	60.4	71.5	0.5	36.2	0.3
rd53	17	18	19	4	6	25	25.1	0.3	11.6	0.1
sao2	59	59	61	13	13	89.2	90.7	1.2	48.4	0.4
bw	84	88	118	24	23	112.4	116.9	7	48	0.6
misexl	29	28		10	9	37.6	36.13	0.3	13.8	

(1) : in s on a SUN 3/60
(2) : in s on a VAX 8650

Table.1

CONCLUSION

The technology mapper presented here allows to reach in a very short time a dense circuit having good timing performance. It is based on a depth decreasing approach. If applied to an adequate factorized expressions , it leads immediately to a circuit exhibiting good performance characteristics. If the performance requirements are very severe, it constitutes an excellent starting point for further critical path optimization. This last step is presently under development.

REFERENCES

[1] R. A. Bergamaschi. "Automatic Synthesis and Technology Mapping of Combinational Logic". ICCAD-88, Santa Clara, California, pp. 466-473.

[2] E. Detjeus, G. Gannot. "Technology Mapping in MIS". ICCAD, Santa Clara (USA), November 1987, pp. 116-119.

[3] K. Keutzer. "DAGON: Technology Binding and Local Optimization by DAG Matching ". 24th DAC, Miami (USA), June 87, pp 341-347.

[4] M. C. Lega. "Mapping Properties of Multi-level Logic Synthesis Operations", ICCD-88, October 3-5, 1988, pp. 257-261

[5] R. Lisanke, F. Berglez, G. Kedem. "McMAP: A Fast Technology Mapping Procedure for Multi-Level Logic Synthesis", ICCD-88, pp 329-333.

[6] Benchmark set, Microelectronic Center of North Carolina 1989.

[7] P. Paulin, F. Poirot, "Logic Decomposition Algorithms for the Timing Optimization of Multi-level Logic", ICCD-89, October 2-4 1989, pp. 257-261

[8] G. Saucier. "Un Algorithme Efficace Recherchant l'Isomorphisme de 2 Graphes".Revue Française d'Informatique et de Recherche opérationnelle n° R-3, 1971, pp. 39-51.

Case Studies

Logic and Architecture Synthesis
P. Michel and G. Saucier (Editors)
Elsevier Science Publishers B.V. (North-Holland)

Evaluation of high-level design decisions using the Cathedral-II silicon compiler to prototype a DSP ASIC

Jan Vanhoof, Ivo Bolsens, Stefan De Troch, Eddy Blokken, Hugo De Man*

IMEC Laboratory, B-3030 Leuven, Belgium

Abstract

This paper describes the design of an ASIC chip set implementing a 512 kBit/s modem. Major design decisions are taken by the designer and evaluated in depth by generating a detailed layout with the Cathedral-II silicon compiler.

1 Introduction

In order to evaluate the performance of silicon compilation for complex system design, a number of design exercises featuring industrial complexity have been done using the Cathedral-II V3.0 compiler. The consequences of high-level design decisions on the overall chip area of a prototype ASIC will be illustrated.

First a brief overview of the current state of Cathedral-II is presented, primarily from a user-practical point of view. Next we will introduce the specifications of a design example which is representative for the current state of Cathedral-II.

Cathedral-II combines a large number of optimising transformations. The relative importance of these optimisations will be evaluated, using the introduced ASIC example. We will demonstrate that some optimisations may have unexpected side-effects on the performance or on the chip area. The importance of user intervention during the compilation will be stressed.

In order to draw relevant conclusions, a detailed chip layout has been generated for each design alternative, including pads and hardware for testability. Finally, the overhead of built-in testability in terms of chip area will be evaluated.

2 Birds view of Cathedral-II

Before presenting the application, an overview of Cathedral-II [1][2] from a user-practical point of view, is given below.

2.1 Target applications

Cathedral-II compiles digital signal processing (DSP) algorithms into application-specific DSP processors. The behaviour of the DSP algorithm is described with bit-level semantics in the functional language Silage (Fig. 1) [3].

*Professor at K.U.Leuven, Belgium

```
#define WT1 fix<16,12>
#define IN_CODE0  3.5
...
#define IN_CODE7 -3.5

FUNC biquad(in,b0,b1,c0,c1:WT1):WT1 =
BEGIN
    state@@1 = 0;
    state@@2 = 0;
    state   = in      - (WT1(state@1 * c1)
                      +  WT1(state@2 * c0));
    return  = (state  + (WT1(state@1 * b1)
                      +  WT1(state@2 * b0)))>>1;
END;

FUNC coder(in:bool) out:WT1 =
count,code:WT1;
BEGIN
  count@@1 = 0;
  (out,count) = IF (count@1 == 2)
                -> (code,0) || (0,count@1+1)
                FI;
  code = IF (!in@2 & !in@1 & !in) -> IN_CODE0
         || (!in@2 & !in@1 &  in) -> IN_CODE1
         || (!in@2 &  in@1 & !in) -> IN_CODE2
         || (!in@2 &  in@1 &  in) -> IN_CODE3
         || ( in@2 & !in@1 & !in) -> IN_CODE4
         || ( in@2 & !in@1 &  in) -> IN_CODE5
         || ( in@2 &  in@1 & !in) -> IN_CODE6
         || ( in@2 &  in@1 &  in) -> IN_CODE7
         FI;
END;
```

Fig. 1. Part of the Silage code describing the transmitter of a 512 kBit/s modem. The complete modem description covers about two pages.

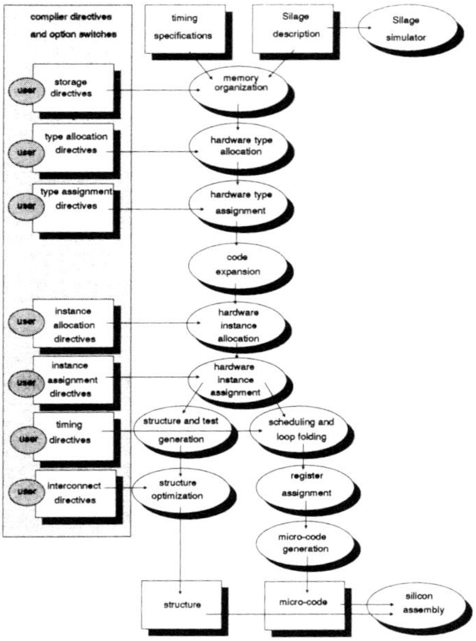

Fig. 2. Architectural synthesis design script of Cathedral-II. Design alternatives are generated by user-supplied compiler directives.

Cathedral-II applications currently feature fixed input and output sample rates. By no means the application domain is restricted to filters. Hierarchy, repetition and conditional operations support the design of large and complex non-linear systems. Starting from a Silage description and a number of compiler directives, a complete and testable chip layout is produced automatically. For a multi-chip design, system partitioning is done manually by splitting the Silage description into several files.

2.2 Target architecture

Cathedral-II generates synchronous DSP processors, with a dedicated micro-coded controller and a data path consisting of several parallel arithmetic and/or memory modules, called execution units. The micro-programme to control these programmable execution units is stored on-chip in ROM. The chips may be initialized, started, stopped or put in a test mode by a host processor. The two-phase clock is to be provided externally. The execution units are generated using a 3μ CMOS cell library, offering ROM, RAM, ALUs, multipliers, address computation units (ACUs), comparators and several types of I/O pads.

2.3 Cathedral-II from a user-practical viewpoint

In a first phase (architecture synthesis), the compiler generates the parameters of each execution unit, a dedicated interconnection network and the micro-programme. In a second phase (silicon assembly), module binders and a floorplanner generate the layout. In this text we will concentrate on the architectural synthesis.

Architecture synthesis is a sequence of several subtasks, combined by a design script (Fig. 2). Tasks with a large impact on the design are scheduled first. A task performs some transformation in view of its built-in optimisation criterion, such as minimising the number of machine cycles, busses or registers. Each task separately is deterministic. For a certain Silage description and a particular set of directives, only one solution is obtained. Influencing the outcome of each task by means of compiler directives generates design alternatives. With each task a specific type of directives is associated. All types of directives share a common syntax. They are specified in a separate input file to ensure reproducible designs, while design alternatives may be explored without re-simulating the Silage description.

2.3.1 Compiler directives

Following directives are currently supported.

- Storage directives suggest the memory organisation and the addressing schemes. This determines the amount of registers, ROM and RAM, as well as the complexity of the address computations.
- Type allocation directives serve as a declaration of the execution unit types the compiler may use. Important speed/area trade-offs are made here, as the additional area cost of an accelerator execution unit compared to a general-purpose ALU is considerable.
- Type assignment directives decide which execution unit type will implement a class of Silage constructs, such as loops, multiplications or delay-lines. Advanced users remove operations from the critical path by means of this feature.
- Instance allocation directives set an upperbound to the number of instances of any execution unit type that may be used. This allows to explore the potential parallelism in the signal flow graph.
- Instance assignment directives influence the assignment of individual operations to particular execution unit instances. Basically this trades machine cycles for interconnection complexity and register cost.
- Timing directives may change the instruction scheduling. They allow both absolute and relative timing constraints on I/O operations.
- Interconnect directives control the transformation of a dedicated multiplexer-based crossbar switch into a bus-oriented interconnection network. Again this is a speed/area trade-off.

For most tasks, the designer may omit directives completely and the compiler comes up with a solution of its own. On the other hand directives allow the designer to specify the final solution in great detail. Any situation in between is feasible. Specifying all details of a solution is undesirable, since this may require a large number of directives. Some tasks do not require or allow user interaction. Directives are to be used for major design decisions only, i.e. decisions the designer may imagine the effect of. The details of the final implementation are left to the compiler.

2.3.2 Design script

It is up to the designer to judge the quality of the solution. The design script is responsable of starting tasks in a meaningful order, providing filtered feedback to the designer, suggesting improvements and alternative solutions. Examples of feedback are the occupation of the execution units and the busses as a function of time. While scanning design alternatives, consistency is maintained by a make-utility based on file generation dates, both during architecture synthesis and silicon assembly.

3 A 512 kBit/s modem as test case

As representative test vehicles we selected a set of chips taken from a state-of-the-art DSP system (Fig. 3). The chip set implements the digital part of a 512 kBit/s modem, crucial part of a baseband PAM communication system. In order to transmit 8 channels of 64 kBit/s binary data simultaneously over a dedicated transmission line with a bandwidth of somewhat less than 125 kHz, a binary transmission scheme is not sufficient [4]. Instead a more complex 8-ary transmission scheme was adopted.

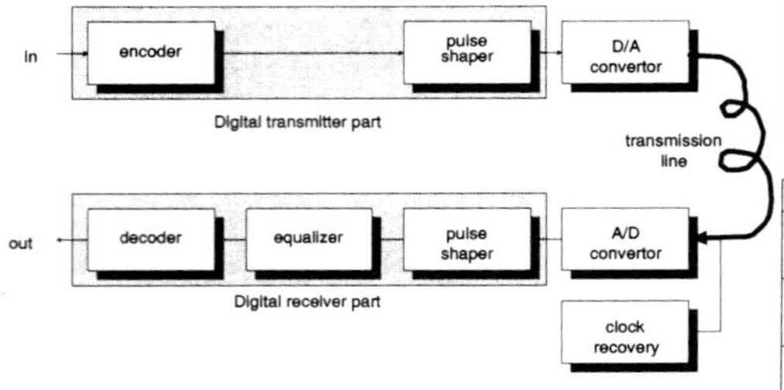

Fig. 3. Block diagram of a 512 kBit/s modem.

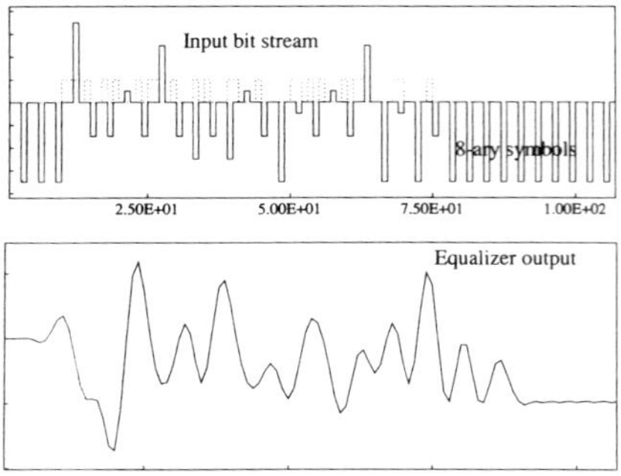

Fig. 4. Time-domain simulation of the complete modem.

3.1 Transmitter

The transmitter chip implements the down-sampling binary to 8-ary encoder and the transmitter pulse shaper. The encoder transforms the incoming bitstream into 8-ary symbols, represented as -3.5, -2.5, ..., 3.5. This decimates the data rate from 512 kBit/s to 170 kHz. In between two consecutive symbols, the encoder inserts two zeroes to again obtain a 512 kHz output (Fig. 4). To minimise inter-symbol interference at the receiver side, the transmitter pulse shaper applies a rooted raised-cosine transformation to the 512 kHz pulse train [4]. Classical filter synthesis techniques show that a cascade of three biquad filter sections is one way to implement this transformation. The output of the pulse shaper is a 512 kHz discrete waveform suited for D/A conversion and transmission.

3.2 Receiver

At the receiver side the incoming analog waveform is sampled at 512 kHz. The receiver chip implements the reception pulse shaper, featuring the same frequency response as the transmission shaper [4], but with a different dynamic range. A 2nd-order line equaliser is included in the receiver chip to compensate the frequency distortion caused by the transmission line. The digital signal is then ready for down sampling to 170 kHz and decoding into a binary data stream of 512 kBit/s.

4 Implementation alternatives

Due to the timing constraints, which are tight for a micro-coded processor realisation, the implementation is not straightforward. Numerous design decisions are hard to evaluate without generating at least a detailed floorplan. In the next paragraphs we will highlight the most important design decisions in detail. Layouts, area and timing figures are provided, whenever relevant.

4.1 Quantisation and simulation

To investigate quantisation effects in filters, analysis techniques are preferred over simulation, but analysis is only applicable to linear systems. Most Cathedral-II applications include non-linear parts, for which extensive time-domain simulation is the only resource left. Therefore we included simulations in this text as illustrative examples (Fig. 4 and Fig. 5).

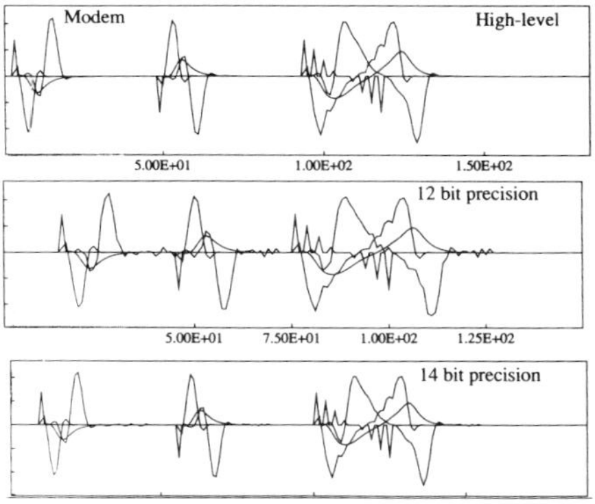

Fig. 5. Quantisation effects can be investigatec by means of time-domain simulations.

Fig. 6. Layout of the separate encoder chip. The control logic consumes most of the area.

Minimising the word length of intermediate signals yields smaller data paths, but only if operations with the same word length are grouped on the same operator. Otherwise the area of each operator is determined by the largest operation it executes, and significant area is additionally lost for type adjusting hardware. If all signals have the same type, the area lost because of using wider data paths is compensated by the absence of cast multiplexers.

4.2 System partitioning

The design is partitioned into one transmitter chip and one receiver chip. Implementing the encoder on a separate chip indicates a very large controller and I/O overhead (Fig. 6). Integrating the encoder and the transmitter pulse shaper on one chip shows that the area and machine cycle overhead, due to the encoder, is neglectable compared to the pulse shaper, if a separate 4-bit ALU is allocated to accomodate the encoder (Fig. 16).

4.3 Memory organisation

The organisation of the memory has a large impact both on the area and on the sample rate.

4.3.1 Storing delay lines

The amount of memory occupied by delay lines is very limited. In the receiver only 5 scalar delay lines of length 2 are to be implemented. The amount of registers per execution unit input is limited by performance to a bank of 8 register fields, but an implementation with 2 or 3 ALUs contains sufficient registers (maximally 32 resp. 48) to accomodate all delay lines. Therefore the delay lines are stored in registers for fast access. A small RAM is not necessarily more area efficient than a small register bank, and at least one address computation unit (ACU) capable of modulo counting is required to address it. A solution with one ACU (Fig. 7) suffers from an address computation bottleneck, slowing down the whole system. The delay lines are too small to split up the RAM in parallel parts for faster access. The area overhead of several small RAMs and ACUs is obviously not acceptable.

4.3.2 Storing constants

The filter coefficients may be stored in register files for fast access or in a data path ROM. Immediate addressing avoids the use of an ACU, since all addresses are known at compile time. The same holds for delay-line initialization constants. Iterative multiplications on an ALU eliminate the need to store the filter coefficients, as the multiplication sequence is controlled by the micro-code rather than stored constants.

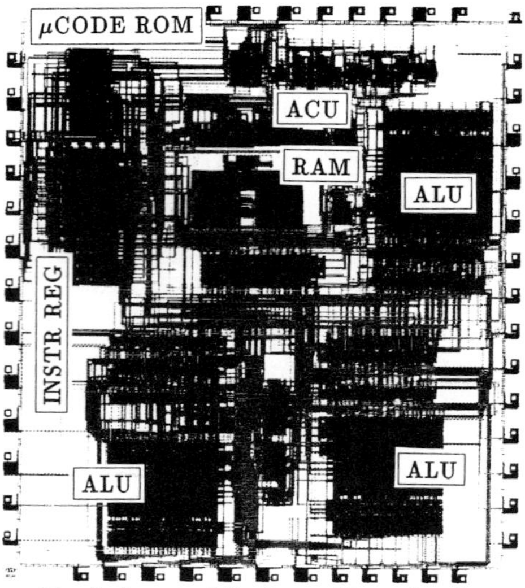

Fig. 7. Layout of the transmitter pulse shaper with delay lines stored in RAM.

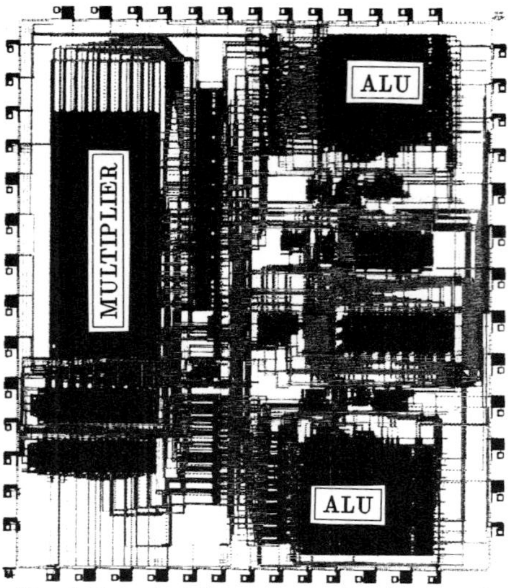

Fig. 8. Layout of the receiver pulse shaper and equaliser with 2 ALUs and a multiplier.

4.3.3 Storing intermediate signals

Signals consumed by more than one operation may either be stored in a single register, or broadcasted to different registers for each consuming operator. Avoiding storage duplication of such signals usually saves registers, often at the cost of machine cycles. Since each execution unit reads its inputs from its own input registers, two potentially parallel operations cannot read a signal from the same register at the same time, because this would imply the same execution unit executing two different operations at the same moment. Therefore one of the operations must be postponed, usually causing an extra machine cycle. If the input signal is broadcasted to two different execution unit input registers, potential parallelism is never lost. In general, signals with non-overlapping lifetimes or signals alive under disjoint conditions can share the same register. This reduces the number of registers significantly. Surprisingly, avoiding storage duplication by *not* broadcasting signals does not always allow to save *additional* registers, due to particular signal lifetime intervals. The designer may switch broadcasting ON or OFF to evaluate a particular design.

4.4 Hardware type allocation

The most complex arithmetic operations in the receiver are multiplications. These multiplications may either be implemented on array multipliers or computed serially on ALUs, depending on the data rate. The high data rate suggests to allocate array multipliers. This proves to be a wrong decision. Since the CSD coded coefficients contain very little non-zero bits, most serial multiplications take only 2 or 3 machine cycles on an ALU. Loading the registers of the multiplier consumes cycles, so a processor with 3 ALUs (Fig. 15) is almost as fast as a processor with 2 ALUs and 1 multiplier (Fig. 8). The area overhead of a multiplier can therefore not be justified (Fig. 16).

Several design alternatives exist to implement I/O operations. Pads may be 3-state or just plain buffers. Either separate pads are used for input and output, or combined bidirectional I/O pads are selected. In general this decision depends on I/O bandwidth, external I/O-timing constraints and on the chip area and perimeter. In all design alternatives, the perimeter ring was spacious enough to accomodate separate pads for input and output. As no bidirectional pads had to be allocated, off-chip communication is simplified.

4.5 Hardware instance allocation

The remaining arithmetic operations, such as additions, subtractions, shifts and delay-line updates, may be executed by one or several ALUs, due to a significant amount of parallelism in the signal flow graph. Fig. 16 indicates data rates for the receiver pulse shaper and equaliser implemented with 2 and 3 ALUs. The decoder demands its own 14-bit ALU in order not to influence the data rate of the receiver chip.

```
/* Allocation directives */

ALLOC(alu,3);
ALLOC(inpad,1);
ALLOC(outpad,1);

/* Assignment directives */

ASSIGN(main(_)\biquad( in,_,_,_,_)\(_),alu,0);
ASSIGN(main(_)\biquad(tb1,_,_,_,_)\(_),alu,1);
ASSIGN(main(_)\biquad(tb2,_,_,_,_)\(_),alu,2);
```

Fig. 9. Directives for 3 ALUs, one set of input pads and one set of output pads. A minimum number of well-chosen assignment directives reduces the routing on the transmitter pulse shaper chip to straightforward abutment.

Fig. 10. Layout of the transmitter pulse shaper with manual distribution of operations over the executions units. The controller is distributed.

4.6 Hardware instance assignment

The number of allocated execution units and the distribution of the operations over the execution units, heavily affect the interconnection requirements, including busses, (3-state) output buffers and input multiplexers. The automatically computed distribution for a 3 ALU transmitter pulse shaper minimises the number of machine cycles, but connects the output of every ALU to every input of every other ALU. This yields a chip looking almost identical to the receiver chip in Fig.15. A manual distribution, simply grouping all operations of one biquad on the same ALU (Fig. 9), reduces the routing area considerably (Fig. 10), as the execution units can simply be abutted. Moreover manual assignment groups all signals of the same type on the same ALU, reducing the word length of the ALUs and the amount of type casting multiplexers. For the total chip, this means 10% area less, with no increase of machine cycles. For the same reason it is advantageous to implement the encoder on its own 4-bit ALU, rather than using the 16-bit ALUs of the shaper. A similar strategy for the receiver chip is not applicable, since the 4 biquads can be implemented on 3 ALUs (Fig. 15) without violating the data rate specifications.

4.7 Micro-code generation

The designer may decide to use a central controller, featuring a programme counter, one micro-code ROM and one instruction register, or a distributed controller, featuring one programme counter but separate instruction registers and micro-code ROMs for each execution unit. This decision depends on the resource allocation, the floorplan strategy and the overhead for controller testing. The best fitting controller was selected for each design alternative. For the transmitter chip the area gain due to a distributed controller was 25% (compare Fig. 10 to Fig. 12), while for the receiver it makes no sense to distribute the controller (compare Fig. 15 to Fig. 13).

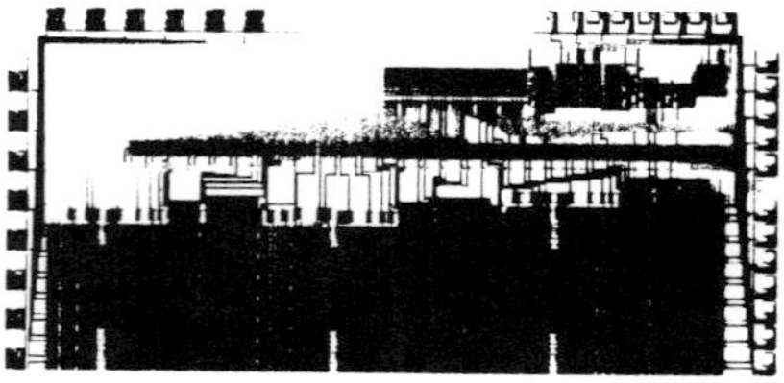

Fig. 11. Layout of the manual design of the transmitter pulse shaper without testability hardware.

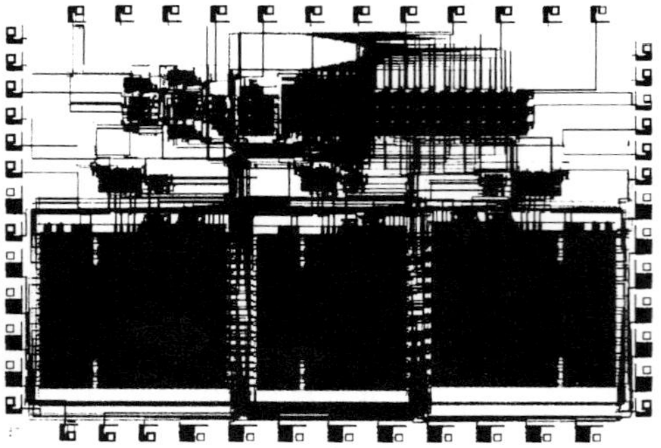

Fig. 12. Layout of the transmitter pulse shaper with manual distribution of operations over executions units. The centralised controller does not allow for an efficient floorplan.

4.8 Structure and testability generation

A combination of scan path testing and functional testing is currently being used in Cathedral-II. Testing overhead is demonstrated for the most area efficient transmission chip (Fig. 10), because this is the solution the designer ultimately selects. To compare, the layout of the transmitter pulse shaper has been generated with test facilities and compared to a manual design (Fig. 11) according to the same specs, and using the same silicon library, but without test facilities. The difference in area indicates a testing and initialization overhead of 30%.

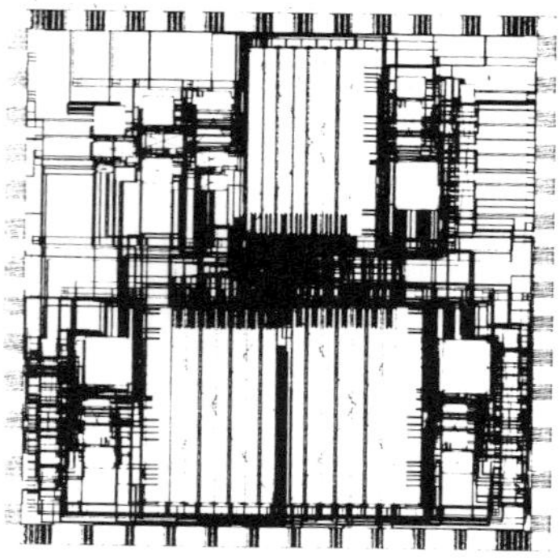

Fig. 13. Floorplan of the receiver pulse shaper and equaliser with distributed controller. The distributed controller does not allow for an efficient floorplan.

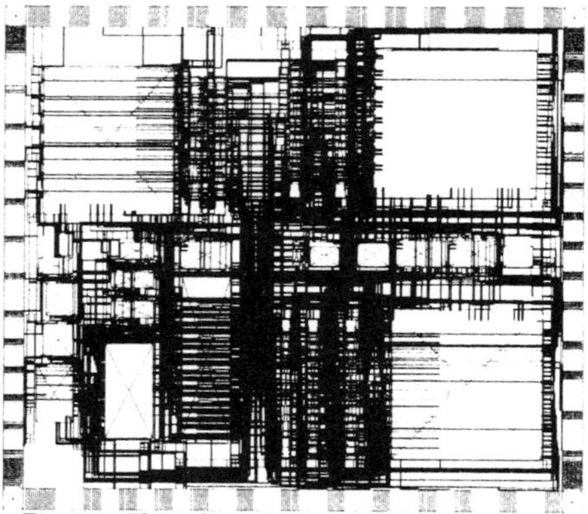

Fig. 14. Floorplan of the receiver pulse shaper. The input multiplexers of the execution units are routed separately.

4.9 Floorplanning, placement and routing

After disappointing experiments with automatic placement, the placement of the execution units and the controller was done manually. A good placement is the result of several iterations including detailed routing. The routing *strategy* is equally important for the overall chip area. Evaluating over-cell routing, switching equivalent pins, power-line tapering and layer switching per channel in order to avoid vias proved to be time-consuming experiments. The gain in area stays below 10%. The conclusion is that these features are useful, but only for optimising the final chip version. The absence of abutment and the overhead of power and ground routing increase the area significantly when a finer cell granularity is used in the floorplan. Simply handling input multiplexers of execution units as separate cells, already causes an area penalty of 15% (Fig. 14).

5 Conclusions

Following conclusions may be drawn for the elaborated examples.

- Abstract optimisation criteria for silicon compiler subtasks have to be selected very carefully. Often reasonable-looking criteria cause unexpected and unwanted side-effects on the area or the performance of a chip. For example optimal instance assignment with respect to clock cycles, creates an important routing area overhead. The use of acceleration execution units does not guarantee a faster design. Minimal signal word lengths do not imply minimal execution unit area.
- Manual intervention is a substantial part of the design, even after completion of the system design at the bit-level. It was felt that compilers such as Cathedral-II do provide an efficient workbench to investigate the consequences of high-level design decisions in a reasonable amount of time.
- Design turn-around time from system description to detailed layout takes about 1 hour on an Apollo DN-3500 workstation. Obtaining switch-level simulation results of the transistor netlist extracted from the layout, takes one additional day on a VAX-780. A dozen design iterations have been done to explore the design alternatives for the modem in depth. The manual design (Fig. 11) took about 1 man-year to complete, as the library was available. The receiver has been processed for performance verification (Fig. 15) .
- A considerable, however unavoidable, area overhead of 30% for testing was encountered.
- Scaling to a competitive technology makes the automatic design of systems such as 2 MHz ISDN modems with Cathedral-II definitely feasible.

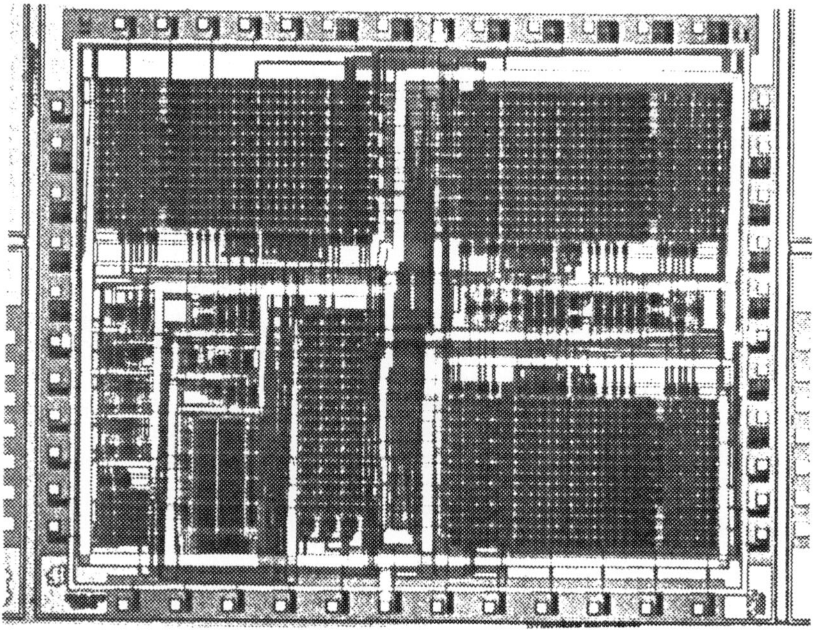

Fig. 15. Micro-photograph of the receiver pulse shaper and equaliser, containing 3 ALUs of 14 bit (38 mm^2 in 3μ CMOS).

TRANSMITTER	layout	resources	area (mm^2)	cycles	data rate(*)
down-sampling encoder	Fig.6	1 ALU (4 bit)	13.5	11	900
transmitter pulse shaper	Fig.10	1 ALU (16 bit) 2 ALU (14 bit)	35.0	16	625
transmitter pulse shaper	Fig.12	1 ALU (16 bit) 2 ALU (14 bit)	46.5	16	625
encoder and transmitter pulse shaper	—	1 ALU (4 bit) 1 ALU (16 bit) 2 ALU (14 bit)	37.5	16	625
transmitter pulse shaper with RAM	Fig.7	1 ALU (16 bit) 2 ALU (14 bit) 1 RAM (12×16 bit) 1 ACU (4 bit)	67.0	18	550
pulse shaper manual design	Fig.11	1 ALU (16 bit) 2 ALU (14 bit)	25.5	12	830
RECEIVER					
pulse shaper and equaliser	—	2 ALU (14 bit)	27.5	27	370
pulse shaper and equaliser	Fig.15	3 ALU (14 bit)	38.0	19	525
pulse shaper and equaliser	—	1 ALU (18 bit) 1 MPY (18 bit)	—	30	330
pulse shaper and equaliser	Fig.8	2 ALU (18 bit) 1 MPY (18 bit)	78.5	17	580
distributed control	Fig.13	3 ALU (14 bit)	51.5	19	525
auto-placed	—	3 ALU (14 bit)	54.5	19	525
separate muxes	Fig.14	3 ALU (14 bit)	46.5	19	525
up-sampling decoder	—	1 ALU (14 bit)	—	17	580

*Fig. 16. Design alternatives overview. * Worst case data rate estimation in kHz, based on a clock period of 100 ns for a 24 bit data path.*

References

[1] H. De Man et al. "*CATHEDRAL-II: A Synthesis System for Multiprocessor DSP Systems*", in Silicon Compilation, Ed. by D. Gajski, Addison-Wesley Publishing Company Inc., 1988, pp. 311-360.

[2] J. Vanhoof et al., "*A knowledge-based CAD system for synthesis of multi-processor digital signal processing chips*", Proc. IFIP Internat. Conf. VLSI, Vancouver, Aug. 1987, pp. 41-56.

[3] P. Hilfinger "*Silage, a high-level language and silicon compiler for digital signal processing*", Proc. IEEE-CICC 1985, Portland, May 1985, pp. 213-216.

[4] K. Sam Shanmugam "*Digital and analog communication systems*", John Wiley & Sons, Inc., New York, 1979, pp. 189-250.

Logic and Architecture Synthesis
P. Michel and G. Saucier (Editors)
Elsevier Science Publishers B.V. (North-Holland)
IFIP, 1991

EXPERIENCES WITH A HIGHLY AUTOMATED DESIGN ENVIRONMENT FOR VLSI

A. ENGBERSEN

IBM Research Division, Zurich Research Laboratory, 8803 Rüschlikon, Switzerland

The design of large, complex VLSI chips requires a different set of tools than the common gate-level entry tools used today. Hardware design languages, logic synthesis and timing analysis are much more appropriate to attack the problems of complex VLSI design. Thorough simulation and verification become mandatory when time-to-market is crucial to the success or failure of a product. A complete design methodology will be discussed with these considerations in mind, based on the practical experience of designing a complex VLSI chip with this environment. Problems with and improvements of the existing tool concepts will be outlined.

1. Introduction

Continuous advances in silicon technology presently allow the integration of complex systems of 100 K or more equivalent 2-input NAND gates, or about 1 million transistors, on a single chip, using a standard cell approach. This number is expected to increase significantly in the near future.

These advances promise new opportunities for those who can master the design of such complex systems and still achieve a short time-to-market, which largely determines the success or failure of a product.

In the past, chip design technology has steadily progressed approximately in line with advances in silicon technology. The first chips were built with hand-drawn masks, a job that was soon delegated to mask-generating software. Today, from placement and wiring all the way to manufacturing, the process has been largely automated. Quite naturally, detailed gate-level design will be the next level to be automated, and the current interest in logic synthesis tools [1] testifies to this trend.

Error-free chip operation after the first design pass is crucial in determining the time-to-market. Significant improvements in the designer's productivity and the reduction of the probability of errors are necessary to design these complex circuits within shorter times. Key instruments used to achieve this are design partitioning, high-level design description with specification- level simulation and logic synthesis. Increasingly, timing issues will be handled by the logic synthesis process, allowing the designer to concentrate on the accuracy of the specification. The use of delayless simulation, which is inherently faster then delay simulation, will add to the designer's productivity.

In the following sections we will describe experiences gathered during the design of a 100 K standard cell VLSI chip. The design methodology was completely based on the above-mentioned instruments. The simulation methodology, which contributed significantly to the achievement of a first-pass successful chip design, is outlined in detail.

2. Design Process

Increasing complexity of VLSI designs dictate the use of user-friendly "natural language"-like design entry tools aimed at increasing designer productivity. Experience shows that productivity is largely determined by the turnaround time of the design-entry to design-verification iteration cycle. Reducing this turnaround time is crucial for improving designer productivity. Register transfer (RT) design entry using a hardware design language (HDL) fulfills this requirement. The RT level guarantees fast compilation and fast simulation, while HDL introduces easy-to-use, high-level programming language constructs to the specification phase. This reduces the time and effort required to modify the design specification in comparison, for example, to a gate-level description.

The bulk of the combinatorial logic can be efficiently designed using high-level programming language constructs like IF-THEN-ELSE and CASE statements, followed by logic synthesis. However, having used an HDL [3], we also consider it important to have the option of specifying gate-level constructs directly at the HDL level. In dense and complex chip designs there will always be opportunities for an experienced designer to use little 'tweaks' to improve performance and save silicon real-estate. The logic synthesis process must subsequently imbed these special designs without modification in order to retain the improvements intended by the designer.

Aside from easy specification, a second key feature of HDL is the ease with which the specification can be changed. The design-entry to design-verification iteration cycle mentioned above facilitates a change of specification. The speed at which a specification can be changed – and simulated again – can motivate a designer to try alternative and innovative approaches to improve the quality of a design. Our experience shows that compiling and starting the simulation of the design must take less then a few tens of seconds for a designer to remain sufficiently motivated to experiment with design alternatives.

An HDL description should also be regarded as a design document that allows, for instance, specifications to be annotated with comments explaining certain constructs and, possibly, the creation of some technology independence.

Design Capture

Design capture is a series of partitioning and specification steps, starting from the architectural description, and gradually moving towards the gate-level specification. Partitioning of the design should result in functionally coherent partitions that humans can design. A functionally coherent partition is characterized by the fact that most of the interconnections between logic gates are contained within the partition, and well-defined interfaces exist to other partitions. The size and complexity of partitions are determined by functional completeness and by the abilities and performance of the available CAD tools. If the partitions are too small, the total design will contain too many partitions and, therefore, require too many interface definitions to be practical.

If the partitions are too large, CAD tools will either not be able to process a partition at all, or will consume prohibitive amounts of processing time. Our experience has shown that partitions are typically in the range of 1000-3000 equivalent 2-way NAND gates for random (control) logic. When strongly regular macros (e.g. memory) are involved, such a partition may typically contain this macro and a minimum of supporting logic (clock drivers, special test multiplexers etc.). Functionally coherent partitions allow designers to identify themselves with these parts of the design, which results in a greater job satisfaction and thus productivity. During the partitioning phase it is possible to identify identical parts of the design which are therefore candidates for design reuse. Partitioning and design reuse are the driving forces for adopting hierarchical design approaches. At each level of the hierarchy it is possible to use lower level designs (vassals) more then once, each instantiation having unique interconnects to the higher levels.

Figure 1 shows a logical view of our recent chip design [2]. This design has 16 input circuits, a memory array, a control structure and 4 output circuits. The chip design was partitioned into 18 functionally coherent partitions, each with equivalent gate counts between 800 and 3500 gates. There was one unique design for the input circuitry and another for the output circuitry, which were re-used 16 and 4 times, respectively. Other partitions that are much less evident from the logical view (and the description of which is beyond the scope of this paper) were also reused. The sum of netlists of all the **different** partitions amounted to a total of 25,000 lines, whereas the final 'flat' netlist of the **entire chip** design contained 64,000 lines. In this case, adopting a hierarchical design methodology exploiting partition reuse enabled us to reduce the design entry effort by a factor of 2. Experiments with CAD tool turnaround time showed that logic synthesis runs have a typical linear behavior with respect to the

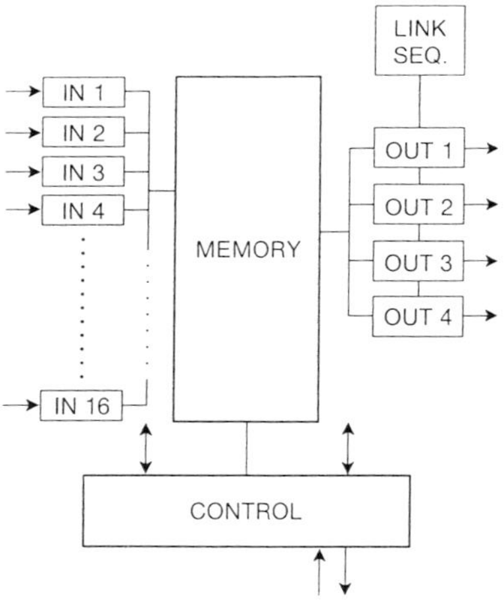

Fig. 1 High-level view of chip

number of gates. Our average partition size of 1000 gates has a turnaround time of approximately 15 minutes, whereas a cell design 15 times larger requires over four hours for a synthesis run on the same machine. In light of the required hand-tuning of certain nets, this is a very important difference, as will become clear in the following section.

3. Design Synthesis

Design synthesis is the generation of a netlist-level, technology-dependent implementa- tion of the HDL-level specification of a design. Although this can be done manually from the HDL specification, it is a very error-prone task, and requires additional design verification afterwards. Advanced automatic logic synthesis tools [1, 4] promise a drastic reduction of time and effort required for VLSI design. However logic synthesis is still not a fully automatic process. For faster turnaround and to reduce complexity, the partitioning of the logic should be maintained throughout the logic synthesis phase. This in turn places the burden on the designer to specify timing requirements for all interfaces between partitions. The inability of the logic synthesis process to synthesize across partition boundaries will also lead to designs which may be less than optimal in terms of gate count and cycle time. In some cases manual intervention is required and should therefore be supported as mentioned earlier.

In CMOS designs, special attention must be paid to capacitive loading due to the wiring, and its effect on the timing conditions. Therefore an indication of wire lengths is required **before** logic synthesis with timing optimization can be executed. In our design we used a two-step approach. First, logic synthesis is performed without timing requirements and with a minimum gate count objective. This process will produce a first approximation of the partition area size and allow a rough floor plan of the chip to be sketched. From this rough floor plan, wire lengths can be estimated. They serve as input to the timing optimization step of the second logic synthesis run.

In addition to these estimates, the timing specification must convey information about noncritical and multipath nets. Noncritical nets have only very slowly varying or even quasi-static signals, like power-on initialization, chip address or chip configuration signals. Multipath nets are those which are allowed several clock cycles to propagate their signals between source and sink. Unless these nets are properly specified, the logic synthesis process will also optimize logic circuits which are timing insensitive. As a consequence, silicon real estate may be wasted. Unfortunately the only method of discovering noncritical and multipath nets is the careful inspection of the design at various levels, and by repeated timing analyses. We are not aware of systems presently available that allow an annotation at the HDL level for these types of nets. This ability would be most desirable for future HDL-synthesis design systems.

In summary, logic synthesis is a two-cycle process. The first cycle produces rough geographic information which is used in the second cycle to specify sufficiently precise timing information. In the second cycle, meeting all timing requirements is the optimization objective. Timing requirements not met after these two cycles require hand-specification at the gate level in the HDL description. Note, however, that any change to the HDL-level requires, in principle, an additional verification cycle in order that design correctness be maintained.

Once all partitions are available in a gate level description, together with the layout-based timing information, a timing analysis of the complete design is required. Owing to their sheer size, these timing analysis runs consume a significant turnaround time. The success of this final timing analysis is largely determined by the quality of the specified timing requirements of the individual partitions. Therefore the quality of these specifications largely determines the turnaround time of the entire synthesis/timing analysis process.

4. Design Verification

Design verification has to be an on-going activity at every stage of the specification. It should not be a single task, only started when the design specification has already been completed.

In order to make design verification as pervasive as possible, an HDL language with a fast, simple path to verification (simulation) is required together with the creation of a flexible, automatic verification environment. The construction of such an environment is described next. The interaction of the HDL language with a verification system has been previously reported [5].

The objective of a design verification environment is to create a system which exhaustively exercises and tests all possible modes of operation of a particular design. Special emphasis must be placed on including exceptional conditions which may rarely occur in the system in practice.

It is important that this environment be created by people not directly involved with the actual hardware design, i.e. the HDL specification. This strategy prevents identical conceptual errors from being made in both. The system described here is not designed to detect such errors, but the method described will reduce the probability of such errors occurring.

From a project planning point of view, it is important that the test environment is ready and debugged prior to, or at least concurrently with the completion of the final HDL specification. A schematic timetable of a VLSI design project in this case is given in Figure 2.

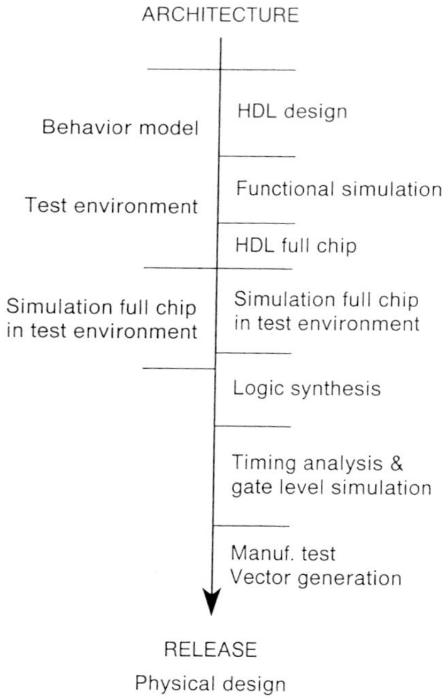

Fig. 2 VLSI Design project schedule

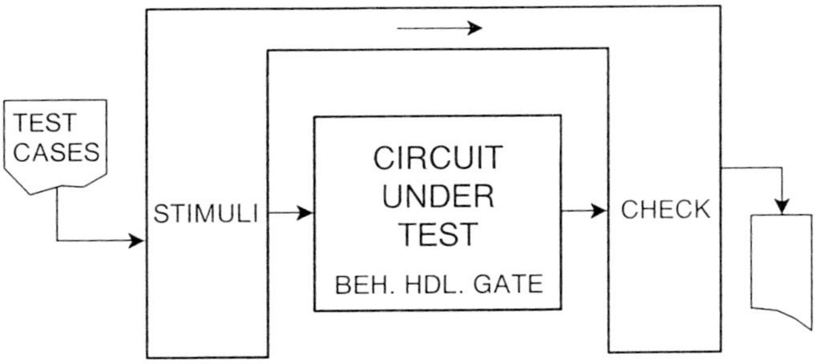

Fig. 3 Simulation environment

Figure 3 is a schematic view of a typical verification environment which has three basic parts:

1. a model of the circuit under test (CUT),
2. a stimulator/checker system,
3. a large set of adequate test cases.

Typically, the CUT is imbedded in the stimulator/checker system as shown. Stimuli for the CUT are generated by the stimulator based on commands in the test cases. Pertinent information is passed from the stimulator to the checker, which computes and compares the actual CUT responses.

For the CUT model, three variations will be used throughout the project's lifetime. In the first phase a behavior model, provided by people not involved in the actual HDL design work, will be required to build the verification environment and help debug the stimulator/checker system and the test cases. The behavior model should be a pin-for-pin replacement of the final expected HDL specification of the chip. In the second phase, the behavior model is replaced by the (compiled) HDL specification CUT. In the third phase, the HDL model is replaced by a (compiled) gate-level model, which is obtained by combining the synthesis results of the partitions. The simulation of the CUT is finished when all test cases run successfully. A gate-level simulation is carried out as a check on the synthesis process (where manual changes may have been introduced). Ideally replacing the CUT with another phase should incur no changes to the stimulator/checker or to the set of test cases.

The stimulator/checker system has three functions. It is typically written in the behavior language provided by a mixed-level simulation system in order to:

1. provide an interface between the test cases and the actual CUT,

2. compute the responses of the CUT,

3. compare the actual CUT responses with the computed ones and generate appropriate error messages.

In most cases, actual stimuli sequences on the bit level for a VLSI chip (the CUT) are very detailed and complex. A higher level of interface is provided for the test case programmer to avoid repetitive coding of these complex sequences. Transformation into actual sequences of bits is then performed by the stimuli generator. A significant reduction of test case development time, and improved test case readability will result. With this kind of stimuli generator, more compact data can be passed to the actual part of the environment which computes the CUT's responses based on the functional design objectives of the VLSI chip. This results in an increase in simulation speed because it is not necessary to interpret the detailed bit sequences within this part of the model.

Finally, the computed results are compared with the actual output sequences of the CUT in the checker part of the test environment, and detailed error messages are generated which indicate the point of failure. Construction of the system such that the occurrence of an error will stop the simulation and save pertinent tracking data proved to be very useful. This allows simulations to be run in batches and over weekends without manual intervention, while still providing detailed information on errors. The availability of a behavior model of the CUT has the advantage that coding and debugging of the test cases can be finished even before the HDL specification is available.

A design verification system organized along these guidelines guarantees that fast re-simulation is possible after any small design change made in the HDL, or even in the gate-level description, during the synthesis process. It is evident that every designer is responsible for verifying the correct operation of his partition ("functional simulation" in Figure 3), during the HDL specification phase. When the HDL specification of the entire chip is ultimately assembled, the probability of arriving at an operational model is maximized.

5. Conclusion

A design methodology for highly complex VLSI designs has been presented. Special attention must be paid to the cycle between design-entry and design-verification to achieve high designer productivity. The key design-verification principles required to maximize the probability of operational first-pass designs of this complexity in a timely manner have been explained. Finally, initial experiences with a fully synthesized design are reported. Automatic logic synthesis increases designer productivity, but may still benefit from certain manual interventions to achieve, for example, the fastest possible circuit operation. With the availability of timing **analysis** tools, which are an integral part of the logic synthesis process, **simulation** can be delayless – a simplification which will reduce simulation turnaround time.

Acknowledgements

I sincerely wish to thank all those involved in the design of this chip. It is due to their individual contributions that this project has become such a success. Special thanks go to the chip designers R. Luijten and R. Mueller, to K. Quinlan and W. Denzel for their contributions to the simulation environment, and to the IBM Laboratory for Research and Development in Boeblingen, Germany, for its continuous support.

References

1. A.J. de Geus, "Logic Synthesis Speeds ASIC Design", *IEEE Spectrum,* Vol. 26, No. 8, August 1989, pp. 27-31.
2. R.P. Luijten, "An Integrated Circuit for Self-Routing Data and Voice Switching," *submitted to EURO ASIC '90, Paris, France, May 28 – June 1, 1990.*
3. W. Roesner, "A Hardware Design Language for Logic Simulation and Synthesis in VLSI", *Proc. IEEE Compeuro '87 Conference, Hamburg, W. Germany, May 11-15, 1987,* pp. 311-314.
4. J.A. Darringer, D. Brand, J.V. Gerbi, W.H. Joyner, L. Trevillyan, "LSS: A System for Production Logic Synthesis," *IBM J. Res. Develop.,* Vol. 28, No. 5, Sept. 1984.
5. W. Roesner, "A Mixed Level Simulation System for VLSI Logic Designs", *Proc. IEEE Compeuro '87 Conference, Hamburg, W. Germany, May 11-15, 1987*, pp. 196-199.